D1429312

CASTI HANDBOOK OF

STAINLESS STEELS & NICKEL ALLOYS

Second Edition

Stephen Lamb
Technical Editor

Executive Editor
John E. Bringas, P.Eng.

CASTI Publishing Inc.
Suite 210, 10544-106 Street
Edmonton, Alberta, T5H 2X6, Canada
Tel: (780) 424-2552 Fax: (780) 421-1308
casti@casti.ca
www.casti.ca

The Materials
Information Society

ASM International
9639 Kinsman Road
Materials Park, Ohio 44073-0002 USA
Tel: (440) 338-5151 Fax: (440) 338-4634
www.asminternational.org

ISBN 1-894038-66-5
Printed in Canada

ii

National Library of Canada Cataloguing in Publication Data

Main entry under title:

CASTI handbook of stainless steels & nickel alloys

Includes index.
ISBN 1-894038-66-5 (bound). -- ISBN 1-894038-67-3 (CD-ROM)

1. Iron-nickel alloys. 2. Corrosion resistant alloys. 3. Steel alloys. I.
Lamb, Stephen. II. Bringas, John E., 1953- III. Title: Handbook of
stainless steels & nickel alloys. IV. Title: Handbook of stainless steels
and nickel alloys.
TA479.S7C37 2002   620.1'6   C2002-910254-5

CASTI Handbook of Stainless Steels & Nickel Alloys – Second Edition

CASTI PUBLICATIONS

CASTI CORROSION SERIES™

Vol. 1 - *CASTI* Handbook of Cladding Technology
Vol. 2 - *CASTI* Handbook of Stainless Steels & Nickel Alloys
Vol. 3 - *CASTI* Handbook of Corrosion in Soils
Vol. 4 - Corrosion Control

CASTI GUIDEBOOK SERIES™

Vol. 1 - *CASTI* Guidebook to ASME Section II, B31.1 & B31.3 - Materials Index
Vol. 2 - *CASTI* Guidebook to ASME Section IX - Welding Qualifications
Vol. 3 - *CASTI* Guidebook to ASME B31.3 - Process Piping
Vol. 4 - *CASTI* Guidebook to ASME Section VIII Div. 1 - Pressure Vessels
Vol. 5 - Plant Project Engineering Guidebook: for Mechanical and Civil Engineers
Vol. 6 - 2001 ASME Section VIII and 2002 Addenda Code Revisions Explained:
Div. 1 & 2 and Selected Code Cases

CASTI METALS DATA BOOK SERIES™

CASTI Metals Black Book™ - North American Ferrous Data
CASTI Metals Black Book™ - European Ferrous Data
CASTI Metals Red Book™ - Nonferrous Metals
CASTI Metals Blue Book™ - Welding Filler Metals

First printing, March 2002
Second printing, October 2002
Third printing, March 2003
Fourth printing, October 2003
ISBN 1-894038-66-5   Copyright © 1999, 2000, 2001, 2002, 2003

FROM THE PUBLISHER

IMPORTANT NOTICE

Our Mission

CASTI Publishing Inc.

Our mission at the *CASTI* Group of Companies is to provide the latest technical information to engineers, scientists, technologists, technicians, inspectors, and other technical hungry people. We strive to be your choice to find technical information in print, on CD-ROM, on the web and beyond.

We would like to hear from you. Your comments and suggestions help us keep our commitment to the continuing quality of all our products. All correspondence should be sent to the author in care of:

CASTI Publishing Inc.,
Suite 210, 10544-106 Street
Edmonton, Alberta, T5H 2X6, Canada
tel: (780) 424-2552   fax: (780) 421-1308
e-mail: casti@casti.ca   website: www.casti.ca

ASM International

ASM International is a society whose mission is to gather, process and disseminate technical information. ASM fosters the understanding and application of engineered materials and their research, design, reliable manufacture, use and economic and social benefits. This is accomplished via a unique global information-sharing network of interaction among members in forums and meetings, education programs, and through publications and electronic media.

Formerly named the American Society for Metals, the society name was changed to ASM International in 1985 to reflect a global membership with interests in metallurgy and engineering materials. For more information on services and membership, call (440) 338-5151 or visit www.asm-intl.org.

Chapter	Author	Peer Review
1	W. I. Pollock (Consultant)	G. Kobrin (Consultant) F. G. Hodge (Haynes International)
2	M. Blair (Steel Founder's Society) R. Pankiw (Duraloy Technologies Inc.)	C. S. Nalbone (Consultant) J. L. Gossett (Fisher Controls) A. Paris, J. Echlin and D. Driggers (Duraloy Technologies Inc.)
3	A. Sabata (Armco, Inc.) W. J. Schumacher (Armco, Inc.)	J. Ziemianski (Consultant)
4	J. D. Redmond, C. W. Kovach (Technical Marketing Resources, Inc.)	J. Fritz (Allegheny Ludlum)
5	G. E. Coates (Nickel Development Institute)	C. Reid (Consultant)
6	I. A. Franson (Consultant) J. F. Grubb (Allegheny Ludlum)	R. Davison (Technical Marketing Resources Inc.) D. E. Bardsley (Beloit Piping)
7	J. R. Crum, E. Hibner (Specialty Metals Corp.) D. R. Munasinghe (Specialty Metals Corp.) N. C. Farr (Specialty Metals Corp.)	P. Crook (Haynes International) P. Elliott (Consultant)
8	D. J. Tillack (Consultant)	A. Lesnewich (Consultant) S. D. Kiser (INCO Welding Products)
9	L. M. Smith (Intetech Ltd.)	

PREFACE

It is our intent to preserve the technical experience and knowledge gained over the past fifty years by material engineers in selecting corrosion resistant alloys for the handling of a wide range of corrosive environments. This has become especially important with the continual downsizing in industry today, consolidation of technical staff through corporate mergers, and retirement of senior engineers.

Industry continues to optimize its processes, often demanding higher temperatures and pressures to yield higher productivity, which requires the use of more corrosion resistant engineering materials.

Stainless steels and nickel alloys provide solutions to many of the problems encountered in aggressive environmental conditions and meet particular corrosive conditions related to stress corrosion cracking, reducing and oxidizing environments, halogenation, salts, hydrogen sulfide, and high temperatures, to name just a few.

Likewise, this *CASTI* Handbook was written to provide the practicing engineer, inspector, designer, or plant operator with guidelines and up-to-date information to help prevent or minimize mistakes that have been made in the past.

The information presented in this book was prepared by 30 authors and peer reviewers, who's total combined work experience is more than 700 years. These individuals are either involved with manufacturing products, developing operating processes, or are practicing engineers and consultants working in industry today. Our sincere thanks, appreciation and recognition is extended to them.

Stephen Lamb (Consultant)
Technical Editor

TABLE OF CONTENTS

Chapter

1

SOME HISTORICAL NOTES

Dr. Warren I. Pollock

Wilmington, Delaware

Historians studying the far reaching technological impacts of metallic materials of construction can rightly call the 20th century the *Corrosion-Resistant Alloy Age* or maybe more precisely the *Stainless Steel and Nickel-Based Alloy Age.* Although for thousands of years precious metals, copper, lead, tin, and zinc, and some alloys based on these elements, were known for resistance to various corrosive media, utilization of these materials was mostly for coinage, jewelry, cooking and other so-called utilitarian necessities. As essential materials of construction for relatively large scale equipment, their use was limited.

Carbon steel, iron castings, and wrought iron were the primary metals and alloys that gave the industrial era of the 19th century its solid foundation and remarkable success. During the 20th century, it was the development and production of corrosion-resistant iron- and nickel-based alloys that permitted the new industrial complexes of chemical, petrochemical, refining, pulp and paper, and other process industries, as well as dairy, food, and pharmaceutical companies, to design and fabricate critical equipment, piping, and storage units handling the widest range of hazardous environments, often at high temperatures and pressures. It is these important operating facilities – so often taken for granted by the general public – that produce the multitude of chemicals, plastics, drugs, foods, beverages, and many other types of products, even silicon chips, that are at the very core of today's industrial advances. And certainly it will be these types of important operating facilities that will be at the very core of tomorrow's multifaceted innovative developments.

Monel®: Ni-Cu Alloy

An alloy-application history of many of these corrosion-resistant iron- and nickel-based alloys is given in Figure 1.1. Compiled by Stephen Lamb, it starts with the discovery of Monel, an alloy of nominal 67% nickel, balance copper, now designated with the Unified Numbering System (UNS) designation N04400 (alloy 400).

Monel was created by Robert C. Stanley of The International Nickel Co. (Inco), formed by the consolidation of Canadian Copper Company at Copper Cliff, Ontario, and Orford Copper Company, in New Jersey in 1902. The alloy's origin was derived from the ores of the Sudbury basin in Ontario, Canada that contain nickel and copper in just about the same two-to-one ratio. 1905 is the date given for its discovery, although the request for the first car of matte to be used experimentally for direct production of the nickel-copper alloy was written by Stanley in December 1904.

The alloy was first designated "Monell Metal" in honor of Ambrose Monell, then Inco president. In 1906, to register a trademark, the name was changed to Monel. Stanley, at the time, was the young company's assistant superintendent. He later became president and chairman of the board of Inco.

To help evaluate the new alloy, Inco engaged Columbia University. One finding was it resisted sulfuric acid so well it could not be pickled without the addition of ferric chloride or nitric acid to the pickling liquor. Two early applications were as a shaft in a pump in the company's plant handling water from New York and roofing for the new Pennsylvania Railroad Station in New York City.

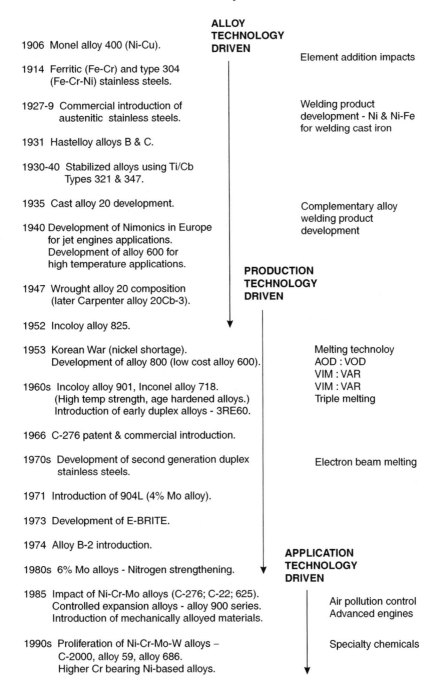

ALLOY TECHNOLOGY DRIVEN

1906 Monel alloy 400 (Ni-Cu).

Element addition impacts

1914 Ferritic (Fe-Cr) and type 304 (Fe-Cr-Ni) stainless steels.

1927-9 Commercial introduction of austenitic stainless steels.

Welding product development - Ni & Ni-Fe for welding cast iron

1931 Hastelloy alloys B & C.

1930-40 Stabilized alloys using Ti/Cb Types 321 & 347.

1935 Cast alloy 20 development.

Complementary alloy welding product development

1940 Development of Nimonics in Europe for jet engines applications. Development of alloy 600 for high temperature applications.

PRODUCTION TECHNOLOGY DRIVEN

1947 Wrought alloy 20 composition (later Carpenter alloy 20Cb-3).

1952 Incoloy alloy 825.

1953 Korean War (nickel shortage). Development of alloy 800 (low cost alloy 600).

Melting technoloy AOD : VOD VIM : VAR

1960s Incoloy alloy 901, Inconel alloy 718. (High temp strength, age hardened alloys.) Introduction of early duplex alloys - 3RE60.

VIM : VAR Triple melting

1966 C-276 patent & commercial introduction.

1970s Development of second generation duplex stainless steels.

Electron beam melting

1971 Introduction of 904L (4% Mo alloy).

1973 Development of E-BRITE.

1974 Alloy B-2 introduction.

APPLICATION TECHNOLOGY DRIVEN

1980s 6% Mo alloys - Nitrogen strengthening.

1985 Impact of Ni-Cr-Mo alloys (C-276; C-22; 625). Controlled expansion alloys - alloy 900 series. Introduction of mechanically alloyed materials.

Air pollution control Advanced engines

1990s Proliferation of Ni-Cr-Mo-W alloys – C-2000, alloy 59, alloy 686. Higher Cr bearing Ni-based alloys.

Specialty chemicals

Figure 1.1 Alloy/Application History

Needing plate and sheet to develop the anticipated sizable market for the alloy, Inco built a melting facility and rolling mill in Huntington, West Virginia, which went into operation in 1922. The site today is the headquarters of Inco Alloys International, earlier called Huntington Alloys, now Special Metals Corporation.

In 1924, based on the work of William A. Mudge and the significant discovery of Paul D. Merica that aluminum and titanium led to precipitation hardening of nickel-based alloys, Inco introduced K-Monel® containing about 3% aluminum and 0.5% titanium. Today's UNS designation is N05500 (alloy 500).

Ferritic (Fe-Cr) Stainless Steel

The discovery of iron-chromium stainless steel is usually attributed to the Englishman Harry Brearley. Working at the Brown Firth Research Laboratories in Sheffield, England, a joint research and development laboratory of Thomas Firth and Sons and John Brown and Co., Brearly became interested in the manufacture of steels with high chromium contents. Trying to prevent erosion and fouling in rifle barrels, he concluded that a steel with upward of 10% chromium could possibly be of advantage. The date of record is June 4, 1912.

The first commercial cast was made in August of the following year. Part of the heat was later fabricated into "rustless" table cutlery blades. Its composition, 12.86% chromium and 0.24% carbon, a plain chromium martensitic steel, is the forerunner of type 420 stainless steel (S42000). Brearley is credited with the name "stainless," yet his own account of the discovery in 1924 states Brearley first heard the knives described as stainless by Ernest Stuart, a cutlery company manager, after Stuart had attempted to stain them with food acids.

The first commercial application was knife making. During the First World War the stainless steel was used in aero-engine exhaust valves. Subsequently, there was a demand for a softer steel, and this was met in 1920 by a steel containing 12-14% Cr and 0.1% C which became known as "stainless iron."

In 1915, the first stainless steel ingot cast in America was made at Firth Sterling Ltd. in Pittsburgh, Pennsylvania. Brearley applied for a U.S. patent the same year, and it was granted in September 1916. Curiously, no patent application was made in Great Britain.

Not often mentioned is that Elwood Haynes, founder of the Haynes Stellite Company, Kokomo, Indiana, now Haynes International, previously Cabot Corporation, also discovered that stainlessness could be imparted to steel by the addition of sufficient quantities of chromium in a series of experiments made between November 1911 and April 1912. Although Haynes' U.S. stainless steel patent application antedated by nearly a year the application submitted by Brearley, it was denied. Haynes filed for interference, which he eventually obtained, and received a patent in April 1919. By that time, however, Haynes and Brearley had merged their interests. Both assigned patents to an American patent-holding company which issued licenses to steel companies desiring to manufacture stainless steel.

Austenitic (Fe-Cr-Ni) Stainless Steel

Meanwhile in Germany, austenitic iron-chromium-nickel stainless steel was being developed. In October of 1912, German authorities granted Fried. Krupp Werke (Friedrich A. Krupp Works) the first patent for the manufacture of components in stainless chromium-nickel steels.

A 1909 Krupp laboratory report mentions that high-alloy chrome and chrome-nickel steels showed no signs of rust even after months of exposure to humid laboratory air. The discovery is attributed to Eduard Maurer who noticed that some alloys that Benno Strauss had made were impervious to attack after months of exposure to acid fumes in Maurer's laboratory. Maurer had joined Kruppschen Forschungsanstalt (Krupp Research Institute) at Essen that year as its first metallurgist.

The 1912 patent for austenitic stainless steels filed by Strauss and Maurer was followed by another application in 1913, with "full patents" granted in 1918. These steels were introduced into the U. K. under license around 1923. They were called "super stainless steels" in England to distinguish them from the plain chromium steels.

Strauss was director of the Krupp research facility. His name is perhaps best known for the copper sulfate/sulfuric acid test for detection of intergranular attack (IGA) of austenitic stainless steels. Although often called the Strauss Test or the Moneypenny-Strauss Test, a literature study by Michael A. Streicher showed that according to Strauss' own publication, the test solution was first used by William H. Hatfield in England.

In 1911, also in Germany, P. Monnartz published his doctoral research observing that chloride salts impair the passive state of iron-chromium alloys and that molybdenum additions increased the resistance in such environments. With his thesis advisor, W. Borchers, Monnartz patented a 30-40% Cr alloy with 2-3% Mo. The alloy, however, proved to be too brittle for fabrication.

Strauss and Maurer no doubt were aware of Monnartz' research and probably initiated studies on molybdenum addition to Fe-Cr-Ni alloys in 1911-12, perhaps even earlier; the exact date seems not to be recorded. The invention of molybdenum-containing type 316 stainless steel (S31600) is credited to Krupp. Molybdenum is a ferrite former and the nickel content must be increased to about 12% to maintain the austenitic structure. Molybdenum, in addition to improving resistance to reducing acids, also makes type 316 superior to type 304 in chloride pitting and crevice corrosion environments.

Industrial uses of austenitic stainless steel began soon after its discovery. By 1914, rapidly increasing quantities of Krupp V2A (V stands for "Versuch," the German word for test, and A for austenite), originally a 20% Cr-7% Ni-0.25% C composition which was later balanced at 18% Cr and 8-10% Ni, today's type 304, and V4A, which may have been molybdenum bearing, were supplied to Badische Anlin und Sodafabrik (now BASF) in Ludwigshafen,

In 1974, Haynes International announced an improved version of alloy B with low-iron and low-carbon contents. Called alloy B-2 (N10665), it contains nominal 28% Mo, 0.8% Fe, and 0.002% C. The introduction of alloy B-2 made "wrought" alloy B obsolete. "Cast" Ni-Mo alloys designated N-7M (N30007) and N-12MV (N30012) are considered to have an alloy B, not alloy B-2, composition.

Twenty years later, several improved Ni-Mo variants were introduced: alloy B-3 (N10675) by Haynes International with nominal 28.5% Mo, 1.5% Cr, 1.5% Fe, and 0.003% C and alloy B-4 (N10629) by Krupp VDM, formerly Vereinigte Deutsche Metallwerke (VDM), with nominal 28% Mo, 1.3% Cr, 3% Fe, and 0.005% C. These alloys provide a level of thermal stability superior to alloy B-2.

Hastelloy alloy C, a Ni-Cr-Mo alloy, was patented by stainless steel specialist Russell Franks of UCC Research Laboratories in 1931. Studies on substituting chromium for some of the molybdenum in the already patented alloy B started in the late 1920s, perhaps as early as 1926. A ratio of 16% Cr and 16% Mo gave a high resistance to oxidizing media. Alloy C also contained about 4% W and 6% Fe.

Early technology limited the product range of alloy C to castings, the high carbon content limiting ductility and weldability. Later, wrought products could be manufactured due to developments in melting and processing practices, in particular, oxygen blowing of heats which made possible the reduction of carbon content to nominal 0.05%. 1937 is the year cited for when Hastelloy alloy C launched the "Hastelloy family of Ni-Cr-Mo alloys."

In the late 1940s, Haynes Stellite developed an alloy, designated Hastelloy alloy F, containing copper and with higher iron content than alloy C, intended to be a lower cost alloy C and reportedly a competitor to alloy 20 (see discussion later). Projected costs, however, placed it too close to alloy C to make marketing attractive.

Alloy C served the needs of the chemical industry for about 30 years. However, it had several serious limitations. In many applications, vessels fabricated from alloy C had to be solution heat-treated to remove detrimental weld heat-affected zone (HAZ) precipitates. When used in the as-welded condition, alloy C was often susceptible to serious intergranular corrosion attack in many oxidizing and chloride-containing environments.

Recognizing the importance of the relatively high carbon and silicon levels, BASF in Germany, based on studies by I. Class, H. Grafen, and E. Scheil published in 1962, modified the composition of alloy C. Both the carbon and silicon contents were reduced more than 10 fold, typically 0.005% C and 0.04% Si. This alloy came to be known as alloy C-276 (N10276).

A U. S. patent for the invention of alloy C-276 was awarded to BASF in 1965. The patent was granted only after BASF and Haynes agreed to work together with Haynes dropping its own application. Alloy C-276 was then produced by Haynes in the U. S. under license from BASF. Robert B. Leonard and others at Haynes Stellite division of Union Carbide were instrumental in the commercial development of alloy C-276.

The introduction of alloy C-276 made "wrought" alloy C obsolete. "Cast" Ni-Cr-Mo with the designation CW-12MW (N30002) is considered to have an alloy C, not alloy C-276, composition.

Several variants of alloy C-276 have been introduced. In 1973, Haynes International announced alloy C-4 (N06455) with improved thermal stability; the U.S. patent was granted in 1978. Tungsten was totally omitted and iron content reduced to about 1%. The alloy also contained about 0.3% Ti. The casting grade CW-2M (N26455) is equivalent to wrought alloy C-4.

In 1982, alloy C-22 (N06022) was developed by Haynes International with about 3% W and 4% Fe, a U.S. patent being awarded three years later. Chromium content was raised to about 21% and molybdenum content reduced to about 13%. The alloy also contained about 0.2% V and 1.7% Co. It offered better resistance to chloride-induced localized corrosion and stress corrosion cracking (SCC). The casting equivalent for this alloy is CX-2MW (N26022).

In 1990, Krupp VDM introduced alloy 59 (N06059) with chromium content about 23%, iron less than 1%, and about 0.2% Al for very severe corrosive applications. And in 1995, alloy C-2000 (N06200) was announced by Haynes International with about 1.6% copper added which improves corrosion resistance in sulfuric acid and reducing media.

In 1964, Inco Alloys International (now Special Metals) introduced Ni-Cr-Mo alloy 625 (N06625), containing nominal 20% Cr, 8-9% Mo, 4% Fe, stabilized with about 3.5% Cb. Derived from alloy 718 (N07718), it was intended for aerospace applications. This alloy is usually, although not always, somewhat less resistant than the 15-16% Mo grades in chemical services.

Melting operations for the Hastelloy alloys initially were at Niagara Falls, gradually moving to the Kokomo site in the late 1930s. Forging and rolling were at the Simmonds Saw and Steel plant in nearby Lockport, New York, and then at the Huntington, West Virginia Inco plant.

With a price at the time of $2.00 to $2.50 per pound for wrought products, compared to 35 to 50 cents per pound for stainless steel, it was difficult for Haynes Stellite to get customers to accept its alloys. One of the earliest installations of any size was at the DuPont Company's Chambers Works in Deepwater, New Jersey. It was for alloy D, a nickel-silicon-copper alloy, the fourth in Haynes Stellite's 1930s' alphabet-designated alloys, to replace stainless steel which was lasting only about one year.

DuPont would not use the highly corrosion-resistant alloy unless the "net" to it for making an autoclave of alloy D would not cost more than stainless. The cost was roughly five times more fabricated from alloy D. To get the order, Haynes Stellite gave a written warranty that this unit would last five years. If not, Haynes Stellite would reimburse DuPont for the unfilled life. The autoclave lasted well beyond five years.

Interestingly, where Ni-Cr-Mo alloys are today associated mostly with process industry equipment and piping systems handling very harsh chemicals, the first large-scale use of alloy C was for a military wartime use: 12-inch diameter reflectors for Navy searchlights which were required by the thousands. The experience gained in making this product was invaluable later for the manufacture and fabrication of plate and sheet for process needs.

The first application of alloy C in the chemical industry is reported to have been made by C. P. Dillon at the Texas City, Texas plant of Union Carbide in about 1948. It was for a bucket liner in a newly developed weak-acid isopropanol process.

Alloy 20

In the mid-1930s, alloy 20, a Fe-Cr-Ni-Mo-Cu composition, was invented. In 1934, Mars A. Fontana, after doctorate studies at the University of Michigan (his Ph.D. degree was awarded in 1935), joined E. I. du Pont de Nemours Company at its Experimental Station in Wilmington, Delaware, and started a comprehensive research program on materials of construction for sulfuric acid service. The following year he found that an iron-based casting containing 20% Cr, 29% Ni, 2.25% Mo, and 3.25% Cu had superior resistance. The composition became known as "alloy 20," named, it is said, for the casting being the 20th that Fontana made or perhaps tested. Another explanation is the alloy contains 20% chromium.

Fontana left DuPont in 1945 to become professor of metallurgical engineering at The Ohio State University where he established one of the first courses in corrosion in 1946. He continued to work on cast corrosion resistant alloys, developing with colleagues alloy CD-4MCu (J93370), a duplex stainless steel containing about 25% Cr, 5% Ni, 2% Mo, and 3% Cu.

Based on Fontana's discovery, the DuPont Company began seeking castings of alloy 20 for several troublesome components, valves in particular. Suitable foundry practices needed to be developed and the composition modified with silicon in order to cast complex shapes. The Duriron Company, now Flowserve Corp., in Dayton, Ohio, a leader in specialty corrosion-resistant cast alloys, developed its own proprietary technology, naming the alloy Durimet® 20. The designation for "cast" alloy 20 is CN-7M (N08007, formerly J95150).

The need for wrought products was also recognized, and Carpenter Steel Company, now Carpenter Technology Corporation (CarTech), in Reading, Pennsylvania initiated development studies. But it was not until 1947 that improved melting techniques made it possible to produce wrought products, designated alloy Carpenter® 20. The blowing of oxygen through the melt resulted in reduction of carbon content. (This is also the technology that reduced carbon content in stainless steels to 0.02-0.03%, resulting in introduction the same year of the low-carbon "L" grades, namely, types 304L [S30403] and 316L [S31603]).

In 1948, columbium (niobium) was added to alloy 20 to combine with carbon and thereby minimize chromium carbide precipitation and susceptibility to intergranular corrosion. In 1960, it was found that this alloy was subject to a unique form of stress corrosion cracking in sulfuric solutions in the range of 20 to 80%. The problem was overcome in 1965 by increasing the nickel content from nominal 29 to 34%, which is the nominal nickel composition in today's alloy Carpenter 20Cb-3.

Although alloy 20 is considered to be a stainless steel, it does not contain 50% or greater iron and, therefore, its UNS prefix is not "S" for stainless steels (listed as "heat and corrosion-resistant steels" in the United Numbering System) but instead "N" for nickel and nickel-based alloys: N08020. Accordingly, it was not included in American Society for Testing Materials (ASTM) A specifications (covering irons and steels). Instead it was listed in ASTM B specifications (covering nickel- and cobalt-based alloys). A recent definition by ASTM now includes alloy 20 in A specifications.

Carpenter Steel, principally a bar manufacturing mill, melted and forged alloy 20Cb-3 ingots in-house, but tolled out rolling of flat products. Until about the mid-1980s it operated a welded stainless steel tubing mill in Union, NJ. After considerable processing studies, satisfactory procedures for manufacturing welded tubular products of this alloy were developed. Alloy 20Cb-3 tubing quickly found important applications in replacing type 300 series stainless steels in environments causing chloride stress corrosion cracking (Cl⁻SCC).

The story is told that DuPont corrosion and materials engineers "standardized" on a three-tier-level of corrosion-resistant alloys: The first tier was type 304 or type 316 if pitting resistance was needed. If neither was satisfactory, especially for chloride stress corrosion cracking (Cl⁻SCC) problems, the automatic next choice was the second tier, principally alloy 20, that is, alloy 20Cb-3. And if this alloy would not perform satisfactorily, then the automatic choice was the third tier, alloy C (alloy C-276) or perhaps alloy B (alloy B-2) or titanium. Obviously, going to a higher corrosion-resistant level meant higher materials of construction cost, but this was justified economically by the need for trouble-free manufacture of specialty products to meet market demands and to maintain and increase market share.

Alloy G

Haynes Stellite introduced alloy F containing nominal 22% Cr, 22% Fe, 6% Mo, and 2% Cb, which evolved into alloy G (N06007) primarily intended for handling phosphoric acid. Alloy G had nominal composition 22% Cr, 20% Fe, 6.5% Mo and 2% Cu, with about 2% Cb added for stabilization. In the 1970s, alloy G-3 (N06985) was introduced with considerably less columbium content, 0.3%, and with less iron, 18%. Only 0.3% Cb was added because of much lower carbon content.

In 1982, alloy G-30 (N06030) was introduced with chromium content increased to about 30%, iron content reduced further to about 15%, and molybdenum content reduced to about 5.5%, with 2.5% W and 0.7% Cb added, giving an alloy with a broader range of applications. The U. S. patent for alloy G-30 was granted in 1983.

Alloy 825

After the Second World War, Inco increased its research and development efforts with a program aimed at corrosion-resistant alloys. In 1952, it introduced Ni-O-Nel, today called alloy 825 (N08825), containing nominal 42% Ni, 22% Cr, 32% Fe, 2.5% Mo, and 2% Cu, stabilized with about 0.8% Ti. Alloy 825 became one of Inco's important "workhorse" alloys. Test coupons and spool pieces were widely distributed to obtain a large bank of corrosion information, especially in-service data.

AOD and New Refining Technology

Starting about 1970, a dramatic improvement in melting corrosion-resistant alloys was made possible by the argon-oxygen decarburization (AOD) refining process developed by W. A. Krivsky and patented in 1966. It enabled alloys to be produced routinely with extra-low carbon levels and close control of critical compositions. For developers of new corrosion-resistant alloys, it permitted compositions to be explored and commercially specified with deliberately added nitrogen contents as high as 0.55% and carbon contents as low as 0.003%, along with precise control of major metallic alloying elements and sharp reduction of tramp elements. Contributing to making the process an industrial success was the concurrent drop in the price of argon. Where ELC (extra-low carbon) stainless steels had been premium grades, they soon became standard products for many mills. Titanium and columbium did not need to be added for stabilization. AOD melting technology was the most influential factor in producing the newer nickel-based alloys as well as the nitrogen-containing, duplex and "6-Mo" stainless steels.

Another important development was electroslag refining in which a stainless steel ingot is remelted under carefully controlled conditions. The resulting product has fewer non-metallic inclusions and a lower tendency toward segregation of inclusions. Vacuum melting and degassing technologies were also introduced to meet the more demanding specifications of the aerospace, nuclear, and power-generation industries. In addition, continuous casting processes markedly improved the economics of stainless steel manufacturing.

Duplex Stainless Steels

Duplex stainless steels (DSS), comprising a two-phase structure of austenite and ferrite, usually a 50/50 ratio, have been used since the 1930s. Nevertheless, it is only since the 1970s that these alloys became popular by the increased use of AOD refining and continuous casting of slabs.

Avesta Jernverks (Ironworks) in Sweden developed two duplex alloys in the early 1930s. They were designed to reduce the problem of intergranular attack of austenitic type 300 series stainless steels from heat treatment and welding. At the time, the minimum carbon level that could be reached with existing furnaces and refining techniques was about 0.08%. Grade 453E contained nominal 26% Cr, 4% Ni, and 0.1% C. Grade 453S had an addition of about 1.5% molybdenum for increased corrosion resistance. To compensate for the Mo addition, Ni was increased to about 5%.

These ferritic-austenitic alloys showed improved resistance to intergranular corrosion and chloride stress corrosion, and had equal or better resistance to uniform corrosion than austenitic stainless steel grades. They also had higher mechanical strengths. Further, the comparatively high chromium content gave good oxidation (scaling) resistance.

Grade 453E was used extensively for high temperature applications in the 1930s. The Mo-containing grade 453S was used mainly in the pulp industry as castings, bars, and plate. It was not until 1947 that grade 453S was included in the Swedish standard as SIS 2324. Later, this duplex steel was designated type 329 (S32900) by the American Iron and Steel Institute (AISI). Type 329 stainless contains nominal 25% Cr, 3-4% Ni, and 1.5% Mo, with about an 85% ferrite content.

French workers were also active in development of ferritic-austenitic stainless steels in the 1930s. The J. Holtzer steel works in France was granted a patent in 1936 for alloys containing 16-23% Cr, 1.5-6.5% Ni, up to 3% Mo and W, and up to 2.5% copper. From this patent, alloy UR® 50 was developed and introduced in about 1970. Containing nominal 21% Cr, 7.4% Ni, 2.4% Mo, 1.5% Cu, and 0.07%N, with UNS designation S32404, it was a pioneer duplex alloy of today's Creusot-Loire Industrie (CLI), the large French steel conglomerate. UR is the abbreviation for Uranus®. The story is told that CLI's principal metallurgist at the time, J. Hochmann, was an amateur astronomer who convinced company management to use the planet's name as the trade designation.

Duplex stainless steels have been classified by generation: first, second, and third. Type 329 and alloy UR 50, along with the previously mentioned cast alloy CD-4MCu (J93370), are considered first generation. The prime distinction between first and second generation is second generation duplex steels contain deliberate additions of nitrogen, about 0.10 to 0.25%. Third generation duplex stainless steels have a pitting index number >40; see discussion later.

Sandvik Steel in Sweden introduced alloy 3RE60 (S31500) in the 1960s. Applications in the pulp and paper industry, from tubing and sheet in preheaters to Kraft and sulfite digesters, started about 1970. This alloy was particularly designed to resist stress corrosion cracking (SCC), replacing types 304 and 316 stainless steels, and found many such applications as tubular products. For its time, it had a relatively low carbon content, 0.03% maximum, made possible by improvements in refining techniques in the electric furnace. Nominally 20% Cr and 5% Ni, the composition of alloy 3RE60 is lean compared to other first generation duplex stainless steels and grades developed in later years. The alloy also contains about 1.7% Si.

Originally, duplex alloy 3RE60 was made without a deliberate nitrogen addition. But in the 1970s about 0.10% nitrogen was added– and, accordingly, this composition would be considered a second generation duplex. The UNS designation S31500 for alloy 3RE60, however, does not specify a nitrogen range.

Perhaps the first superduplex was alloy 255 (Ferralium® 255), introduced initially as a casting alloy by Bonar Langley Alloys, later Langley Alloys, of England. (The Ferralium trademark is now owned by Meighs, Ltd.) Its nominal composition is 25% Cr, 6.5% Ni, 3% Mo, and 1.6% Cu with 0.20% nitrogen. The UNS designation for the wrought alloy is S32550. This DSS has about 55% ferritic phase.

With the development and introduction of so many duplex austenitic-ferritic alloys in the last 20 or so years, classification by generations is not always particularly useful, and other terminology is fashionable. "Duplex" now often refers to grades with nominal 22% Cr content, while "super duplex" or "superduplex" indicates grades with nominal 25% Cr.

The development of alloy 2205 (S31803), with nominal 22% Cr, 5.5% Ni, 2.7% Mo, and 0.08 to 0.20% N, can be traced to a 1972 German patent application. Compared to the earlier ferritic-austenitic stainless steels, this alloy had a high Mo content and, importantly, also a nitrogen addition. It was designed to have better intergranular corrosion resistance after welding, a problem with the first duplex stainless steels. The proposed composition range, particularly that of nitrogen, in the patent was very wide allowing for high levels of ferrite in heat-affected zones. Furthermore, S31803 has a wide composition range.

Other Europeans steelmakers also were involved with a 22% alloy. The development of duplex alloy 2205 (S31803) with high nitrogen content can be credited to efforts of Sandvik Steel and Avesta, now Avesta Sheffield, both in Sweden, and reportedly also of Mannesmann in Duisburg, Germany. Today, alloy 2205, sometimes called alloy 22/5 in Europe, is considered the DSS workhorse and is produced worldwide. It has a 50/50 austenitic-ferritic structure.

Early duplex and superduplex grades were both characterized by an intermediate level of nitrogen, generally about 0.12%, and low molybdenum content, about 2.7%. Although these steels had good mechanical properties and corrosion resistance in the unwelded condition, in the as-welded condition the low phase stability led to very high ferrite contents in heat-affect zones, decreasing toughness at low temperature and also corrosion resistance. Higher nitrogen additions in the base material–and also in the filler metal–improved the stability of the two phases, as well as the mechanical and corrosion-resistant properties.

To avoid these problems, today's alloy 2205 normally contains 0.15 to 0.20% N. In 1996, a new grade of alloy 2205 with minimum nitrogen content of 0.14% was included in ASTM Specification A 240, "Heat-Resisting Chromium and Chromium-Nickel Stainless Steel Plate, Sheet, and Strip for Pressure Vessels," along with a new UNS designation. Cleverly, it was assigned UNS designation S32205.

Increasing PREN (pitting resistance equivalent number) has been a major driving force in duplex stainless steel development. PREN, an index of pitting resistance in chloride-containing environments, is based on the stainless steel's composition:

$$\text{PREN} = \text{Cr\%} + 3.3 \times \text{Mo\%} + 16 \times \text{N\%}^{*}$$

Alloy 2205 has a PREN 35. Several superduplex grades have PREN greater than 40. (At least one company marketing a superduplex with PREN >40 not surprisingly states this PREN value is "the" defining characteristic for a superduplex stainless steel!)

For over 50 years, the French have been developing austenitic-ferritic alloys. Alloy UR 45N was introduced about 1990 with the same nominal metallic alloying content as alloy 2205 but containing about 0.17% N. It meets requirements of S32205. To obtain more pitting resistance, in 1992 the Cr, Ni, and Mo contents were increased and the N level increased to 0.18%, and the grade given the designation alloy UR 45+. It too meets the requirements of S32205. Alloy UR 45+ has about 55% austenite phase.

---

* Many PREN formulas have been proposed. The "%C + 3.3 x %Mo" portion is generally accepted. The nitrogen coefficient, however, is more contentious, with values of 22 and 30 often widely cited.

In the late 1980s, Sandvik introduced superduplex alloy 2507 (S32750) with nominal 7% Ni. 4% Mo, and 0.28% N. Weir Metals in England has introduced superduplex alloy Zeron® 100 (S32760) with nominal 25% Cr, 7% Ni, 3.7% Mo, and 0.26% N, plus about 0.7% Cu and 0.7% W.

6-Mo Alloys

The "6 Moly" austenitic stainless steels have become a very important group of corrosion-resistance alloys. A forerunner containing about 4.5% Mo was alloy 904L (N08904), invented by Sweden's Nyby Uddeholm, now incorporated within Avesta Sheffield. Its composition is about 25% Ni, 20% Cr, 4.5% Mo, and 1% Cu. Tubular applications of alloy 904L date from 1971.

Alloys with 6% Mo are often called "super austenitics" or "superaustenitics," though this terminology is also used for other highly alloyed stainless steels usually with copper, e.g., alloys 904L, 20Cb-3 and 28 (N08028), the latter developed by Sandvik in the late 1970s containing nominal 31% Ni, 27% Cr, 3.5% Mo, and 1% Cu. "High-performance stainless steels" is also used. By themselves, the terms "high performance alloys" and "high performance materials" can be misleading as they often also comprise nickel-based alloys.

Alloy 254 SMO® (S31254), developed by Avesta Jernverks, today Avesta Sheffield, was commercially introduced in about 1977. Its nominal composition is 54% Fe, 18% Ni, 20% Cr, 6% Mo, 0.7% Cu, and 0.20% N. This alloy evolved from the standard austenitic stainless steel grades, with the addition of nitrogen. Nitrogen is economical because it decreases the need for nickel to stabilize the austenitic structure, while providing higher strength and improved pitting resistance.

Allegheny Ludlum's alloy AL-6X (N08366), designed for power plant condensers, was installed in service applications starting about 1973. Its nominal composition, 20% Cr, 24% Ni, and 6% Mo, evolved from Inco's alloy IN-748. This nickel-based alloy containing over 9% Mo was designed for cast rather than wrought products. Nitrogen was not enhanced in alloy AL-6X which limited products to light gage strip and tubing.

Alloy AL-6XN®, the nitrogen-enhanced version of alloy AL-6X requiring a different UNS designation, N08367, was introduced in 1983 and has replaced alloy AL-6X. The nitrogen content of about 0.2% allows alloy AL-6XN to be produced over a full range of product forms and sizes.

In 1998, AL-6XN PLUS was introduced with Cr content increased to about 22%; Ni, to about 25%; and nitrogen, to 0.24%. This modified composition, which is within the existing specification for N08367 and thereby carries the UNS designation N08367 also, has a PREN of 50 minimum, giving it improved resistance to localized chloride attack compared to AL-6XN.

The 6-Mo grade with the UNS designation N08926, called alloys 926 and 1925hMo by Krupp VDM and alloy 25-6MO by Inco Alloys International, is a modification of the lower molybdenum-containing alloy 904L. N08926 has nominal composition 20% Cr, 25% Ni, 6.5% Mo, 1% Cu, and 0.20% N. When first introduced it had UNS designation N08925 with a lower N range, 0.10-0.20%. When the nitrogen specification was changed to 0.15-0.25%, a different UNS designation was required. The so-called nitrogen-modified N08926 has replaced N08925.

Carpenter Technology modified its alloy 20Cb-3 and introduced a 6-Mo grade, alloy 20Mo-6® (N08026). It has nominal 24% Cr, 36% Ni, 5.5% Mo, and 3% Cu. This alloy is not nitrogen enhanced. In the late 1980s, Krupp VDM introduced alloy 31 (N08031) containing about 31% Ni, 27% Cr, 6.5% Mo, 1% Cu, and 0.20% N.

Alloy 654 SMO® (S32654), the highest molybdenum- and nitrogen-containing superaustenitic grade, was introduced by Avesta Sheffield in the early 1990s. It contains nominal 22% Ni, 24% Cr, 7.3% Mo, 0.5% Cu, and 0.50% N.

Ferritic (Fe-Cr-Mo) Stainless Steels

In the last thirty years, there also has been considerable development of ferritic stainless steels, in particular Fe-Cr-Mo compositions. Ferritics of more than nominal 26% Cr and/or nominal 2% Mo are commonly called "super ferritics" or "superferritics."

In the early 1970s, alloy E-BRITE® (S44625), also called alloys 26-1 and XM-27, containing nominal 26% Cr and 1% Mo, was introduced by Airco Vacuum Metals, then a division of Airco based in New Jersey. This alloy was developed primarily as a replacement for type 300 austenitic stainless steels in applications requiring chloride stress corrosion cracking (Cl⁻SCC) resistance.

Type 446 (S44600), the 26% ferritic stainless steel, has good Cl⁻SCC characteristics but poor toughness. Alloy 26-1 was an effort to improve the properties of type 446, 1% molybdenum being added for better pitting resistance.

The generally good toughness of alloy E-BRITE results from its extra-low carbon and nitrogen interstitial levels, less than about 200 ppm, achieved by a new melting technology, electron-beam continuous-hearth refining in a high vacuum. However, even this high-tech process did not give low enough concentrations of nitrogen to avoid intergranular attack (IGA) of weldments, and several years later columbium was added in concentrations from 13 to 29 times nitrogen content – with S44625 still retained as the UNS designation.

In 1977, Allegheny Ludlum, now an Allegheny Teledyne Company, acquired the E-BRITE trademark from Airco and started producing alloy 26% Cr-1% Mo alloy by vacuum induction melting. This melting process required additions of stabilizing elements to tie up carbon and nitrogen. A new grade was introduced, designated S44627, yet still called alloys 26-1 and XM-27, having columbium addition of 0.05 to 0.20%, with carbon and nitrogen maximum contents of 0.10% and 0.015%, respectively. This was followed by another stabilized grade, designated S44426 and called alloys 26-1S and XM-33, having a minimum titanium addition of seven times the C+N, with carbon and nitrogen maximum contents of 0.10% and 0.015%, respectively.

In 1966, DuPont Company's Michael A. Streicher initiated studies on Fe-Cr-Mo-Ni ferritic alloys and developed a high purity 29% Cr and 4% Mo alloy (S44700). Patent applications were filed in 1970. The alloy manufactured today by Allegheny Ludlum with Cb+Ti added to stabilize C+N is designated alloy AL-29-4C® (S44735).

Streicher also invented high purity alloy 29-4-2 (S44800) containing nominal 2% nickel. Allegheny Ludlum modified the composition by adding Cb+Ti, developing stabilized alloy AL-29-4-2C® (S44736).

In the 1970s, Trent Tube Division of Colt Industries (now Trent Tube Division of Crucible Materials Corp.), East Troy, Wisconsin introduced alloy Sea-Cure® (S44660) containing nominal 26% Cr, 3% Mo, and 2.5% Ni stabilized with Cb+Ti.

Also in the 1970s, the Swedish stainless steelmaker Granges Nyby, later Nyby Uddeholm and now incorporated into Avesta Sheffield, sought markets for titanium-stabilized alloy 18-2, a ferritic stainless steel with nominal 18% Cr, 2% Mo, and 0% Ni. An important application was cooling water systems using river waters with high chloride content. In these applications, types 304 and 316 austenitic stainless steels failed by chloride stress corrosion cracking (Cl⁻SCC). The economic incentive was alloy 18-2 priced comparable to type 304. Duplex alloy 2205 is now widely used in Europe for many services requiring resistance to Cl⁻SCC.

Nyby Uddeholm also introduced alloy MONIT®, a titanium-stabilized extra-low interstitial ferritic with nominal 25% Cr, 4% Mo, and 4% Ni in the 1970s. In 1971, Deutsche Edelstahlwerke, later Thyssen Edelstahlwerke (TEW) and now Krupp Thyssen Nirosta, in Krefeld, Germany, announced SUPERFERRIT®, a columbium-stabilized alloy of nominal 28% Cr, 4% Ni, and 2% Mo composition.

Cr-Fe-Ni Alloy

Alloy 33, recently introduced by Krupp VDM, is included in these historical notes because it is unique among the corrosion-resistance austenitic materials. It is a *chromium*-based alloy with nominal composition of 33% Cr, 32% Fe, 31% Ni, 1.6% Mo and 0.4% N. The UNS designation is R20033.

Alloy 33 resulted from a cooperative program between Krupp VDM and Bayer in Germany, starting in 1992, to develop a metallic material with the highest possible corrosion resistance to oxidizing media such as highly concentrated sulfuric acid. Chromium additions to stainless steels increase their resistance in oxidizing acids. Commercially available stainless steels and nickel-based alloys usually have a limit of about 29% Cr. Alloy 33 has 33% Cr.

It was important for the alloy to have an austenitic microstructure and thereby good mechanical and fabrication properties. This was achieved by the substantial addition of nitrogen which stabilizes the austenitic structure.

The Future?

These historical notes highlight a large number–but certainly not all (e.g., titanium, zirconium, refractory, wear resistant, and high temperature alloy are not mentioned)–of the major corrosion-resistant materials advances in this century, the *Corrosion-Resistant Alloy Age*. Looking to the 21st century, this era might earn the name of *Technological Corrosion-Resistant Alloy Age*. For specific and limited niche markets, new materials of construction will be required with still greater corrosion, oxidation, and carburization resistance.

Alloys with higher mechanical strength and improved integrity at high temperatures also will be needed. Produced in relatively small quantities, these materials most likely will find uses in specialized process, for example, manufacturing high-value-end products. Such alloy advancements will no doubt include dispersion strengthening, namely mechanical alloying, and super clean and metallographically controlled microstructures. And with these alloy developments will come challenges in processing, welding and fabrication technologies.

Sources

1.   *Harry Brearley: Stainless Pioneer.* Sheffield, U.K.: British Stainless Steel (now Avesta Sheffield), 1989. p. 105.

2.   *Materials Science and Engineering for the 1990s.* Washington, D.C.: National Academy Press, 1989.

3.   "Stainless Steel: The Inventor, Harry Brearley, and His Invention." *Materials Performance* 20, 3 (1990): pp. 64-68.

4.   "Celebrating Ninety Years of the 'Monel' Alloys." Huntington, WV: Inco Alloys International, 1995.

5.   "It took a century of experience to develop 'Hastelloy' 'C-2000' alloy...." Kokomo, IN: Haynes International, 1996.

6.   Agarwal, D. C., W. R. Herda, "The 'C' family of Ni-Cr-Mo alloys' partnership with the Chemical Process Industry; the last 70 years," Materials and Corrosion 48 (1997): pp. 542-548.

7.   Agarwal, D. C., U. Heubner, R. Kirchheiner, M. Koehler. "Cost-Effective Solutions to CPI Corrosion Problems with a New Ni-Cr-Mo Alloy." Corrosion/91, paper no. 179. Houston, TX: NACE International, 1991.

8.   Agarwal, D. C. Private communication.

9.   Amato, I., Stuff: the materials the world is made of (New York: Basic Books/HarperCollins, 1997).

10.   Bates, J. M. Private communication.

11. Brady, G. S., H. R. Clauser, J. A. Vaccari. *Materials Handbook*. Fourteenth Edition. New York: McGraw-Hill, 1997. p. 841.

12. Charles, J. "Duplex stainless steels: from the theory to the practice." Presented at Process and Materials Innovation Stainless Steels, Florence, Italy, October 1993.

13. Coates, G. Private communication.

14. Colombier, L., *Molybdenum in Stainless Steels and Alloys*. London, U.K.: Climax Molybdenum Company Ltd., no date. p. 9.

15. Crum, J. R. Private communication.

16. Davidson, R. M., H. E. Deverall, J. D. Redmond. "Ferritic and Duplex Stainless Steels." In *Process Industries Corrosion – The Theory and Practice*. eds. B. J. Moniz, W. I. Pollock. Houston, TX: NACE International, 1986. pp. 427-443.

17. Davison, R. M., J. D. Redmond. "Practical Guide to Using 6 Mo Austenitic Stainless Steels." *Materials Performance* 27, 12 (1988): pp. 39-43.

18. Davison, R. M., J. D. Redmond. "Practical Guide to Using Duplex Stainless Steels." *Materials Performance* 29, 1 (1990): pp. 57-62.

19. Davison, R. M. Private communication.

20. Degnan, T. F. Private communication.

21. Dillon, C. P., *Corrosion Control in the Chemical Process Industries*. Second Edition. St. Louis, MO: Materials Technology Institute of the Chemical Process Industries, 1994.

22. Dillon, C. P. Private communication.

23.   Dupoiron, F., B. Bonnefois, M. Verneau, J. Charles. "Superduplex Stainless Steels for Offshore Applications." In *Duplex Stainless Steel '94*. Proceedings of 1994 Conference, Glasgow, Scotland. Abington, Cambridge, UK: TWI, 1994.

24.   Eiselstein, H. Private Communication.

25.   Elliott, D., S. M. Tupholme. *An Introduction to Steel Selection: Part 2, Stainless Steels*. London, UK: Design Council, British Standards Institution and Council of Engineering Institutions, 1981.

26.   Fontana, M. A. *Corrosion Engineering*. Third Edition. New York: McGraw-Hill, 1986.

27.   Franson, I.A. Private communication.

28.   Garvin, R. C. Private communication.

29.   Gemmel, G., S. Nordin. "MONIT and 904L – Two High Alloy Stainless Steels for Seawater Applications." In *Advanced Stainless Steels for Seawater Applications*. Proceedings of 1980 Symposium, Piacenze, Italy. Greenwich, CN: Climax Molybdenum Company, 1981. pp. 69-80.

30.   Gray, R. D. *Stellite: A History of The Haynes Stellite Company, 1912-1972*. Kokomo, IN: Cabot Corporation (now Haynes International), 1974.

31.   Grubb, J.F. Private communication.

32.   Heubner, U. *Nickelwerkstoffe und hochlegierte Sonderedelstahle*. Ehningen bei Boblingen, Germany: expert verlag, 1985.

33.   Hodge, F. G. Private communication.

34.   Kohler, M., U. Heubner, K.-W. Eichenhofer, M. Renner. "Alloy 33." *Stainless Steel World* 11, 4 (May 1999): pp. 38-49.

35.   Lamb, S. Private communication.

36.   Liljas, M. "The welding of duplex stainless steels." In *Duplex Stainless Steels '94*. Proceedings of 1994 Conference, Glasgow, Scotland. Abington, Cambridge, UK: TWI, 1994.

37.   Mankins, W. L., S. Lamb. "Nickel and Nickel Alloys." In *ASM Handbook*. Volume 2, *Properties and Selection: Nonferrous Alloys and Special-Purpose Materials*. Materials Park, OH: ASM International, 1991.

38.   Manning, P. E., J. D. Smith, J. L. Nickerson. "New Versatile Alloys for the Chemical Industry." *Materials Performance* 27, 6 (1988): pp. 67-73.

39.   Maurer, J. R. "Development and Application of High Technology Stainless Steels for Marine Exposures," In *Advanced Stainless Steels for Seawater Applications*. Proceedings of 1980 Symposium, Piacenza, Italy. Greenwich, CN: Climax Molybdenum Company, 1981. pp. 11-29.

40.   Olsson, J, M. Liljas, "60 Years of Duplex Stainless Steel Applications." *Acom (Avesta Corrosion Management)* 2, 1996.

41.   Portisch, H. "Corrosion fighters." *Krupp VDM Metal Times* 10 (1993): pp. 26-29.

42.   Redmond, R. D. Private communication.

43.   Streicher, M. A. "Stainless Steels: Past, Present and Future." In *Stainless Steel '77*. Proceedings of 1977 Conference, London, England. Greenwich, CN: Climax Molybdenum Company, 1978.

44.   Streicher, M. A.. "The Discovery of Austenitic Stainless Steels." *Materials Performance* 20, 4 (1990): p. 64.

45.   Streicher, M. A. Private communication.

46.   Visser, J. N. "The 'Hastelloy' family." *Stainless Steel Europe* October 1992: pp. 54-57.

47.   Zapffe, C. A. *Stainless Steels*. Materials Park, OH: ASM (now ASM International), 1949.

Chapter

2

CAST CORROSION- AND HEAT-RESISTANT ALLOYS

Malcom Blair
Steel Founders' Society of America, Des Plaines, Illinois

Roman Pankiw
Duraloy Technologies, Inc., Scottdale, Pensylvania

Overview

The broad category of *corrosion resistant alloys* consists of high alloy steels and nickel-base alloys. Subcommittee A01.18 of the American Society for Testing and Materials (ASTM) has responsibility for developing the standards for this broad category, unlike the wrought products where different subcommittees develop standards for ferrous and non-ferrous alloys. The primary alloying elements in the high alloys are chromium, nickel, iron, and carbon. The high alloy group is usually divided into two broad groups: corrosion-resistant and heat-resistant. The heat resistant grades are generally used in conditions where oxidation and carburization resistance are important as well as high temperature strength. In steel castings the difference between corrosion-resistant and heat-resistant grades is reflected in the differences in the carbon levels of these types.

Because of the dominance of wrought products, in tonnage terms, cast versions of wrought grades are commonly available. Tables 2.1a, 2.1b, and 2.1c list the various corrosion and heat resistant cast grades, including their ACI and UNS designations and chemical compositions. It should be recognized that because wrought and cast products are manufactured by different processes there are compositional requirements which aid the processing and manufacture of these products. Consequently, since the wrought materials require a capability for hot or cold deformation and castings require greater fluidity, then the compositions are often different and each product may also have some microstructural differences. For example, cast materials may contain higher levels of silicon to aid fluidity, and the microstructures of the austenitic grades may contain ferrite. The mechanical properties of cast and wrought materials also differ. There are the effects of larger grains in the cast materials and the directionality effects in wrought materials.

Stainless steel castings form part of the category of alloys known as *high alloy steels* within the corrosion-resistant alloys. Stainless steels are characterized by their resistance to corrosion in aqueous and elevated temperature environments. They are distinguished from other steel grades by their higher chromium levels which impart oxidation resistance or passivity when present in amounts greater than 12%. These alloys are commonly termed *stainless steels*.

Table 2.1a ACI Alloy Designations and Chemical Composition Ranges for Corrosion-Resistant Castings

CHEMICAL COMPOSITION OF CORROSION-RESISTANT CAST STAINLESS STEELS[a]

Grade	UNS No.	ASTM Specification	%C	%Mn	%Si	%Cr	%Ni	Others (%)	Similar[b] Wrought
CA-15	J91540	A 217, A 487, A 743	0.15	1.00	1.50	11.5-14.0	1.0	0.50 Mo	410
CA-15M	J91151	A 487, A743	0.15	1.00	0.65	11.5-14.0	1.0	0.15-1.00 Mo	410Mo
CA-28MWV	---	A 743	0.20-0.28	0.50-1.00	1.00	11.0-12.5	0.50-1.00	0.9-1.25 Mo, 0.9-1.25 W, 0.2-0.3 V	422
CA-40	J91153	A 743	0.20-0.40	1.00	1.50	11.5-14.0	1.0	0.5 Mo	420
CA-40F	---	A 743	0.20-0.40	1.00	1.50	11.5-14.0	1.0	0.5 Mo, 0.20-0.40 S	---
CA-6N	---	A 743	0.06	0.50	1.00	10.5-12.5	6.0-8.0	---	IN833
CA-6NM	J91540	A 352, A 356, A 487, A 743	0.06	1.00	1.00	11.5-14.0	3.5-4.5	0.4-1.0 Mo	F6NM
CB-6	J91804	A 743	0.06	1.00	1.00	15.5-17.5	3.5-5.5	0.5 Mo	---
CB-7Cu-1	J92180	A 747	0.07	0.70	1.00	15.50-17.70	3.60-4.60	2.50-3.20 Cu, 0.15-0.35 Cb, 0.05 N	17-4 PH
CB-7Cu-2	J92110	A 747	0.07	0.70	1.00	14.0-15.50	4.50-5.50	2.50-3.20 Cu, 0.20-0.35 Cb, 0.05 N	15-5 PH
CB-30	J91803	A 743	0.30	1.50	1.50	18.0-21.0	2.0	0.90-1.20 Cu optional	431, 442
CC-50	J92615	A 743	0.50	1.00	1.50	26.0-30.0	4.0	---	446
CD-3MWCuN	J93380	A 351, A 890 Gr 6A	0.03	1.00	1.00	24.0-26.0	6.5-8.5	3.0-4.0 Mo, 0.5-1.0 Cu, 0.5-1.0 W, 0.20-0.30 N	Zeron 100
CD-4MCu	J93370	A 351, A 744, A 890 Gr 1A	0.04	1.00	1.00	24.5-26.5	4.75-6.00	1.75-2.25 Mo, 2.75-3.25 Cu	---
CE-8N	J92805	---	0.08	1.00	1.50	23.0-26.0	8.0-11.0	0.50, 0.20-0.30 N	---
CE-8MN	J93345	A 890 Gr 2A	0.08	1.00	1.50	22.5-25.5	8.00-11.00	3.00-4.50 Mo, 0.10-0.30 N	---
CE-30	J93423	A 743	0.30	1.50	2.00	26.0-30.0	8.0-11.0	---	312
CF-3	J92500	A 351, A 743, A 744	0.03	1.50	2.00	17.0-21.0	8.0-12.0	---	304L
CF-3M	J92800	A 351, A 743, A 744	0.03	1.50	1.50	17.0-21.0	9.0-13.0	2.0-3.0 Mo	316L
CF-3MN	J92700	A 743	0.03	1.50	1.50	17.0-22.0	9.0-13.0	2.0-3.0 Mo, 0.10-0.20 N	---
CF-8	J92600	A 351, A 743, A 744	0.08	1.50	2.00	18.0-21.0	8.0-11.0	---	304
CF-8C	J92710	A 351, A 743, A 744	0.08	1.50	2.00	18.0-21.0	9.0-12.0	(8 x C) Cb min to 1.00	347

Table 2.1a ACI Alloy Designations and Chemical Composition Ranges for Corrosion-Resistant Castings (continued)

CHEMICAL COMPOSITION OF CORROSION-RESISTANT CAST STAINLESS STEELS[a] (continued)

Grade	UNS No.	ASTM Specification	%C	%Mn	%Si	%Cr	%Ni	Others (%)	Similar[b] Wrought
CF-8M	J92900	A 351, A 743, A 744	0.08	1.50	2.00	18.0-21.0	9.0-12.0	2.0-3.0 Mo	316
CF-10	J92590	A 351	0.04-0.10		2.00	18.0-21.0	8.0-11.0	0.50 Mo	---
CF-10M	J92901	A 351	0.04-0.10		1.50	18.0-21.0	9.0-12.0	2.0-3.0 Mo	---
CF-10MC	J92971	A 351	0.10		1.50	15.0-18.0	13.0-16.0	1.75-2.25 Mo, (10 x C) Cb min to 1.20	---
CF10SMnN	J92972	A 351, A 743	0.10	7.00-9.00	3.50-4.50	16.0-18.0	8.0-9.0	7.00-9.00 Mn, 0.08-0.18 N	Nitronic 60
CF-16F	J92701	A 743	0.16	1.50	2.00	18.0-21.0	9.0-12.0	1.50 Mo, 0.20-0.35 Se	303Se
CF-16Fa	---	A 743	0.16		2.00	18.0-21.0	9.0-12.0	0.40-0.80 Mo, 0.20-0.40 S	
CF-20	J92602	A 743	0.20		2.00	18.0-21.0	8.0-11.0	---	302
CG-3M	---	A 351, A 743, A 744	0.03		1.50	18.0-21.0	9.0-13.0	3.0-4.0 Mo	317L
CG-6MMN	J93790	A 743	0.06	4.00-6.00	1.00	20.5-23.5	11.5-13.5	1.50-3.00 Mo, 4.00-.00Mn, 0.10-0.30 Cb, 0.10-0.30 V, 0.20-0.40 N	Nitronic 50
CG-8M	---	A 351, A 743, A 744	0.08	1.50	1.50	18.0-21.0	9.0-13.0	3.0-4.0 Mo	317
CG-12	J93001	A 743	0.12	1.50	2.00	20.0-23.0	10.0-13.0	---	---
CH-8	J93400	A 351	0.08		1.50	22.0-26.0	12.0-15.0	0.50 Mo	---
CH-10	J93401	A 351	0.04-0.10		2.00	22.0-26.0	12.0-15.0	0.50 Mo	---
CH-20	J93402	A 351, A 743	0.04-0.20	1.50	2.00	22.0-26.0	12.0-15.0	0.50 Mo	309
CK-3MCuN	J93254	A 351, A 743, A 744	0.025	1.20	1.00	19.5-20.5	17.5-19.5	6.0-7.0 Mo, 0.50-1.00 Cu, 0.180-0.240 N	2545MO
CK-20	J94202	A 743	0.20	2.00	2.00	23.0-27.0	19.0-22.0	---	310
CN-3M	J94652	A 743	0.03		1.00	20.0-22.0	23.0-27.0	4.5-5.5 Mo	IN862
CN-3MN	J94651	A 743, A 744	0.03	2.00	1.00	20.0-22.0	23.5-25.5	6.00-7.00 Mo, 0.75 Cu, 0.18-0.26 N	AL-6XN
CN-7M	N08007	A 351, A 743, A 744	0.07	1.50	1.50	19.0-22.0	27.5-30.5	2.0-3.0 Mo, 3.0-4.0 Cu	Alloy 20
CE3MN	J93404	A 890 Gr 5A	0.03		1.00	24.0-26.0	6.0-8.0	4.0-5.0 Mo, 0.10-0.30 N	2507

Table 2.1a ACI Alloy Designations and Chemical Composition Ranges for Corrosion-Resistant Castings (continued)

CHEMICAL COMPOSITION OF CORROSION-RESISTANT CAST STAINLESS STEELS[a] (continued)								
Grade	UNS No.	ASTM Specification	%C	%Mn	%Si	%Cr	%Ni	Others (%)	Similar[b] Wrought
CN-7MS	J94650	A 743, A744	0.07	1.00	2.50-3.50	18.0-20.0	22.0-25.0	2.5-3.0 Mo, 1.5-2.0 Cu	---
CD6MN	---	A 890 Gr 3A	0.06		1.00	24.0-27.0	4.00-6.00	1.75-2.50 Mo, 0.15-0.25 N	---
CD3MN	J92205	A 890 Gr 4A	0.03		1.00	21.0-23.5	4.5-6.5	2.5-3.5Mo, 1.00 Cu, 0.10-0.30N	2205
CD4MCuN	J93372	A890 Gr 1B	0.04	1.00	1.00	24.5-26.5	4.7-6.0	1.7-2.3 Mo, 2.7-3.3 Cu, 0.10-.25N	---

a. Although manganese, sulfur and phosphorous contents are not listed in this table due to limited space, they are specified; see appropriate ASTM standard for more details.

b. Similar wrought designations are listed only as a guide for comparison to cast grades; they are not equivalent.

All values are maximums, unless otherwise specified.

Most of the standard grades listed are covered by the ASTM specifications A 217, A 351, A 447, A 451, A 452, A 494, A 608, A 743, A 744, A 757 and, A 890.

Table 2.1b ACI Alloy Designations and Chemical Composition Ranges for Heat-Resistant Castings

CHEMICAL COMPOSITION OF HEAT-RESISTANT CAST STAINLESS STEELS[a]

Grade	UNS No.	ASTM Specification	%C	%Mn	%P	%S	%Si	%Cr	%Ni
HC	J92605	A 297	0.50	1.00	0.04	0.04	2.00	26.0-30.0	4.00
HC30	J92613	A 608	0.25-0.35	0.5-1.0	0.04	0.04	0.50-2.00	26-30	4.0
HD	J93005	A 297	0.50	1.50	0.04	0.04	2.00	26.-30.0	4.0-7.0
HD50	J93015	A 608	0.45-0.55	1.50	0.04	0.04	0.50-2.00	26-30	4-7
HE	J93403	A 297	0.20-0.50	2.00	0.04	0.04	2.00	26.0-30.0	8.0-11.0
HE35	J93413	A 608	0.30-0.40	1.50	0.04	0.04	0.50-2.00	26-30	8-11
HF	J92603	A 297	0.20-0.40	2.00	0.04	0.04	2.00	18.0-23.0	8.0-12.0
HF30	J92803	A 608	0.25-0.35	1.50	0.04	0.04	0.50-2.00	19-23	9-12
HH	J93503	A 297, A 447	0.20-0.50	2.00	0.04	0.04	2.00	24.0-28.0	11.0-14.0
HH30	J93513	A 608	0.25-0.35	1.50	0.04	0.04	0.50-2.00	24-28	11-14
HH33	J93633	A 608	0.28-0.38	1.50	0.04	0.04	0.50-2.00	24-26	12-14
HI	J94003	A 297	0.20-0.50	2.00	0.04	0.04	2.00	26.0-30.0	14.0-18.0
HI35	J94013	A 608	0.30-0.40	1.50	0.04	0.04	0.50-2.00	26-30	14-18
HK	J94224	A 297	0.20-0.60	2.00	0.04	0.04	2.00	24.0-28.0	18.0-22.0
HK30	J94203	A 351, A 608	0.25-0.35	1.50	0.04	0.04	1.75	23.0-27.0	19.0-22.0
HK40	J94204	A 351, A 608	0.35-0.45	1.50	0.04	0.04	1.75	23.0-27.0	19.0-22.0
HL	N08604	A 297	0.20-0.60	2.00	0.04	0.04	2.00	28.0-32.0	18.0-22.0
HL30	N08613	A 608	0.25-0.35	1.50	0.04	0.04	0.50-2.00	28-32	18-22
HL40	N08614	A 608	0.35-0.45	1.50	0.04	0.04	0.50-2.00	28-32	18-22
HN	J94213	A 297	0.20-0.50	2.00	0.04	0.04	2.00	19.0-23.0	23.0-27.0
HN40	J94214	A 608	0.35-0.45	1.50	0.04	0.04	0.50-2.00	19-23	23-27

Table 2.1b ACI Alloy Designations and Chemical Composition Ranges for Heat-Resistant Castings (continued)

CHEMICAL COMPOSITION OF HEAT-RESISTANT CAST STAINLESS STEELS[a] (continued)

Grade	UNS No.	ASTM Specification	%C	%Mn	%P	%S	%Si	%Cr	%Ni
HP	N08705	A 297	0.35-0.75	2.00	0.04	0.04	2.00	24-28	33-37
HT	N08605	A 297	0.35-0.75	2.00	0.04	0.04	2.50	15.0-19.0	33.0-37.0
HT30	N08603	A 351	0.25-0.35	2.00	0.04	0.04	2.50	13.0-17.0	33.0-37.0
HT50	N08050	A 608	0.40-0.60	1.50	0.04	0.04	0.50-2.00	15-19	33-37
HU	N08004	A 297	0.35-0.75	2.00	0.04	0.04	2.50	17.0-21.0	37.0-41.0
HU50	N08005	A 608	0.40-0.60	1.50	0.04	0.04	0.50-2.00	17-21	37-41
HW	N08001	A 297	0.35-0.75	2.00	0.04	0.04	2.50	10.0-14.0	58.0-62.0
HW50	N08006	A 608	0.40-0.60	1.50	0.04	0.04	0.50-2.00	10-14	58-62
HX	N06006	A 297	0.35-0.75	2.00	0.04	0.04	2.50	15.0-19.0	64.0-68.0
HX50	N06050	A 608	0.40-0.60	1.50	0.04	0.04	0.50-2.00	15-19	64-68

a. Molybdenum is for all alloys 0.5% maximum.
All values are maximums, unless otherwise specified.

Table 2.1c ASTM A 494 Chemical Composition Ranges for Heat-Resistant Castings

ASTM A 494 - CHEMICAL COMPOSITION OF NICKEL & NICKEL ALLOY CASTINGS		
Grade	UNS	Chemical Composition (Weight %)
CZ-100	N02100	Ni remainder C 1.00 max. Cu 1.25 max. Fe 3.00 max. Mn 1.50 max. Si 2.00 max.
M-35-1	N24135	Ni remainder C 0.35 max. Cu 26.0-33.0 Fe 3.50 max. Mn 1.50 max. P 0.03 max. S 0.03 max. Si 1.25 max.
M-35-2	---	Ni remainder C 0.35 max. Cb 0.5 max. Cu 26.0-33.0 Fe 3.50 max. Mn 1.50 max. P 0.03 max. S 0.03 max. Si 2.00 max.
M-30H	N24030	Ni remainder C 0.30 max. Cu 27.0-33.0 Fe 3.50 max. Mn 1.50 max. P 0.03 max. S 0.03 max. Si 2.7-3.7
M-25S[a]	N24025	Ni remainder C 0.25 max. Cu 27.0-33.0 Fe 3.50 max. Mn 1.50 max. P 0.03 max. S 0.03 max. Si 3.5-4.5
M-30C	N24130	Ni remainder C 0.30 max. Cb 1.0-3.0 Cu 26.0-33.0 Fe 3.50 max. Mn 1.50 max. P 0.03 max. Si 1.0-2.0
N-12MV[a]	N30012	Ni remainder C 0.12 max. Cr 1.00 max. Fe 4.0-6.0 Mn 1.00 max. Mo 26.0-30.0 P 0.040 max. S 0.030 max. Si 1.00 max. V 0.20-0.60
N-7M[a]	N30007	Ni remainder C 0.07 max. Cr 1.0 max. Fe 3.00 max. Mn 1.00 max. Mo 30.0-33.0 P 0.040 max. S 0.030 max. Si 1.00 max.
CY-40[a]	N06040	Ni remainder C 0.40 max. Cr 14.0-17.0 Fe 11.0 max. Mn 1.50 max. Si 3.00 max.
CW-12MW[a]	N30002	Ni remainder C 0.12 max. Cr 15.5-17.5 Fe 4.5-7.5 Mn 1.00 max. Mo 16.0-18.0 P 0.040 max. S 0.030 max. Si 1.00 max. V 0.20-0.40 W 3.75-5.25
CW-6M[a]	N30107	Ni remainder C 0.07 max. Cr 17.0-20.0 Fe 3.0 max. Mn 1.00 max. Mo 17.0-20.0 P 0.040 max. S 0.030 max. Si 1.00 max.
CW-2M[a]	N26455	Ni remainder C 0.02 max. Cr 15.0-17.5 Fe 2.0 max. Mn 1.00 max. Mo 15.0-17.5 P 0.03 max. S 0.03 max. Si 0.80 max. W 1.0 max.
CW-6MC	N26625	Ni remainder C 0.06 max. Cb 3.15-4.50 Cr 20.0-23.0 Fe 5.0 max. Mn 1.00 max. Mo 8.0-10.0 P 0.015 max. S 0.015 max.
CY5SnBiM	N26055	Ni remainder Bi 3.0-5.0 C 0.05 max. Cr 11.0-14.0 Fe 2.0 max. Mn 1.5 max. Mo 2.0-3.5 P 0.03 max. S 0.03 max. Si 0.5 max. Sn 3.0-5.0
CX2MW[a]	N26022	Ni remainder C 0.02 max. Cr 20.0-22.5 Fe 2.0-6.0 Mn 1.00 max. Mo 12.5-14.5 P 0.025 max. S 0.025 max. Si 0.80 max. V 0.35 max. W 2.5-3.5

a. See ASTM A 494 for details regarding class designations and heat treat requirements.

Description of Alloy Designation System (ACI)

Cast high alloy steels and nickel-based alloys generally use grade designations that are different from their wrought counterparts. This system can make recognition or identification of the cast grades more difficult. Cast stainless steels and nickel-base alloys are split into two categories–corrosion-resistant and heat-resistant–and, as mentioned earlier, the principal difference between these two types is reflected in the carbon levels and how the carbon level is referenced in the alloy designation. The system used is known as the *ACI system* and was developed some years ago by the Alloy Casting Institute which was merged into the Steel Founders' Society of America (SFSA). The ACI system is, in principle, very simple and gives some indication of the principal alloying elements contained in the material. For example, the alloy designations for two common grades, CF3M and HK40, are:

> CF3M   C = corrosion resistant
> F = nickel content
> 3 = 0.03% maximum carbon content
> M = molybdenum is a principal alloying element

> HK40   H = heat resistant
> K = nickel content
> 40 = the mid range of the carbon content

The ACI designations for all corrosion-resistant stainless steel grades and some nickel-base alloy grades have the prefix "C," while all heat-resistant grades have the prefix "H." Immediately following the "C" or "H" designation is a series of numbers and letters.

The letter after the "C" ("F" in CF3M) indicates the nickel content, the letters in use ranging from A to Z, although each letter of the alphabet is not used. It is important to note that the letter after the "C," in this case "F," is not assigned a particular value, but is merely an indicator of relative nickel content; for example, the letter "A" indicates that there is less nickel than an alloy with the letter "F," and the letter "X" indicates that the grade contains more nickel than an alloy with a letter "F." The relationship between the chromium and nickel levels with respect to the second letter designation is shown in figure below.

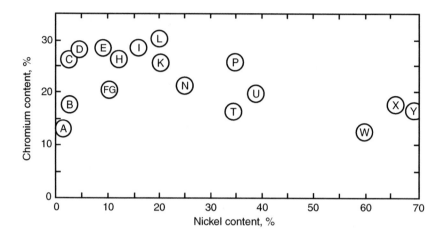

Figure 2.1 Ranges of chromium and nickel in standard grades of heat- and corrosion-resistant castings.[5]

The number after the second letter indicates the carbon content. In corrosion-resistant "C" materials, this indicates the maximum carbon limit in one-hundredths of a percent. For heat-resistant "H" materials, this number indicates the middle of the carbon range in one-hundredths of a percent, and it is not uncommon for this range to be ±0.05%.

Letters following the number indicate the principal alloying elements which will have a range assigned to them in the alloy specification.

Examining the composition of grade CF3M (J92800) shows that it is similar to the wrought type 316L (S31603); however, the chromium and nickel levels in the cast and wrought specifications are different as shown in Table 2.2.

Table 2.2 Comparison of Grade CF3M and Type 316 Compositions

	ASTM A 351 Gr. CF3M (UNS J92800)	Type 316 (S31603)
%C	0.03 max.	0.03 max.
%Mn	1.5 max.	2.00 max.
%Si	2.00 max.	1.00 max.
%Cr	17.0-21.0	16.0-18.0
%Ni	8.0-12.0	10.0-14.0
%Mo	2.0-3.0	2.0-3.0

This difference in composition also brings about differences in the microstructure. It is common for the cast grades to contain ferrite, unlike the fully austenitic wrought grades. The effect of ferrite is (1) it makes the material magnetic, (2) it increases the strength, (3) it improves the weldability and also increases the resistance to stress corrosion cracking. Further information on the effect of ferrite is detailed earlier in this chapter.

HK40 is similar to S31000 (type 310) as shown in Table 2.3.

Table 2.3 Comparison of HK40 and Type 310 Compositions

	ASTM A 351 Gr. HK40 (UNS J94204)	Type 310 (S31000)
%C	0.35-0.45	0.25 max.
%Mn	1.50 max.	2.00 max.
%Si	1.75 max.	1.50 max.
%Cr	23.0-27.0	24.0-26.0
%Ni	19.0-22.0	19.0-22.0

Metallurgy

To understand the nature of these materials it is necessary to review some of the basic metallurgy of iron. In particular, it is important to understand the differences between austenitic and ferritic phases and the reasons for the differences in mechanical properties of these phases. On cooling from the melting point or casting temperatures (Figure 2.2), these materials initially form delta ferrite (δ), which has a body centered cubic (bcc) structure (Figure 2.3), where atoms are located at the corners of the unit cube and one is located at the center of the cube. As the structure continues to cool, it transforms to austenite. The crystal structure of austenite (γ) is a face centered cubic (fcc) (Figure 2.4), where atoms are positioned at the corners of the unit cube and also in the center of each face of the unit cube. With further cooling, the structure then changes to ferrite, a bcc (α) structure, again.

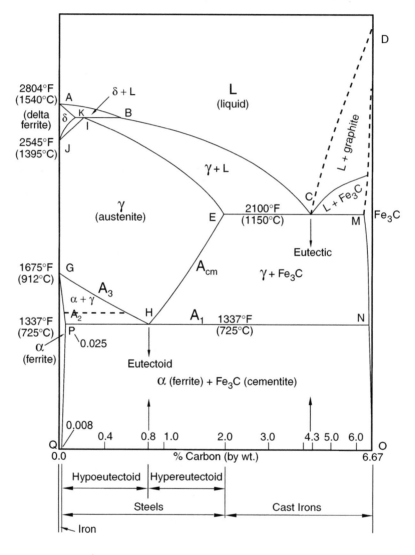

Figure 2.2 Fe-F$_3$C phase diagram.

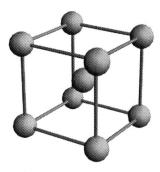

Figure 2.3 Body centered cubic structure, bcc α.

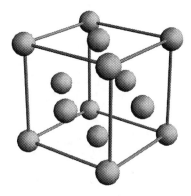

Figure 2.4 Face centered cubic structure, fcc γ.

The transformation from delta ferrite (δ) to austenite (γ) to ferrite (α) and vice versa is affected by the composition and cooling rate of the alloy involved. The most notable effect is the γ to α transformation which may be prevented from happening when the cooling rate and composition of the alloy may cause a transformation to martensite, γ′. This transformation is caused by a shearing action of the lattice where the fcc structure transforms to a body-centered tetragonal structure (bct) (Figure 2.5). The transformation is known as the martensitic transformation, and because it occurs by a shearing action, a highly stressed structure is produced which has a very high hardness and poor ductility. This martensitic transformation is one of the most important transformations in the metallurgy of steels and is largely responsible for the wide range of strength levels achieved in carbon and low alloy steels. It is also extremely important in high alloys and is responsible for the development of the martensitic and precipitation hardening grades.

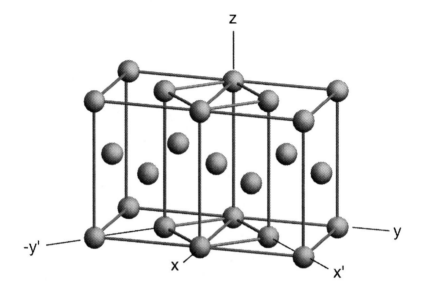

Figure 2.5 Martensite–body centered tetragonal (bct γ′).

In addition to the strengthening effects of the martensitic reaction, there are differences in the strengths of the α, γ, and δ phases and, as a consequence, their presence and their proportions can have a significant effect on the strength levels of the alloy produced. These different strength levels can be largely rationalized by considering the effect of subjecting their structures to loading. When a load is applied to a fcc structure it can deform when one or more planes slide over each other. In the fcc structure there are many more slip planes than in the bcc structure; consequently, it is more difficult to deform the bcc structure giving it higher strength.

Another effect, which has to be considered when producing castings, is the effect of alloy type on the solidification shrinkage which takes place. When alloys solidify, they undergo a number of shrinkage stages (Figure 2.6). The first is the contraction of the liquid due to cooling. The second is the volumetric change associated with solidification. And the third is the shrinkage occurring during the cooling of the solid. The volumetric change is the phenomenon which most strongly affects the dimensions to which the pattern will be built. The volumetric change that occurs is approximately 6% in steels. The prediction of the volumetric shrinkage is not straightforward and depends on the geometry of the part and the material being cast. As a general rule, ferritic and martensitic materials use a shrinkage allowance of ¼ inch per foot, while the austenitic grades use a larger allowance of ⁵⁄₁₆ inch to ⅜ inch per foot. This difference can be explained by the fact that the bcc structure is less dense than the fcc structure. The use of patterns manufactured for the production of carbon and low alloy steel castings will give significant dimensional problems if they are used for high alloy castings without any modification.

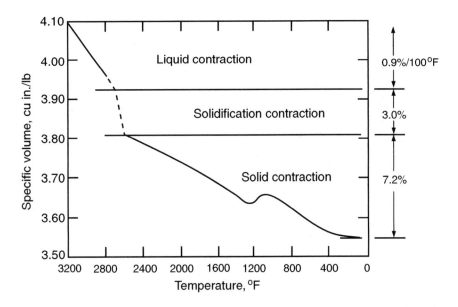

Figure 2.6 The change in specific volume of
solidifying and cooling steel.

Ferrite

Because of its importance, it is worthwhile spending some time discussing the effects of ferrite on the performance and processing of corrosion-resistant alloys.

Ferrite affects strength, weldability, corrosion resistance, and toughness. Ferrite is not unique to cast materials. It also occurs in wrought and weld metals, and therefore the comments made here are relevant to all product forms. Ferrite can be both beneficial and detrimental. Ferrite, both δ and α, have largely similar effects, but their origin is different. However, for the purposes of this discussion, they will be treated as the same under the generic term *ferrite*.

Cast austenitic stainless steels can have from 5% to 20% ferrite present by volume. In the austenitic materials, the ferrite is discontinuous and appears as pools in an austenite matrix (Figure 2.7).

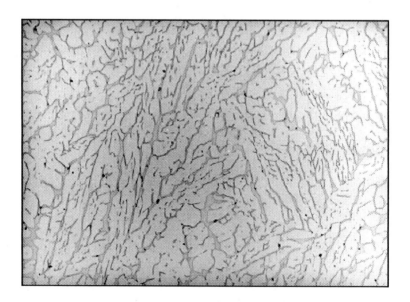

Figure 2.7 Grade CF3A microstructure: austenite with ferrite;
Condition: solution treated 1 hr at 2050°F (1120°C), water quenched.
Ferrite number: 13. Mag. 200X, Glyceregia.

Ferrite improves the weldability of CF (similar to wrought 300 series) austenitic alloys. Austenite is susceptible to hot cracking which may occur in the heat affected zone and/or the weld metal. This hot cracking can be avoided if the filler metal is alloyed to produce a minimum ferrite level of about 4%. In the cast grades, ferrite is intentionally present and therefore these products are less likely to be subject to this problem.

The presence of ferrite in the CF grades also has the effect of improving the resistance to stress corrosion cracking (SCC) and intergranular corrosion (IGC) attack. It is thought that the presence of ferrite in the form of pools surrounded by austenite increases the resistance to crack propagation. The effect of ferrite with respect to IGC is associated with the preferential precipitation of carbides in sensitized castings in the ferrite rather than at the grain boundaries in fully austenitic materials. The effect of ferrite in corrosive environments depends on the media to which the ferrite is exposed. Consequently, it is not possible to make a general statement that ferrite is always beneficial in all corrosive media. Clearly, each service application should be considered on its merits.

Generally, ferrite does not have a detrimental effect on toughness of the austenitic grades. However, ferrite does have detrimental effects on the mechanical properties of all product forms used at temperatures around 600°F (360°C) and above. At these temperatures carbides precipitate at the edges of the ferrite pools. When these materials are heated to higher temperatures, in the region of 1000°F (538°C), the ferrite can transform into brittle phases chi (χ) or sigma (σ) (Figure 2.8). Brittle networks of these phases may form providing corrosion paths or metallurgical notches resulting in lower toughness. The most common detrimental effect of high temperature on ferrite is the "885 embrittlement" phenomenon. This occurs by a redistribution or spinodal decomposition of the ferrite (α) to a chromium-rich phase known as alpha prime (α'). The toughness of α' is much lower than α. The American Society of Mechanical Engineers (ASME) developed guidelines regarding the use of materials containing ferrite (Figure 2.9).

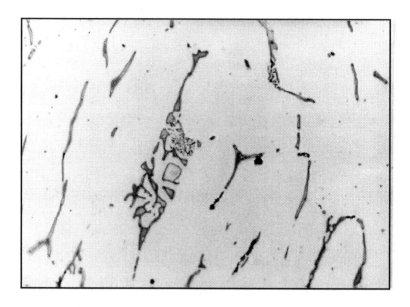

Figure 2.8 Grade CF3M, NaCN microstructure:
austenite with sigma. Mag. 1000X.

							Cautionary Ferrite Guidelines
	Temperature, °F						
% Ferrite	500	600	700	800	900	1000	1100 and higher
0	---	---	---	---	---	---	C
5	---	---	---	---	---	---	C
10	---	---	---	---	C	C	C
15	---	---	---	C	C	C	C
20	---	---	C	C	C	C	C
25	---	C	C	C	C	C	C
30	---	C	C	C	C	C	C
35	C	C	C	C	C	C	C
40	C	C	C	C	C	C	C

a. C stands for caution.
b. At the ferrite levels and temperature identified with the letter C, the subject alloy
will have significant reduction in Charpy V-notch toughness values at room
temperature and below following service exposure. This reduction indicates the
potential for brittle fracture with high rate loading in the presence of sharp notches
or cracks.

Figure 2.9 ASME Table 6-360, cautionary ferrite guidelines.

The development of the duplex stainless steels exploit the strengthening effects of ferrite. In these materials the ferrite level is commonly targeted at 50%. Further details on these alloys can be found in a later section of this chapter, Duplex Stainless Steels.

Clearly, controlling ferrite levels will have a direct effect on mechanical and corrosion performance. Many of the elements commonly present in stainless steels promote the formation of ferrite (α) or austenite (γ). The major α stabilizers are chromium, silicon, molybdenum, and niobium The major γ stabilizers are nickel, carbon, manganese, and nitrogen. By balancing the levels of these elements within the specification limits it is possible to control the ferrite level in the product. The most significant work in developing a method to control ferrite levels was carried out by Schoefer.[1] This work uses an empirical relationship where the major alloying elements are converted into chromium or nickel equivalents:

$$Cr_e = \%Cr + 1.5(\%Si) + 1.4(\%Mo) + \%Nb - 4.99$$
$$Ni_e = \%Ni + 30(\%C) + 0.5(\%Mn) + 26(\%N - 0.02) + 2.77$$

Using the Cr_e and the Ni_e in conjunction with the Schoefer diagram (Figure 2.10), an indication of the ferrite level can be determined from the composition of the product. It is important to recognize that the scatter band in the Schoefer diagram represents 1 σ: therefore, the flexibility in determining the ferrite level required may be much less than the compositional range in the specification. Other methods are available for the determination of ferrite in the product. These include the point count method and the use of secondary standards which determine the ferrite level from the magnetic response. All of these methods and techniques are referenced in ASTM A 799 and ASTM A 800.

The corrosion testing of the cast CF grades is worthy of note here. In ASTM A 262 it is a requirement to sensitize the low carbon (0.03% maximum) grades before conducting the oxalic acid (Practice A) corrosion screening test for intergranular corrosion. Without exception, cast materials with ferrite fail this test, not because they exhibit IGC, but because corrosion occurs at the austenite-ferrite (α-γ)

phase boundaries. It has been shown that even as-cast materials without solution heat treatment and as-welded cast materials will pass the subsequent IGC test.

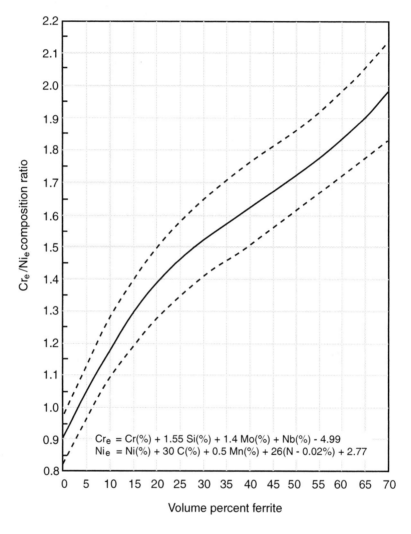

Figure 2.10 Schoefer diagram for estimating the average ferrite content in austenitic iron-chromium-nickel alloy castings. (Reprinted with permission from the Annual Book of ASTM Standards. 100 Barr Harbor Drive, West Conshohoken, PA 194228)

Ferrite in Welds

A method also exists for determining ferrite in welds. It is important to recognize, at this stage, that the level of ferrite in a product is not only a function of the composition, but is also a function of the cooling rate during solidification. Therefore, where welding is carried out without post weld heat treatment, the amount of ferrite will also be affected by the very high cooling rates associated with the heat extraction from the part being welded. Where post weld heat treatment is carried out, such as solution annealing, the cooling rate during welding is not a significant factor, except in instances where cracking of the weld could occur due to very high ferrite levels. The method used for determining ferrite in welds is based on work done by Schneider[2] and Schaeffler.[3] Schneider determined the chromium and nickel equivalents for cooling rates which are typical of fabrication conditions to be:

$$Cr_e = \%Cr + 2(\%Si) + 1.5(\%Mo) + 5(\%V) + 5.5(\%Al) + 1.5(\%Nb) + 0.75(\%W)$$
$$Ni_e = \%Ni + \%Co + 0.5(\%Mn) + 0.3(\%Cu) + 25(\%N) + 30(\%C)$$

Using the Schaeffler diagram (Figure 2.11), the level of ferrite in the weld can be determined.

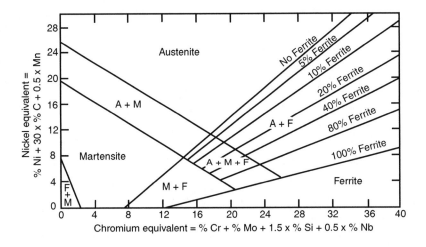

Figure 2.11 Schaeffler diagram showing the amount of ferrite and austenite in weldments as a function of Cr and Ni equivalents.[3]

Heat Treatment

All of the corrosion resistant "C" grades are heat treated (refer to the appropriate ASTM specifications). The heat resistant "H" grades, other than the martensitic grades, are supplied in the as-cast condition.

Heat treatment for the martensitic corrosion-resistant grades usually consists of a normalizing treatment followed by a temper. This treatment imparts the required mechanical properties for these materials. The remainder of the corrosion-resistant grades are usually solution treated. The purpose of this treatment is to reduce carbon segregation, causing any detrimental carbides formed at the grain boundaries during solidification to be dissolved. In addition, the solution treatment dissolves many of the detrimental intermetallic phases, such as σ, which may cause corrosion, mechanical property, or processing problems. The ferritic grades are supplied in the annealed condition.

The heat treatment of the heat-resistant grades is normally restricted to the martensitic materials which are heat treated in the same manner as the corrosion-resistant martensitic grades. Some of the heat-resistant grades such as HU, which is used for heat treatment fixtures, may be heat treated to improve thermal fatigue resistance. It is not clear from the literature what mechanism improves the thermal fatigue life, although this may be associated with the relief stresses in the types of parts produced. Heat treatment is not normally carried out on the heat-resistant grades because these materials are used at high temperatures and rely on the precipitation of carbides during their lifecycle to develop the creep and rupture properties required. Heat treatment as rule does not improve the life of these materials and in some cases may reduce the service life. It is not uncommon for austenitic heat resistant grades which have been in service for some time to be re-solution treated to put intermetallic phases back into solution.

Alloy Types

There are two general categories of cast high alloy steels: corrosion-resistant prefixed by the letter "C" in the ACI designation and heat-resistant high alloy steels prefixed by the letter "H." The first part of this section on high alloy grades will discuss the properties of the "C" type materials, which include the nickel-base alloys, The second section will discuss the "H" series of materials.

Corrosion-Resistant Grades "C"

The corrosion-resistant "C" grades are produced in the following broad types:

> Ferritic
> Martensitic
> Age hardenable
> Austenitic
> Duplex
> Superaustenitic
> Nickel-base alloys

Ferritic Stainless Steels

Table 2.4 Chemical Composition (Weight %)

Grade	UNS No.	C[a]	Mn[a]	Si[a]	Cr	Ni[a]
CB30	J91803	0.30	1.00	1.50	18-22	2.0
CC50	J92615	0.50	1.00	1.50	26-30	4.0

a. Single figures denote maximum.

These materials do not respond to heat treatment. Some improvement in mechanical properties of grade CB30 (J91803) can be achieved by balancing the composition to produce some austenite, although the strength of grade CC50 (J92615) can be improved by solid solution hardening.

Although these fully ferritic grades do not respond to heat treatment they are usually supplied in the annealed conditions.

These materials have Cr contents which offer resistance to many acids, alkaline solutions, and organic chemicals (Figure 2.12). The higher chromium levels in grade CC50 gives it excellent resistance to dilute sulfuric acid in mine waters, mixed nitric and sulfuric acids, and oxidizing acids of all types.

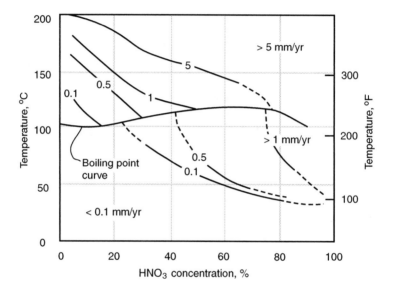

Figure 2.12 Isocorrosion diagram for ACI CB30 in HNO_3.
Castings were annealed at 1450°F (790°C), furnace cooled to 1000°F (540°C), and then air cooled to room temperature.

Typical applications for these grades include:

CB30   Furnace brackets and hangers, pump parts, rabble arms, tube supports, valve bodies, valve parts.

CC50   Bushings, cylinder liners, digester parts, pump casings and impellers, valve bodies, valve seats.

Martensitic Stainless Steels

Table 2.5 Chemical Composition (Weight %)

Grade	UNS No.	C	Mn	Si	Cr	Ni	Mo
CA15	J91150	0.15	1.00	1.50	11.5-14.0	1.0	0.5
CA15	J91151	0.15	1.00	0.65	11.5-14.0	1.0	0.15-1.00
CA40	J91153	0.40	1.00	1.50	11.5-14.0	1.0	0.5
CA40F	J91154	0.2-0.4	1.00	1.50	11.5-14.0	1.0	---
CA6NM	J91540	0.06	1.00	1.00	11.5-14.0	3.5-4.5	0.4-1.0

CA15 (J91150). The microstructure is ferritic, but is usually considered to be a martensitic grade, responding well to heat treatment consisting of normalize and temper. The best mechanical and corrosion resistance is obtained in the hardened condition.

CA15M (J91151). Has improved elevated temperature strength over CA15.

CA40 (J91153). With the higher carbon level, higher strength levels than CA15 can be achieved.

From the definition for stainless steels, CA15 qualifies as the lowest alloyed stainless steel. It has good rural atmospheric corrosion resistance, as well as offering good corrosion resistance to a wide range of organic acid environments. CA15 will rust more rapidly in seacoast and industrial environments.

CA6NM (J91540). One of the most common martensitic alloys. The effect of residual ferrite on this grade is most noticeable. The use of double tempering to meet hardness requirements has become an essential part of the processing procedure for this material.

Precipitation Hardened Stainless Steels

Precipitation hardening is produced by the additions of alloying elements which are put into solution at high temperatures. Rapid cooling causes the alloying elements to remain in solution.

Precipitation is then induced by reheating the material to different temperatures to produce the desired mechanical properties. Time and temperature have a similar effect to each other in terms of the property enhancement. However, the variations in mechanical properties are usually produced by varying the precipitation treatment temperature.

These grades contain more nickel and chromium than the martensitic grades and rely on additions of copper which forms precipitates when heat treated. The most common alloys are listed in Table 2.6.

Table 2.6 Chemical Composition (Weight %)

Grade	UNS No.	C	Mn	Si	Cr	Ni	Cu	Other
CB7Cu-1	J92180	0.07	0.70	1.00	15.5-17.5	3.6-4.6	2.5-3.2	Nb+N
CB7Cu-2	J92110	0.07	0.70	1.00	14.0-15.5	4.5-5.5	2.5-3.2	Nb+N
CD4MCu	J93370	0.04	1.00	1.00	25.0-26.5	4.75-6.0	2.75-3.25	Mo

Grades CB7Cu-1 (J92180) and CB7Cu-2 (J92110) are generally known by their common names 17-4PH and 15-5PH. Their structures are martensitic and they are age hardenable. These are the two most common age hardenable alloys and are usually solution treated at 1925°F (1050°C) followed by rapid cooling. It may be necessary to refrigerate the castings to ensure that transformation has been completed. Incomplete transformation can lead to inferior mechanical properties. Applications for these alloys include airframe components, centrifuge bowls, compressor impellers, food machinery, valve bodies, and discs.

Although CD4MCu (J93370) is not thought of as a precipitation hardened alloy, it was initially developed as such. This alloy is known principally as a duplex stainless steel and is discussed in greater detail in that section.

Austenitic Stainless Steels

The use of the term *austenitic* to describe these alloys should not be taken to mean that they are all fully austenitic. Some of the alloys in this group (summarized in Table 2.7) contain up to 30% ferrite. One of the most commonly used alloys in this group is grade CF8 (J92600), which is similar to the wrought austenitic type 304 (S30400). Although grade CF8 can be produced in the fully austenitic form, it is most commonly produced with some residual ferrite. This residual ferrite is produced by design and imparts the benefits described in the earlier section on ferrite. Grade CF3 (J92500) is the lower carbon version and is much less susceptible to sensitization. It is often used where corrosion resistance in the weld heat affected zone is important.

Table 2.7 Chemical Composition (Weight %)

Grade	UNS	C	Mn	Si	Cr	Ni	Mo
CE30	J93423	0.30	1.50	2.00	26.0-30.0	8.0-11.0	---
CF3	J92500	0.03	1.50	2.00	17.0-21.0	8.0-12.0	---
CF3M	J92900	0.03	1.50	2.00	17.0-21.0	8.0-12.0	2.0-3.0
CF3MN[a]	J92804	0.03	1.50	1.50	17.0-21.0	9.0-13.0	2.0-3.0
CF8	J92600	0.08	1.50	2.00	18.0-21.0	8.0-11.0	---
CF8C	J92710	0.08	1.50	2.00	18.0-21.0	9.0-12.0	---
CF8M	J92900	0.08	1.50	2.00	18.0-21.0	9.0-12.0	2.0-3.0
CF10	J92590	0.04-0.10	1.50	2.00	18.0-21.0	8.0-11.0	---
CF10M	J92901	0.04-0.10	1.50	1.50	18.0-21.0	9.0-12.0	2.0-3.0
CF10MC	J92971	0.10	1.50	1.50	15.0-18.0	13.0-16.0	1.75-2.25
CF10SMnN[a]	J92972	0.10	7.00-9.00	3.50-4.50	16.0-18.0	8.0-9.0	---
CF12M	---	0.12	1.50	2.00	18.0-21.0	9.0-12.0	2.0-3.0
CF16F[b]	J92701	0.16	1.50	2.00	18.0-21.0	9.0-12.0	1.50
CF20	J92602	0.20	1.50	2.00	18.0-21.0	8.0-11.0	---
CG6MMN	J93790	0.06	4.00-6.00	1.00	20.5-23.5	11.5-13.5	1.50-3.00
CG8M	J93000	0.08	1.50	1.50	18.0-21.0	9.0-13.0	3.0-4.0
CG12	J93001	0.12	1.50	2.00	20.0-23.0	10.0-13.0	---
CH8	J93400	0.08	1.50	1.50	22.0-26.0	12.0-15.0	---
CH10	J93401	0.40-0.10	1.50	2.00	22.0-26.0	12.0-15.0	---
CH20	J93402	0.20	1.50	2.00	22.0-26.0	12.0-15.0	---
CK20	J94202	0.20	2.00	2.00	23.0-27.0	19.0-22.0	---
CN7M[c]	N08007	0.07	1.50	1.50	19.0-22.0	27.5-30.5	2.0-3.0

a. Nitrogen addition.
b. Selenium addition.
c. Copper addition.

The molybdenum bearing versions, grades CF8M (J92900) and CF3M (J92900), have increased pitting resistance compared to CF8 and CF3. The corrosion resistance in seawater and chloride-bearing solutions is improved by the addition of molybdenum. The addition of molybdenum has an adverse effect on corrosion resistance in highly oxidizing atmospheres such as boiling nitric acid; however, in weakly oxidizing atmospheres, molybdenum-bearing alloys have superior corrosion resistance. The addition of nitrogen to the CF alloys has some beneficial effects, although these have not been exploited commercially to any great extent (Figure 2.13). In grade CF3MN, the addition of nitrogen increases the yield strength and it has the lowest tendency for carbide precipitation.

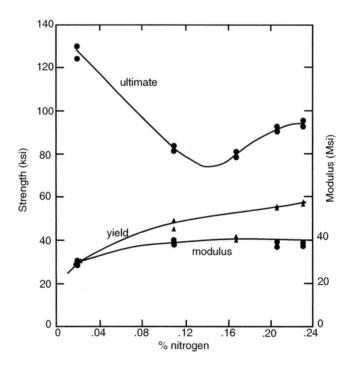

Figure 2.13a Effect of nitrogen on the tensile strength, yield strength and elastic modulus in constant ferrite content CF3 steels. (Series III)

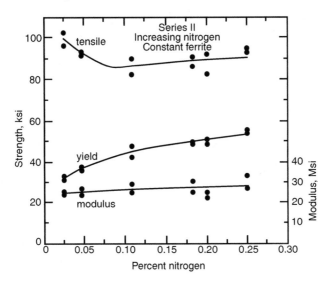

Figure 2.13b Effect of nitrogen on the tensile strength, yield strength
and elastic modulus in constant ferrite content CF8 steels.
(Series II).

Machining of stainless steels frequently presents problems to
manufacturers. Some of these difficulties are associated with an
unfamiliarity with the appropriate machining conditions necessary to
process these materials efficiently. To overcome some of the
machining problems free-cutting grades have been developed. One
such grade is CF16F (J92701) which contains selenium. The corrosion
resistance of this 19% chromium, 10% nickel alloy is adequate, but
the presence of the large number of selenium inclusions reduces the
pitting and corrosion resistance of this material. The effect of ferrite
on machining has not been quantified, although anecdotal evidence
suggests that at ferrite levels lower that 10%, machining becomes
more difficult. Other suggestions for the variation in machining
performance is that the levels of residual elements might be a
significant factor.

Two other grades worthy of mention are CF20 (J92602) and CG8M
(J93000). Grade CF20 is fully austenitic and has relatively high
corrosion resistance to many oxidizing acids. The high carbon content
makes it an absolute requirement that the material be used in the

solution-annealed condition. Grade CG8M has a higher molybdenum content than CF8M and improves corrosion resistance in halide-bearing and reducing acids. The performance in highly oxidizing solutions is not good.

Duplex Stainless Steels

Duplex stainless steels have some of the highest strength levels of any of the corrosion-resistant materials. The improvements in mechanical properties are achieved by the balancing of the structure to be approximately 50% ferrite, 50% austenite–the ferrite being responsible for the improved mechanical properties. The corrosion resistance of these alloys is similar to better than that of CF8M. All of the commercially available alloys listed in ASTM A 890 are alloyed with molybdenum to improve pitting resistance. This group of alloys has seen the greatest amount of development work over the last few years. In addition to the ferrite-balancing requirement in the composition, additional limits are being placed on some of the more recent alloys--essentially requiring a minimum Pitting Resistance Equivalence Number (PREN). The PREN is intended to give an indication of the pitting resistance of the alloy, the higher the PREN the greater the pitting resistance. The calculation of the PREN is as follows:

$$PREN = \%Cr + 3.3\%Mo + 16\%N$$

CD4MCu (J93370). Although grade CD4MCu has been identified as one of the first duplex alloys, it should be remembered that it was developed first as an age-hardening alloy. Grade CD4MCu has a tarnished reputation, and in a number of places is considered to be a problem alloy. This reputation is partly the price of being one of the earliest alloys of its type and the omission of a controlled nitrogen range in the specifications. This problem has been addressed with the introduction of CD4MCuN into ASTM A 890 as Grade 1B. It is recommended that CD4MCuN be ordered in its place since the requirements to control nitrogen levels are clearly stated, and this will ensure that the product will be produced with minimal problems.

CD4MCuN (J93372). This grade was recently added to the ASTM standards and also reflects the need to control nitrogen. Nitrogen has a strong effect on the amount of ferrite present in these alloys.

Because of the relatively high strength of this class of materials and the strong tendency to form intermetallic compounds (Figure 2.14), processing of these materials must be carried out with great care. Such operations as riser size, placement, and removal usually require special procedures. Heat treatment of these alloys is essential before placing in service. The heat treatment is necessary to put the intermetallic compounds back into solution. As mentioned earlier in the section on ferrite, the amount of ferrite formed in an alloy is influenced by composition, but the effect of cooling rate is extremely important when welding these materials. Welds with matching fillers generally produce very high ferrite contents in the weld metal due to high cooling rates. Ferrite levels in excess of 95% have been observed in welds with matching fillers on thick sections. It is common practice for welds to be made with filler metals which have approximately 3 to 4% more nickel than the base material. This balances the weld metal composition so that the weld contains the desired ferrite level.

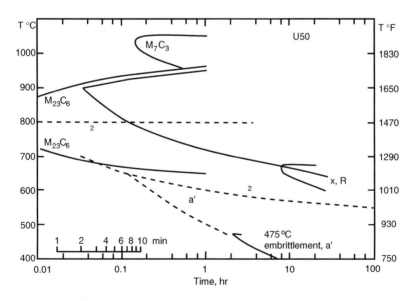

Figure 2.14 Temperature-time precipitation curves
for various phases observed in alloy U50.[7]

The following alloys are also included in ASTM A 890 and respond in a similar manner to CD4MCuN, CE8MN (J93345), CD6MN, CD3MN, CE3MN (J93404), and CD3MWCuN (J93380).

Superaustenitic Stainless Steels

Table 2.8 Chemical Composition (Weight %)

Grade	C	Mn	Si	Cr	Ni	Mo	Cu	N	Others
CK3MCuN	0.025	1.20	1.00	19.5-20.5	17.5-19.5	6.0-7.0	0.50-1.00	0.18-0.24	Cu+N
CN3MN	0.03	2.00	1.00	20.0-22.0	23.5-25.5	6.0-7.0	0.75	0.18-.026	Cu+N

CK3MCuN and other alloys containing 6% molybdenum are more resistant to chloride-bearing solutions. These alloys have the highest molybdenum contents and have the highest pitting resistance in high halide atmospheres. These alloys are used in seawater and equipment where liquids are recirculated with increasing chloride contents. The major alloying elements in the wrought and cast versions of these alloys are the same. It has recently been demonstrated that the effect of welding without post-weld heat treatment (PWHT) causes both cast and wrought versions to have similar corrosion behavior.

Nickel-Based Alloys

Table 2.9 Chemical Composition (Weight %)

Grade	C	Mn	Si	Cr	Ni	Mo	Fe	Others
CU5MCuC	0.05	1.0	1.0	19.5-23.5	38.0-44.0	2.5-3.5	Bal.	0.60-1.20 Cb
CW6M	0.07	1.0	1.0	17.0-20.0	Bal.	17-20	3.0	---
CW6MC	0.06	1.0	1.0	20-23	Bal.	8-10	11.0	3.15-4.50 Cb
CW12MW	0.12	1.0	1.0	15.5-17.5	Bal.	16-18	4.5-7.5	3.75-5.25 W
CX2MW	0.02	1.0	0.8	20-22.5	Bal.	12.5-14.5	2.0-6.0	2.5-3.5 W 0.035 V
CY5SnBiM	0.05	1.5	0.5	11.0-14.0	Bal.	2.0-3.5	2.0	3.0-5.0 Bi 3.0-5.0 Sn
CY40	0.40	1.5	3.0	14-17	Bal.	---	11.0	---
CZ100	1.00	1.5	2.0	---	95.0 min.	---	3.0	1.25 Cu

The development of most of the nickel-base alloys has followed the development of wrought materials. The differences between the wrought and cast materials are frequently associated with the processing requirements in the foundry. For example, fluidity of the molten material in casting is a major concern. However, in the production of wrought products, fluidity is not a concern since the products are produced through mechanical deformation. Nickel-base alloys are utilized for their corrosion resistance, while the mechanical properties of these materials tends to be of secondary concern.

The heat treatment of all of the nickel-base alloys is critical except for CZ100 (N02100) which is generally supplied in the as-cast condition. The heat treatment consists of a solution treatment to minimize the occurrence of intermetallics and molybdenum segregation. This solution treatment improves corrosion performance. Alloys which contain both chromium and molybdenum require solution treating at 2250°F (1232°C). Treatments for shorter times and especially at lower temperatures are not recommended since the rate at which the intermetallics dissolve is slow, and at lower temperatures even more intermetallic product is precipitated. Water quenching is commonly used on these alloys. In cases where distortion of thin sections may be a problem, the parts may be rapidly air cooled.

The alloy CZ100 is known as *cast nickel*. It is very similar to ductile iron where the carbon is treated to produce graphite spheres rather than flakes. Poor mechanical properties result from inadequate spherodizing of the graphite.

Welding of these materials is readily performed. Filler metals which were developed for the wrought materials are commonly used. As with all processes, weldability can be adversely affected by a lack of attention to compositional control, pouring, casting integrity, and heat treatment. It has been observed that most nickel-chromium alloys produced with low carbon (0.01 to 0.02%) produce little precipitation at small welds. It is important to recognize that the full use of the carbon range in the specifications does not ensure the best results. PWHT of small welds is often not considered necessary, but for major welds it should be carried out.

Table 2.10 provides a brief summary of the cast nickel-base grades, their wrought counterparts, and typical uses.

Table 2.10 Nickel-Base Cast and Wrought Alloys
and General Applications

Cast Designation ASTM A 494	Comparable Wrought Grade	General Application
CW12MW	Hastelloy C	Aggressive acids
	Hastelloy C-276	Mixed acids
CW2M	Hastelloy C-4C	High chloride solutions
CX2MW	Hastelloy C-22	As above, intended for both oxidizing and reducing environments
M35-1	Monel 400	Seawater
M35-2	---	Hydrofluoric acids
M30-C	---	---
M35-B	H-Monel	Corrosive wear applications
M35-C	Monel 400	---
M35-D	S-Monel	Machinability
N7M	Hastelloy B modified	---
N12MV	Hastelloy B	---
CY40	Inconel 600	---
CW6MC	Inconel 625	Oxidizing acids
CZ100	Cast nickel	Caustic service

Heat Resistant Alloys "H"

The immediate observation from the composition ranges in Table 2.11 is that the carbon ranges are very wide. At the extreme ends of the carbon ranges, which are rarely used, high carbon gives weldability problems and the lower carbon reduces the creep strength. Silicon is a major alloying element in these materials and significantly improves the oxidation and carburization resistance. It is not uncommon for silicon levels to be specified in the 1.25 to 1.75% range for applications such as ethylene furnaces. Chromium improves oxidation resistance and nickel, an austenite stabilizer, increases resistance to carburization. The creep strength of these materials is developed over time. Nearly all alloys are supplied in the as-cast condition, and the creep strength is developed by the precipitation of carbides in the austenite matrix. Carbon is of primary importance to creep strength (Figure 2.15).

Table 2.11 Chemical Composition (Weight %)

	C	Mn	Si	Cr	Ni	Mo	Nb
Fe-Cr Alloys							
HA	0.20	0.65	1.0	8-10	---	0.90-1.20	---
HC	0.50	1.0	2.0	26-30	4.0	0.5	---
HD	0.50	1.5	2.0	26-30	4-7	0.5	---
Fe-Cr-Ni Alloys							
HE	0.20-0.50	2.0	2.0	26-30	8-11	0.5	---
HF	0.20-0.40	2.0	2.0	18-23	8-12	0.5	---
HH	0.20-0.50	2.0	2.0	24-28	11-14	0.5	---
HI	0.20-0.50	2.0	2.0	26-30	14-18	0.5	---
HK	0.20-0.60	2.0	2.0	24-28	18-22	0.5	---
HL	0.20-0.60	2.0	2.0	28-32	18-22	0.5	---
HN	0.20-0.50	2.0	2.0	19-23	23-27	0.5	---
HP	0.35-0.75	2.0	2.5	24-28	33-37	0.5	---
HP-Nb modified[1]	0.35-0.75	2.0	2.5	24-28	33-37	0.5	0.75-1.51
HT	0.35-0.75	2.0	2.5	15-19	33-37	0.5	---
HU	0.35-0.75	2.0	2.5	17-21	37-41	0.5	---
Ni-Fe-Cr Alloys							
HW	0.35-0.75	2.0	2.5	10-14	58-62	0.5	---
HX	0.35-0.75	2.0	2.0	15-19	64-68	0.5	---
50:50Nb	0.10	0.3	0.5	47-52	Balance	---	1.4-1.7
Low Carbon Fe-Ni-Cr Alloy							
CT15	0.05-0.15	1.5	1.5	19-21	31-34	0.5	0.5-1.5
HP-Nb modifed(LC)[a]	0.10	2.0	2.0	24-28	33-37	0.5	0.75-1.5

a. Composition range proposed in ISO WG 13583-2.

Iron-Chromium Alloys (Fe-Cr) - HA, HC, and HD

These grades are used in conditions where high sulfur-bearing gases
are present and where high strength is not the primary requirement.

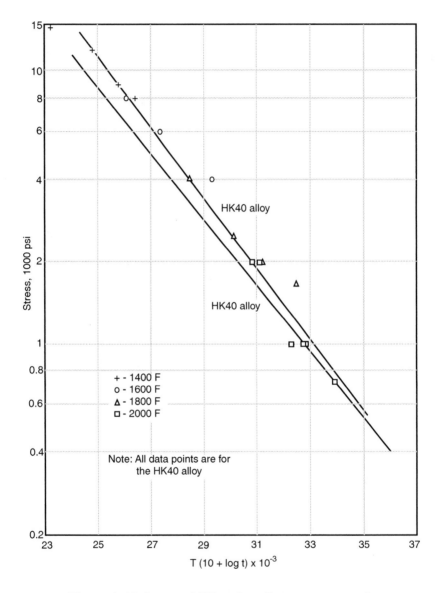

Figure 2.15 Larson-Miller plot of rupture strength
properties for HK40 alloy.[8]

Iron-Chromium-Nickel Alloys (Fe-Cr-Ni) - HE, HF, HH, HI, HK, and HL

Although these alloys are considered to be austenitic, there are two grades of HH: HH1 and HH2. HH1 contains ferrite and HH2 is fully austenitic. Both of these alloys have the same basic specified composition but are produced by balancing the composition to produce the desired structure. HH2 dominates the sales of this type of alloy. It is common to see σ formation in these alloys even though they may be fully austenitic. The high temperature strength of this alloy group is greater than that of the iron-chromium group.

The HK grade has found the greatest application in the chemical and petrochemical industries. It has been replaced to a great extent by the HP alloys but still plays a valuable role. One of the attractions of HK is the lower cost than HP alloys, and some attempts were made to increase its strength through relatively modest changes in the composition. One of the most notable was the increase in the nickel level to 24% (IN519) to minimize the precipitation of σ phase. Other modifications include the addition of tungsten or niobium with microalloys of titanium and zirconium.

Iron-Nickel-Chromium Alloys (Fe-Ni-Cr) - HN, HP, HT, and HU

These materials have enhanced resistance to thermal cycling over the Fe-Cr-Ni alloys with the best resistance in the higher nickel-containing alloys, HT and HU. Confirmation of HT's and HU's resistance to thermal is evidenced by their extensive use in heat treatment furnaces and fixtures where they may be used as trays for the support of steel parts being quenched in oil or water.

HP is rarely used in the unmodified form. HP-Nb modified, which covers a wide group of proprietary materials, dominates the sales of this alloy type. HP is a fully austenitic grade and has replaced HK in many applications and, in particular, in petrochemical furnace tubing and reformer tubing. Unlike HK it does not suffer from the formation of σ. Of all the heat resistant grades, this alloy has seen the greatest development over the last 10 to 15 years. The use of HP-Nb modified is now being superseded by variants which use additions of such elements as titanium, zirconium, and yttrium to produce more stable carbides and, consequently, enhanced stress-to-rupture properties (Figure 2.16). Lower carbon variants of HP-Nb modified and the microalloyed versions have been developed with the carbon level in these versions typically being held to a maximum of 0.10%. These lower carbon versions have been used where large radius tube bends are required. The lower carbon allows these materials to be hot formed into bends without cracking. It is not uncommon for these bends to be machined on the outside surface, as well as the bore, to minimize the tendency to crack during hot forming. HP-modified alloys for use in the hydrocarbon processing industry are discussed later in this chapter.

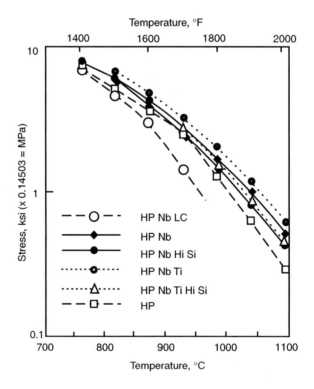

Figure 2.16 Comparison of standard HP grade, niobium-modified
alloys and micro-alloyed compositions–
100,000 hour rupture lives.[9]

Nickel-Iron-Chromium Alloys (Ni-Fe-Cr) - HX and HW

The creep strength of HX and HW is low, and these alloys are used where carburization resistance is required. They might typically be used in heat treatment applications as supports for parts which are being carburized (case hardened).

50:50Nb alloy (IN657) was developed to resist corrosion from fuel ash from low grade fuel oils. It is included in ASTM A 560 along with a 50:50 non-Nb alloy and a 60:40 NiCr alloy. The 50:50Nb alloy is used most frequently, due to its improved resistance to embrittlement. Manufacturing problems associated with surface effects in alloys of this type have been largely overcome. Typical applications for this alloy include furnace fittings/supports and burner cones.

Low Carbon Iron Nickel Chromium Alloy - CT15C

CT15C (N08151) is very similar to alloy 800H (N08810) and has been used in similar applications. Because of its lower carbon level, the creep and stress to rupture properties are lower. This material has been used where higher retained ductility after aging is required. CT15C has been hot formed but has largely been replaced by the low carbon versions of HP-Nb modified in these applications.

Cast Heat Resistant Alloys for the Hydrocarbon Processing Industry

Introduction

Steam methane reformers and ethylene pyrolysis (cracking) furnaces utilize centrifugal cast tubulars for their firebox process piping. These applications require a combination of high temperature strength, oxidation resistance, resistance to carburization and often resistance to other forms of attack. Reformer tubes operate at high temperatures and pressures which induce high stresses, but are not severe applications from a corrosion standpoint. Therefore, stress to rupture properties are the critical factor in reformer tube materials. Ethylene tubes operate at higher temperatures, lower stresses and are subjected to carburization and coke formation.

Alloys for use in petrochemical furnaces are based on specific mixtures of nickel, chromium, and iron, along with additions of modifying elements including niobium, lanthanides, titanium, tungsten, and zirconium. The chromium content is 25% or greater to impart the necessary oxidation resistance. Nickel has to be sufficiently high to produce a fully austenitic structure and offset the tendency for chromium to produce ferrite. Increasing nickel to 35% increases stress-to-rupture properties in chrome-nickel-iron alloys.

Commercial grades most frequently used today are not specified in published standards, such as ASTM, ASME, API, etc. However, ASTM A 297 does list the standard grade HP alloy (26% chromium-35% nickel, balance iron).[11] HP alloy is the base composition for several modified alloys, which are enhanced by the presence of carbide formers such as niobium, molybdenum, and tungsten. HP-modified alloys have displaced the formerly used HK40 alloy (J94224). HP-modified with 1% niobium has a 40% incremental stress rupture strength increase over HK40 (J94224). Thermal fatigue life is at least twice as long.[22]

Limiting service temperatures are typically based on oxidation resistance. HP-modified alloys are commonly used in the service range from 1700 to 2050°F (930 to 1120°C).

For higher operating temperatures from 1900 to 2200°F (1040 to 1205°C) the base alloy typically consists of 26-36% chromium plus 40-50% nickel. This alloy can be improved through additions of solid solution strengtheners, carbide formers, and microalloying elements.

Carbon is an important element common to heat resistant steels. HP-modified alloys contain between 0.30% and 0.50% carbon. Carbon forms carbides that oppose creep (metal deformation) at high temperature. During casting, some of the carbon forms primary carbides that appear in the interdendritic areas of the microstructure (see Figure 2.17). A significant amount of carbon is also retained in solution.

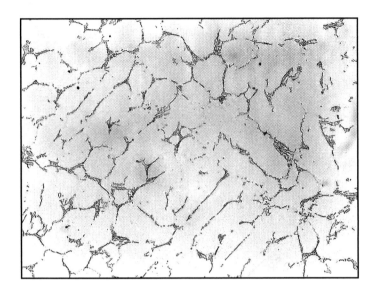

Figure 2.17 Showing a HP-modified as-cast structure of eutectic carbides in an austenite matrix. Aqua Regia Etch. 200X

During exposure above 1200°F (650°C), additional carbon precipitates out in the form of secondary carbides, as shown in Figure 2.18. This process is called aging. Aging of the material can result in a loss of ductility.

Above 1830°F (1000°C) chromium and iron carbides coalesce (grow) quickly at the expense of secondary carbides, which re-dissolve and precipitate out (see Figure 2.19). This causes a sharp decrease in creep rupture strength. Carbide coalescence is retarded by additions of niobium, titanium, tungsten or molybdenum which form dispersed, stable carbides. These elements are required to improve the creep strength in the high temperature range.

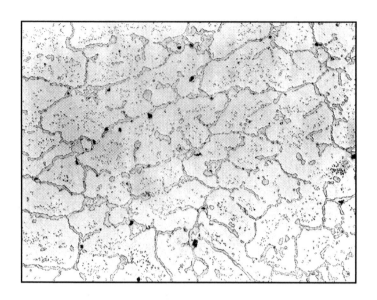

Figure 2.18 Showing precipitated secondary carbides in a
HP-modified cast alloy. Aqua Regia Etch. 200X

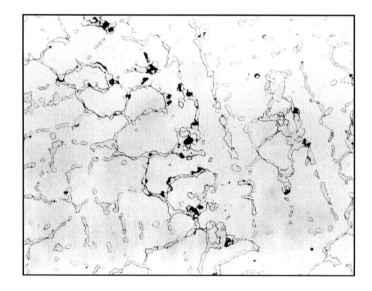

Figure 2.19 Illustrating larger coalesced carbides in a
HP-modified cast alloy. Aqua Regia Etch. 200X

Strengthening additions are dominated by the carbide-forming elements niobium, tungsten, titanium, molybdenum and zirconium. Occasionally, non-carbide formers such as cobalt or nitrogen are used. Tungsten and niobium both lower the cast ductility of HP alloy, with the effect being more pronounced for tungsten. Both elements increase long term rupture strength but the effect is small below 1650°F (900°C). Above 1650°F (900°C) niobium can raise the strength by up to 50% and tungsten by more than 100%. Carburization resistance is not improved by the addition of niobium and only marginally increased by tungsten.[13] Niobium has a detrimental effect on oxidation resistance due to spalling of the protective chromium oxide film and reduces limiting service temperature to 1950°F (1065°C). Tungsten does not exhibit this phenomena and can be used to 2010°F (1100°C).

The major advantage of cast alloys over wrought alloys is creep rupture strength due to higher carbon levels. Grain size control is an important factor in wrought alloys, but is not practical in the case of cast materials. The grain size of cast alloys is normally larger than that obtained in wrought materials thus providing improvements in creep rupture strength. Figure 2.20a shows a typical wrought structure, while a cast structure is shown in Figure 2.20b. Note that the wrought structure has angular grains as opposed to the rounded grains present in the cast specimen.

Centrifugal cast tubes are typically supplied with the inside diameter bored to 125 RMS finish. Pull-bar boring the inside diameter removes the unsound internal shrinkage porosity.

The major mechanical properties of concern in these alloys are: 100,000-hour rupture and the 0.0001%/hour limiting creep stresses; room temperature and elevated temperature tensile properties.

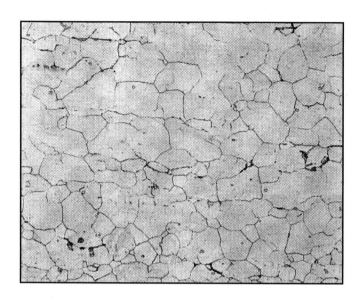

Figure 2.20a Showing a wrought AISI 321 stainless steel.
Electrolytic HNO$_3$ Etch. 250X.

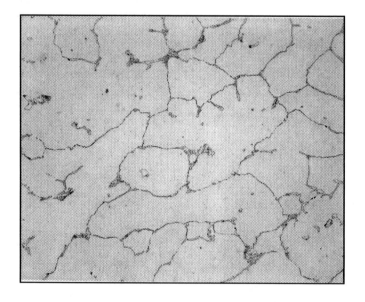

Figure 2.20b Showing a Centrifugal Cast
HP-Modified Alloy Tube. Glyceregia Etch. 200X.

100,000-hour stress-to-rupture data is derived from tests which last less than 10,000 hours. However, the values of interest to designers are primarily stresses to produce a service life of 10,000 to 100,000 hours. Consequently short-term data must be extrapolated to longer times. The Larson-Miller diagram is essential in projecting stress-to-rupture values. Figure 2.21 is a representation of the relative strengths (in 100,000 hour stress-to-rupture values) of the high-temperature alloys discussed. This data was compiled as an average of all of the suppliers listed.

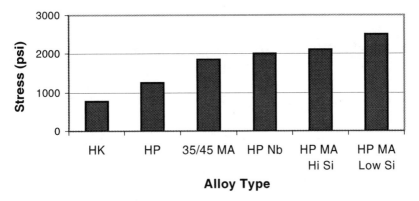

Figure 2.21 Ave. 100,000-hour stress-to-rupture data at 1800°F (980°C).

API Recommended Practice 530 should be used as a design guide.[24]

API attempted to prepare a single standard creep rupture curve for the HP-modified alloy for inclusion in the API RP530 but was not successful. Users must defer to the manufacturer's curves or request raw stress-to-rupture data and produce their own curves.

The standard method used to interpret stress-to-rupture data is the parametric expression developed by Larson and Miller,[15] known as the Larson-Miller Parameter (LMP); an example is shown in Figure 2.15.

The LMP is defined by the following equation.

$$LMP = T (\log t + c)$$
where: T = temperature (Rankine)
t = time (hours)
c = constant

A crucial factor in the use of the LMP is the number assigned to the constant. Different constants applied to the same set of data can give rise to large differences in the predicted stress. Unfortunately users and producers do not always recognize the importance of the constant.

The design and alloy selection for furnace tube coils is generally dependent on whether the coil is destined for ethylene cracking service or reforming service. Ethylene tubes use silicon levels of 1.5-2.0% to improve carburization resistance through the formation of a protective film. Silicon is detrimental to creep resistance and reformer tubes tend to have silicon levels of 1.5% maximum.

Operation of the furnace within the design conditions specified is critical. At temperatures above 1600°F (870°C) small variations in temperature may significantly decrease tube life.

Ethylene Pyrolysis (Cracking) Furnaces

In ethylene production, hydrocarbon feed such as ethane, propane, naphtha, or gas oil is thermally cracked in the presence of steam at low pressure. The radiant section of some of these cracking furnaces operates at end-of-run tube metal temperatures up to 2100°F (1150°C). This is the practical upper limit for most of the weldable, heat resistant alloys.[16]

The radiant coil has a limited life of approximately three years before re-tubing. Tube failure is caused by a variety of factors, but carburization and longitudinal cracking due to high temperature are the most significant causes of failure. Pyrolysis requires materials with high temperature strength and resistance to carburization. Carburization reduces ductility, making the tube much more susceptible to stress damage from either thermal cycles or bending moments. Carburization may result in metal volume changes leading to additional internal stresses.

Silicon is an element that is critical for resistance to carburization, but is detrimental to high temperature strength and fatigue resistance. This reduction in the strength is part of the penalty paid for the improved carburization resistance offered by high silicon.

There has been an evolution in the alloys used in ethylene furnaces. Wrought, heat resistant alloys were displaced by HK40 (J94224), which in turn has been displaced by HP-modified with niobium. Vendors producing HP-modified alloys are listed in Table 2.13. This evolution is continuing with the increasing use of 35% chromium-45% nickel microalloys as shown in Table 2.14.

Table 2.13 Vendors with HP-Modified Analysis

Alloy Name (Manufacturer)	C	Cr	Si	Mn	Nb	Ni	Fe
MO-RE 10 (Duraloy Technologies)[17]	0.4	23/27	2.0 max.	1.5 max.	1.5 max	33/37	Bal.
KHR-35C (Kubota)[18]	0.45	24/28	2.0 max.	2.0 max.	1.0	34/37	Bal.
H39W (Paralloy)[19]	0.4	25	1.5 max.	1.0	1.5 max.	35	Bal.
G-48-52 (Schmidt & Clemens Centracero)[20]	0.4	25	1.5 max.	1.5 max.	1.5 max.	35	Bal.
Wiscalloy 25-35 Nb (Wisconsin Centrifugal)[21]	0.4	23/27	2.0 max.	1.5 max.	1.0 max.	32/37	Bal.
36X (Manoir Ind.)[22]	0.4	23/28	2.0 max	1.5 max.	Add	33/38	Bal.
UCI - 193 (Ultracast)[23]	0.4	24/26	1.75 max.	2.0 max.	1.5 max.	34/36	Bal.
Pyrotherm G 25/35 Nb (Pose. Marre)[24]	0.4	25	1.5 max.	1.5 max.	1.5 max.	35	Bal.

Table 2.14 Vendors with 35% Chromium-45% Nickel Analysis

Alloy Name (Manufacturer)	C	Cr	Si	Mn	Nb	Ni	Other
MO-RE40MA (Duraloy Technologies)[17]	0.45	34/37	2.0 max.	2.0 max.	1.0 max	43/48	Ti, Rare Earths
KHR-45A (Kubota)[18]	0.5	30/35	2.0 max.	2.0 max.	1.5	40/46	Ti, Al
ET45 Micro (Schmidt & Clemens Centracero)[20]	0.45	35	1.6 max.	1.0 max.	1.0 max.	45	Add.
Wiscalloy 35-45 Nb MA (Wisconsin Centrifugal)[21]	0.45	33/36	2.0 max.	2.0 max.	1.0 max.	43/47	Ti & Other
XTM (Manoir Ind.)[22]	0.5	34/37	2.0 max.	2.0 max.	1.0 max.	43/48	Ti, Rare Earths
UCI - 371 (Ultracast)[23]	0.4	---	1.5 max.	1.5 max.	1.5 max.	---	Bal.

Steam Methane Reformers

Steam reforming of hydrocarbons occurs in tubes filled with catalyst. Pressures are typically 350 psi (2450 kPa) with operating temperatures of 1800°F (980°C).

The alloys used for reformer tubing have been going through a transformation from HK-40 to HP-Nb and, more recently, to the microalloyed grades as listed in Table 2.15. The HP-modified category has become more established. During casting, trace quantities of titanium, zirconium and lanthanides (rare earth elements) are added. Adding titanium and zirconium to nearly 0.5% maximum is claimed to improve high temperature stress-rupture resistance by forming more finely dispersed carbides. While some users question the extra stress-rupture performance obtained at temperatures over 1850°F

(1000°C), the risks are low since microalloy additions are usually at little or no extra cost. Process control is very critical in microalloy practice and users are cautioned to deal with reputable foundries. Melting and deoxidation practices must be executed carefully to ensure formation of finely dispersed carbides.

Table 2.15 Vendors with HP Microalloy

Alloy Name (Manufacturer)	C	Cr	Si	Mn	Nb	Ni	Other
TMA 6300 (MO-RE 10MA Duraloy)[17]	0.4	23/27	1.5 max.	1.5 max.	1.5 max.	33-37	Ti, Rare Earths
KHR-35CT (Kubota)[18]	0.45	24/28	2.0 max.	2.0 max.	1.5	34-37	Ti
H39WM (Paralloy)[19]	0.4	25	1.5 max.	1.0	1.5 max.	35	Ti
G-48-52 MICRO (Schmidt & Clemens Centracero)[20]	0.4	25	1.5 max.	1.5 max.	1.5 max.	35	Add.
Wiscalloy 25-35 MA (Wisconsin Centrifugal)[21]	0.4	23/27	1.5 max.	1.5 max.	1.25 max.	33-37	Ti & Other
ManauriteXM (Manoir Ind) [22]	0.4	23/27	1.5 max.	1.5 max.	1.5 max.	33-35	Ti + Other
UCI - 198 (Ultracast)[23]	0.4	23/28	2.0 max.	2.0 max.	1.0 max.	34-36	Ti, Rare Earths

Collectors, Manifolds, and Headers

The transition region from a furnace to the outside atmosphere is a critical portion of any gas processing operation. Process tubes in this area are subject to severe thermal stresses due to temperature gradients and cyclical operation. For this reason specially modified alloys have been developed which emphasize retained ductility and fatigue resistance over strength and corrosion resistance. Alloys for these applications usually have lower carbon and silicon contents compared to their purely heat resistant counterparts.

Initially, many steam methane reformer furnaces had unanticipated failures of risers, manifolds, and transfer headers made of HK, HT, and HU alloys. These alloys lacked ductility and resistance to thermal shock. Failures were due to stresses from thermal cycling during start up and shut down. Welds in positions where thermal stresses and stress concentrations are high should be avoided. Wrought alloy N08810 (alloy 800H), while not as strong as cast alloys, has a greater ductility and thermal shock resistance and consequently became the alloy of choice. Cast 20%Cr-32%Ni+Nb was developed as an alternative to N08810 for outlet header manifolds. This alloy with 0.10% carbon and 0.5 - 1.5% niobium, provides high strength below 1650°F (900°C) and a low tendency for embrittlement. It is less expensive than N08810 because it has higher creep and rupture strength and permits using larger and fewer outlet headers. Elongation is 25% when the alloy is new and over 15% with aging.

Additional Considerations in the use of Heat Resistant Cast Alloys

There is no one "best" alloy which is suitable in all applications. High temperature alloy development is based on a series of compromises to obtain maximum service life in a specific application. Often, an element is beneficial in one application and detrimental in another. An alloy which is excellent at 2200°F (1205°C) may be ill suited for 1600°F (870°C). The major companies in the high-alloy steel casting industry have positioned themselves to interact and respond to the various needs of the petrochemical industry. These companies are continuing to develop new alloys and modify existing alloys for the various stringent conditions and environments found in the applications of their products, to meet the desire for operation at higher temperatures and pressures to improve yield and efficiency, and to meet flexibility in feedstock. These alloys are not listed in published standards. Users should contact the foundries' metallurgical departments for current alloy developments.

Major suppliers also offer "microalloy" modifications of their alloys. Additions of small quantities of titanium and/or zirconium, along with some rare earth metals, enhance the stress-rupture properties through the development of finer, more widely dispersed carbides. They also enhance oxide stability and improve carburization and oxidation resistance. Microalloys offer improved performance with very little cost increase.

Summary

Cast corrosion and heat resistant alloys are available in many shapes and sizes varying in weight from ounces to tons. In general, cast versions are available for all wrought grades. Cast compositions are different from wrought compositions, because of the different manufacturing requirements. It is likely that the majority of new casting alloys will also be derivatives of wrought materials. In the case of materials for use at high temperatures, it is likely that cast grades will be developed independent of wrought materials and continue to be available only as cast grades, due to manufacturing and performance problems with wrought versions The amount of research currently being carried out on cast corrosion and heat resistant ferrous and nickel-based alloys is at the highest level since the late 1970s. This work is dedicated to optimizing performance and manufacture and to the development of new alloys.

References

1.   E. Schoeffer, Appendix to "Mossbauer - Effect examination of ferrite in stainless steel welds and castings," Welding Journal, Research Supplement, 39, Jan. 1974, p.10-S.

2.   H. Schneider, Foundry Trade Journal, 1960, 108, 562.

3.   A. Schaeffler, "Constitution diagram for stainless steel weld metal," Metal Progress, 56, p 680, November 1949.

4.   Flemings M.C., "Lecture V- Origins of Pores and Cavities" The Solidification of Steel Castings, Steel Founders' Society 1965

5.   Peter F. Wieser, Editor, Steel Castings Handbook, 5th Ed., Steel Founders' Society of America, 1980.

6.   Fig 2.13 Bates C.E., "Effects of Nitrogen Additions on CF3 and CF8 Stainless Steels" Steel Founders' Society of American Research Report A78/79, May 1985.

7.   Solomon, Harvey D., and Devine, T.M. Jr., "Duplex stainless steels - a tale of two phases", ASM Metals/Materials Technology Series 8201-089.

8.   Van Echo, Roach and Hall, "Investigation of the Short-Time Tensele and Long-Time Creep-Rupture Properties of the HK35 alloy and A1S1 310 Stainless Steel." Summary Report on Project No. 45 to Alloy Casting Institute, Bottelle Memorial Institute May 20, 1965.Fig 2.15

9.   Malcolm Blair, Thomas L. Stevens, Editors, Steel Castings Handbook, 6th Ed., p. 22-9, Steel Founders' Society, 1995.

10.   Malcolm Blair, Thomas L. Stevens, Editors, Steel Castings Handbook, 6th Ed., pA-6,7, Steel Founders' Society, 1995.

11.   Annual Book of ASTM Standards Vol. 01.02

12.   S.B. Parks & C.M. Schillmoller, Hydrocarbon Processing, October 1997, Improve Alloy Selection for Ammonia Furnaces

13.   C. Steel, Corrosion 83, Paper Number 265, Tungsten and Niobium Modified HP Alloys for High Temperature Service

14. Calculation of Heater-Tube Thickness in Petroleum Refineries API Recommended Practice 530-92

15. G.E. Dieter, Mechanical Metallurgy, Third Edition, McGraw Hill 1986.

16. D.J. Tillack & J.E. Guthrie, NiDi Technical Series 10 071, Wrought and cast heat resistant stainless steels and nickel alloys for the refining and petrochemical industries

17. Duraloy Technologies, Inc. Technical Services Manual

18. Kubota Heat Resistant Alloys Datasheets

19. Paralloy Range of Cast Steels and Alloys

20. Schmidt+Clemens/Centracero Brochure

21. Wisconsin Centrifugal Datasheets

22. Manoir Industries Heat Resistant Alloys for Hydrocarbon Processing

23. Ultracast, Inc. Data sheets

24. Pose-Marre Pyrotherm Brochure

Chapter

3

MARTENSITIC AND FERRITIC STAINLESS STEELS

A. Sabata and W.J. Schumacher

Armco Inc

Middletown, Ohio

Introduction

The years between 1910 and 1915 may be considered the golden years for inventions of stainless steels. The three basic classes of stainless steels were all developed in this period and in three different countries. The man commonly referred to as the inventor of stainless steels is Harry Brearley, an Englishman who developed the 0.35% carbon, 14% chromium alloy which is known today as S42000 (type 420). He was searching for a better alloy for making gun barrels at the time, and was trying unsuccessfully to etch a sample of this composition for metallographic examination. Others had had similar trouble but the spark of genius in Mr. Brearley enabled him to turn this annoyance into a discovery.

At the same time in America, Mr. Dantsizen, working for General Electric Company, was investigating similar alloys primarily for lead-in wires in incandescent lamps. Certain properties of these alloys led him to develop a lower carbon variation, similar to type 410 (S41000), for use as turbine blades. The application is still popular.

Concurrently in Germany two men named Maurer and Strauss were studying iron-nickel alloys for thermocouple protection tubes. They eventually added some chromium which led to an alloy whose composition was very similar to S30200 (type 302).

Historically, stainless steels have been classified by microstructure and are described as martensitic, ferritic, austenitic, or duplex. In addition, a number of precipitation-hardening (PH) martensitic, semiaustenitic stainless steels exist and are normally classified as PH stainless steels. In this chapter, we will discuss ferritic, martensitic, and PH stainless steels.

Martensitic Stainless Steels

The basic martensitic alloy corresponds to S42000 (type 420). This alloy is referred to as *martensitic* because of its structure in the hardened condition. The term *martensite* is derived simply from the name of the man, Martens, who first examined metals under the microscope. The structure is needle-like or acicular and is present in all hardened steels–stainless, carbon, or alloy. See Figure 3.1. It is characterized by high hardness and relatively low ductility. All other hardenable stainless steels may be regarded as modifications of this basic alloy as shown schematically in Figure 3.2.

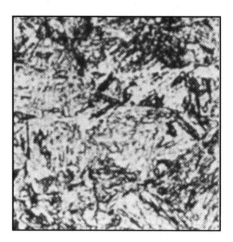

Figure 3.1 Typical microstructure of acicular (or "feathery") martensite found in the hardenable 400 series stainless steels.

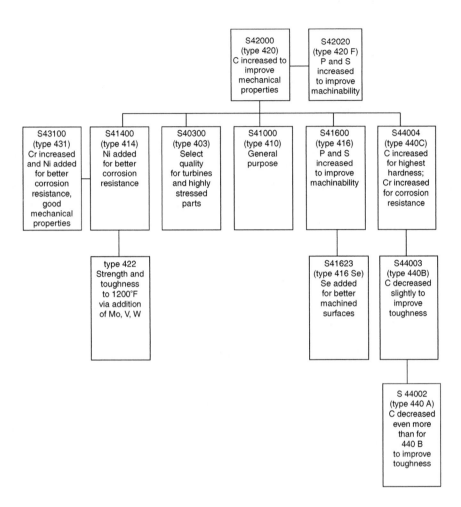

Figure 3.2 Family relationships for
standard martensitic stainless steels.

A description of the most common types of martensitic stainless steels, including corrosion resistance and heat treatment, is found on the following pages. Each martensitic stainless steel is listed by its Unified Numbering System (UNS) designation and AISI type shown in brackets.

UNS S42000 (Type 420)

S42000 (type 420) is corrosion resistant only in the hardened condition. The reason for this is that in the annealed condition the carbon in chromium stainless steels combines with the chromium to form a chromium carbide having a formula of $Cr_{23}C_6$; i.e., about four parts of chromium are combined with one part carbon. Carbon has an atomic weight only one-fourth that of chromium; so in percent weight, one part of carbon combines with close to sixteen parts by weight of chromium. If an alloy has 0.35% carbon, and all the carbon is combined with chromium to form chromium carbide, roughly 5.6% (16 x 0.35%) of the chromium in the alloy is tied up with the carbon. If there is 13% chromium to start with and 5.6% is removed, only 7.4% remains in the matrix of the alloy to give it corrosion resistance, which is not enough. It takes about 10.5% or 11% chromium to give reasonable resistance to atmospheric corrosion.

When type 420 is hardened, the chromium carbides are removed from the structure, and all the chromium is available to provide corrosion resistance. For this reason, the alloy is corrosion resistant in the hardened condition, but not in the annealed condition. Because of this, the usefulness of the alloy is limited. Although it can be tempered or annealed to give quite a range of hardness and tensile strength, it should not be used in those conditions.

UNS S41000 (Type 410)

In order to get around this limitation, it would be logical to reduce the carbon content somewhat, and that is exactly what was done in the case of S41000 (type 410). In this type, the carbon content is dropped from 0.35% down to about 0.10%. This latter amount of carbon does not take too much chromium out of the lattice in the annealed condition, and, therefore, this alloy is resistant to corrosion in the annealed condition. It can develop a very wide range of mechanical properties by different heat-treatments and still be resistant to corrosion. Naturally, this alloy soon became a popular standard stainless grade.

UNS S41040

With S41000, rapid changes in hardness can occur with relatively minor changes in time and temperature. Because of this, it is often difficult to obtain uniform, consistent hardness from part to part–especially when heat treating large furnace loads where temperature variations are likely to exist.

To solve these heat treating problems, S41040 (also known as alloy XM-30) was developed. It has the same composition as S41000, but a small amount of columbium has been added. With S41040 heat treatment is simplified because the effect of temperature and time variations is not so critical. It is easier to obtain desired properties and limit hardness to a narrower range.

In addition to more uniform response to heat treatment, S41040 has higher strength and toughness than regular S41000. S41040 can be used when S41000 is specified.

UNS S44004 (Type 440C)

S42000 was also modified by increasing the carbon content. The objective in this case is to get an alloy which is very hard and abrasion resistant, comparable to a tool steel. The demand for such an alloy resulted in S44004 (type 440C) which has 1% carbon and 17% chromium. Notice that the chromium content has been increased by 4%. The reason is that there is a point in carbon content above which carbides no longer are removed in the hardened condition. The rest of the carbon remains in the structure as free chromium carbides, even in the hardened condition. To compensate for the chromium that is tied up in those free chromium carbides, the chromium is raised to 17%.

UNS S44002/UNS S44003 (Type 440A/Type 440B)

Attempts have been made to arrive at a compromise between ability to harden and brittleness by reducing the carbon contents. This resulted in having two variations of S44004 (type 440C), now designated as S44002 (type 440A) and S44003 (type 440B), which contain 0.60% carbon and a 0.80% carbon, respectively. These alloys harden to a slightly less than S44004 and are a little tougher in the hardened condition.

Early experience with stainless steels showed that they were quite different in machining properties from carbon steels. For one thing, they took more power and were harder on cutting tools. Efforts were made to improve these properties, and it was not long until it was found that additions of sulfur or selenium would appreciably improve their machining properties.

UNS S41600 (Type 416)

A free-machining version of S41000, with an addition of about 0.30% sulfur, is called S41600 (type 416). Through this addition, plus close control of processing and precisely adjusting composition, S41600 has improved cutting rates and tool life as well as providing more consistently uniform machining characteristics.

UNS S42020/UNS S44020 (Type 420F/Type 440F)

S42000 and S44004 also have a free-machining counterpart in alloys S42020 (type 420F) and S44020 (type 440F). In these alloys, selenium is most frequently employed as the free-machining element.

UNS S40300 (Type 403)

There are two other alloys which might be considered as modifications of S41000. The first one, S40300 (type 403), has almost exactly the same composition as S41000. S40300 heats are specially selected to meet the requirements of turbine blade specifications. They must pass rigid specifications in terms of quality in macro etch tests and non-metallic inclusion content as required by turbine blade manufacturers.

UNS S41400/UNS S43100 (Type 414/Type 431)

S41400 (type 414) is a modification of S41000, to which 2% nickel has been added to improve the corrosion resistance, particularly to salt spray and to mild reducing acids. The corrosion resistance, in general, of S41400 is somewhat better than that of S41000. S43100 (type 431) alloy has quite good corrosion resistance as the result of its higher chromium content plus the 2% nickel. It has the highest total alloy content of any of the chromium, hardenable grades and for that reason it is the most corrosion resistant of any of these steels.

As more alloy is added to the composition of S43100, complete transformation during heat treatment becomes more difficult. This leads into the field of semi-austenitic steels which will be covered in the Precipitation Hardening Alloys section.

S43100 has been very popular in England, a good deal more so than in the United States. The alloy is very difficult to process into cold-drawn bars, which has limited its application. When attempts are made to cold-draw this grade in larger size round or hexagon bars, it has a propensity for splitting lengthwise. Many potential applications for S43100 are currently being handled successfully with S17400 (17-4 PH).

Martensitic Stainless Steel Data

Table 3.1 lists chemical compositions of martensitic stainless steels, while mechanical properties can be found in Tables 3.2 through 3.5. Physical properties for martensitic stainless steels are shown in Table 3.6, and American standard cross references are listed in Table 3.7.

Corrosion and oxidation resistance is described for martensitic stainless steels in Table 3.8, along with typical applications for each alloy.

Table 3.1 Chemical Composition of Wrought Martensitic Stainless Steels[a] (Weight %)

UNS No.	Type	C	Mn	Si	Cr	Ni	Mo	Others
S40300	403	0.15	1.00	0.50	11.50-13.00	---	---	---
S41000	410	0.15	1.00	1.00	11.50-13.50	---	---	---
S41400	414	0.15	1.00	1.00	11.50-13.50	1.25-2.50	---	---
S41500	415	0.05	0.5-1.0	0.60	11.50-14.00	3.5-5.5	0.5-1.0	---
S41600	416	0.15	1.25	1.00	12.00-14.00	---	0.6[b]	0.15 S min.
S41623	416Se	0.15	1.25	1.00	12.00-14.00	---	---	0.15 Se min.
S42000	420	0.15 min.	1.00	1.00	12.00-14.00	---	---	---
S42020	420F[b]	0.15 min.	1.25	1.00	12.00-14.00	---	0.6[b]	0.15 S min.
S42200	422	0.20-0.25	1.00	0.75	11.50-13.50	0.5-1.0	0.75-1.25	0.75-1.25 W, 0.15-0.30 V
S43100	431	0.20	1.00	1.00	15.00-17.00	1.25-2.50	---	---
S44002	440A	0.60-0.75	1.00	1.00	16.00-18.00	---	0.75	---
S44003	440B	0.75-0.95	1.00	1.00	16.00-18.00	---	0.75	---
S44004	440C	0.95-1.20	1.00	1.00	16.00-18.00	---	0.75	---
Non-AISI/Common Name								
S41040	410Cb (XM-30)	0.15	1.00	1.00	11.50-13.50	---	---	0.05-0.20 Cb
S41610	416Plus X (XM-6)	0.15	1.5-2.5	1.00	12.00-14.00	---	0.6	---

a. Although sulfur and phosphorous contents are not listed in this table due to limited space, they are specified. See appropriate material standard for more details.
b. Optional.
Single values are maximums, unless otherwise specified.

Table 3.2 Mechanical Properties of ASTM A 276 Martensitic Grades for Bars and Shapes

UNS No. (Type)	Condition[a]	Finish	Diameter or Thickness in. (mm)	Tensile Strength, min. ksi	MPa	Yield Strength, min. ksi	MPa	% El. min.	% RA, min.	HB max.
S40300 (403),	A[a]	Hot-finished	all	70	480	40	275	20	45	---
		Cold-finished	all	70	480	40	275	16	45	---
S4100 (410)	T	Hot-finished	all	100	690	80	550	15	45	---
		Cold-finished	all	100	690	80	550	12	40	---
	H	Hot-finished	all	120	830	90	620	12	40	---
		Cold-finished	all (rounds only)	120	830	90	620	12	40	---
S41040 (XM-30)	T	Hot-finished	all	125	860	100	690	13	45	302
		Cold-finished	all	125	860	100	690	12	35	---
	A	Hot-finished	all	70	480	40	275	13	45	235
		Cold-finished	all	70	480	40	275	12	35	---
S41400 (414)	A	Hot-finished or Cold-finished	all	---	---	---	---	---	---	298
	T	Hot-finished or Cold-finished	all	115	790	90	620	15	45	---
S41500	T	Hot-finished or Cold-finished	all	115	795	90	620	15	45	295
S42000 (420)	A	Hot-finished	all	---	---	---	---	---	---	241
		Cold-finished	all	---	---	---	---	---	---	255
S42010	A	Hot-finished or Cold-finished	---	---	---	---	---	---	---	235
		Cold-finished	all	---	---	---	---	---	---	255
S43100 (431)	A	Hot-finished or Cold-finished	all	---	---	---	---	---	---	285
S44002 (440A), S44003 (440B) S44004 (440C)	A	Hot-finished	all	---	---	---	---	---	---	269
		Cold-finished	all	---	---	---	---	---	---	285

Table 3.2 Mechanical Properties of ASTM A 276 Martensitic Grades for Bars and Shapes (continued)

UNS No. (Type)	Condition[a]	Finish	Diameter or Thickness in. (mm)	Tensile Strength, min. ksi	MPa	Yield Strength, min. ksi	MPa	% El. min.	% RA min.	HB max.
S50400 (9)	A	Hot-finished or Cold-finished	all	60	415	30	207	30	45	179
	T	Hot-finished or Cold-finished	all	100	690	80	550	14	35	241

a. Condition A – Annealed; Condition H - Hardened and tempered at a relatively low temperature; Condition T - Hardened and tempered at a relatively high temperature.

Table 3.3 Nominal Mechanical Properties (Annealed Bar)

UNS No.	Type	Tensile Strength ksi	MPa	Yield Strength ksi	MPa	% Elongation	Hardness Rockwell
S40300	403	75	517	40	276	35	82 HRB
S41000	410	75	517	40	276	35	82 HRB
S41040	---	75	517	40	276	35	80 HRB
S41400	414	120	827	105	724	15	98 HRB
S41600	416	75	517	40	276	30	82 HRB
S41623	416Se	75	517	40	276	30	82 HRB
S42000	420	95	655	50	345	25	92 HRB
S42020	420F	95	655	55	379	22	97 HRB
S42200	422[a]	145	1000	125	862	18	34 HRC
S43100	431	125	862	95	655	20	24 HRC
S44002	440A	105	724	60	414	20	95 HRB
S44003	440B	107	738	62	427	18	96 HRB
S44004	440C	110	758	65	448	14	97 HRB

a. Hardened and tempered.

Table 3.4 Nominal Mechanical Properties
(As Quenched Hardness and Properties After Hardening and Tempering 1 in. (25.4 mm) Diameter Bars)

UNS No.	Type	Hardening Temperature °F (°C)	As Quenched Hardness HB	As Quenched Hardness HRC	Tempering Temperature °F (°C)	Tensile Strength ksi	Tensile Strength MPa	Yield Strength ksi	Yield Strength MPa	% El.	% RA	Izod Impact V-Notch ft-lb (J)	Tempered Hardness HB	Tempered Hardness HRC[b]
S40300	403	1800 (981)	410	43	400 (204)	190	1310	145	1000	15	55	35 (47)	390	41
S41000	410				600 (315)	180	1241	140	965	15	55	35 (47)	375	39
					800[a] (426)	195	1344	150	1034	17	55	---	390	41
					1000[a] (538)	145	1000	115	793	20	65	---	300	31
					1200 (648)	110	758	85	586	23	65	75 (102)	225	97 HRB
					1400 (760)	90	621	60	414	30	70	100 (136)	180	89 HRB
S41040	---	1850 (1010)	410	43	500 (260)	195	1344	161	1110	16	61	65	---	43
					700 (371)	194	1338	162	1117	16	61	56	---	43
					900 (482)	199	1372	156	1076	18	60	44	---	43
					1000 (538)	157	1082	138	951	18	64	37	---	34
					1100 (593)	137	945	121	834	19	64	68	---	28
					1200 (648)	125	862	109	752	20	65	96	---	25
					1300 (704)	114	786	99	683	22	69	110	---	22
					1400 (760)	102	703	90	621	25	70	110	---	98 HRB
S41600	416	1800 (981)	410	43	400 (204)	190	1310	145	1000	12	45	20 (27)	390	41
S41623	416Se				600 (315)	180	1241	140	965	13	45	20 (27)	375	39
					800[a] (426)	195	1344	150	1034	13	50	---	390	41
					1000[a] ((538)	145	1000	115	793	15	50	---	300	31
					1200 (648)	110	758	85	586	18	55	30 (41)	225	97 HRB
					1400 (760)	90	621	60	414	25	60	60 (81)	180	89 HRB

Table 3.4 Nominal Mechanical Properties (continued)

(As Quenched Hardness and Properties After Hardening and Tempering 1 in. (25.4 mm) Diameter Bars)

UNS No.	Type	Hardening Temperature °F (°C)	As Quenched Hardness HB	As Quenched Hardness HRC	Tempering Temperature °F (°C)	Tensile Strength ksi	Tensile Strength MPa	Yield Strength ksi	Yield Strength MPa	% El.	% RA	Izod Impact V-Notch ft-lb (J)	Tempered Hardness HB	Tempered Hardness HRC[b]
S41400	414	1800 (981)	425	44	400 (204)	100	1379	150	1034	15	55	45 (61)	410	43
					600 (315)	190	1310	145	1000	15	55	45 (61)	400	41
					800 (426)	200	1379	150	1034	16	58	---	415	43
					1000a (538)	145	1000	120	837	20	60	---	290	30
					1200 (760)	120	827	105	724	20	65	50 (68)	250	22
S43100	431	1900 (1036)	440	45	400 (204)	205	1413	155	1069	15	55	30 (41)	415	43
					600 (315)	195	1344	150	1034	15	55	45 (61)	400	41
					800a (426)	205	1413	155	1069	15	60	---	415	43
					1000a (538)	150	1034	130	896	18	60	---	325	34
					1200 (760)	125	862	95	655	20	60	50 (68)	260	24
S42000	420	1900 (1036)	540	54	600 (315)	230	1586	195	1344	8	25	10 (14)	500	50
S44002	440A	1900 (1036)	570	56	600 (315)	260	1793	240	1655	5	20	4 (5)	510	51
S44003	440B	1900 (1036)	590	58	600 (315)	280	1931	270	1862	3	15	3 (4)	555	55
S44004	440C	1900 (1036)	610	60	600 (315)	285	1965	275	1896	2	10	2 (3)	580	57

a. Tempering within the range of 750 to 1050°F (400 to 570°C) is not recommended because such treatment will result in low and erratic impact properties and loss of corrosion resistance.

b. Unless otherwise noted.

Note: Variations in chemical composition within the individual type ranges may affect the mechanical properties.

Table 3.5 Mechanical Properties of Wrought Martensitic Stainless Steels

UNS No. (Type) Product Form[a]	ASTM Specification	Heat Treat Condition[b]	Tensile Strength, min. ksi	Tensile Strength, min. MPa	Yield Strength, min. ksi	Yield Strength, min. MPa	% El. min.	% RA min.	Hardness, max.
UNS S41000 (Type 410)									
P, Sh, St	A 240	---	65	450	30	205	20	---	217 HB or 96 HRB
B, shapes	A 276	A, HF	70	480	40	275	20	45	---
	A 479	A	70	485	40	275	20	45	223 HB
		1	70	485	40	275	20	45	223 HB
		2	110	760	85	585	15	45	269 HB
		3	130	900	100	690	12	35	331 HB
		A, CF	70	480	40	275	16	45	---
		T, HF	100	690	80	550	15	45	---
		T, CF	100	690	80	550	12	40	---
		T, HF or CF	120	830	90	620	12	40	---
UNS S41000 (Type 410)									
B, W	A 493	A	82	565	---	---	---	---	---
		LD	85	585	---	---	---	45	---
W	A 580	A	70	485	40	275	20	45	---
		A, CF	70	485	40	275	16	45	---
		IT, CF	100	690	80	550	12	40	---
		HT, CF	120	825	90	620	12	40	---
UNS S41600 (Type 416)									
W	A 581	A	85-125	585-860	---	---	---	---	---
		T	115-145	790-1000	---	---	---	---	---
		HT	140-175	965-1210	---	---	---	---	---
UNS S41000 (Type 410)									
B, P, Sh, St	A 582, A 895	A	---	---	---	---	---	---	262 HB
		T	---	---	---	---	---	---	248-302 HB
		H	---	---	---	---	---	---	293-352 HB

Table 3.5 Mechanical Properties of Wrought Martensitic Stainless Steels (continued)

UNS (AISI Type) Product Form[a]	ASTM Specification	Heat Treat Condition[b]	Tensile Strength, min.		Yield Strength, min.		% El. min.	% RA min.	Hardness, max.
			ksi	MPa	ksi	MPa			
UNS S44002 (Type 440A)									
B, shapes	A 276	A, HF	---	---	---	---	---	---	269 HB
		A, CF	---	---	---	---	---	---	285 HB
B	---	A	105[d]	725[d]	60[d]	415[d]	20[d]	---	95 HRB[d]
		T[c]	260[d]	1790[d]	240[d]	1650[d]	5[d]	20[d]	51 HRC[d]
UNS S41000 (Type 410)									
W	A 580	A, CF	140 max.	965 max.	---	---	---	---	---
UNS S44003 (Type 440B)									
B, shapes	A 276	A, HF	---	---	---	---	---	---	269 HB
		A, CF	---	---	---	---	---	---	285 HB
B	---	A	107[d]	740[d]	62[d]	425[d]	18[d]	---	96 HRB[d]
		T[c]	280[d]	1930[d]	270[d]	1860[d]	3[d]	15[d]	55 HRC[d]
W	A 580	A, CF	140 max.	965 max.	---	---	---	---	---
UNS S44004 (Type 440C)									
B, shapes	A 276	A, HF	---	---	---	---	---	---	269 HB
		A, CF	---	---	---	---	---	---	285 HB
B	---	A	110[d]	760[d]	65[d]	450[d]	14[d]	---	97 HRB[d]
		T[c]	285[d]	1970[d]	275[d]	1900[d]	2[d]	10[d]	57 HRC[d]
W	A 580	A, CF	140 max.	965 max.	---	---	---	---	---

a. B - bar; F - forgings; P - plate; Sh - sheet; St - strip; W - wire.
b. A – annealed; HF - hot finished; CF - cold finished; T - hardened & tempered at relatively low temperature; NT - normalized and tempered; H - hardened and tempered at relatively low temperature; NT - normalized and tempered; LD - lightly drafted.
c. Tempered at 600°F (315°C).
d. Typical values.
Single values are minimums unless otherwise specified.

Table 3.6 Typical Physical Properties of Wrought Martensitic Stainless Steels Annealed Condition

UNS No.	Type	Density g/cm³ (lb/in.³)	Elastic Modulus GPa (10⁶ psi)	Mean CTE[a] from 0°C (32°F) to: 315°C (600°F) μm/m°C (μin./in.°F)	538°C (1000°F) μm/m°C (μin./in.°F)	Thermal Conductivity at 100°C (212°F) W/m°K (Btu/ft h°F)	at 500°C (932°F) W/m°K (Btu/ft h°F)	Specific Heat at 0-100°C (32-212°F) J/kg°K (Btu/lb°F)	Electrical Resistivity ηΩm	Magnetic Permeability
S40300	403	7.8 (0.28)	200 (29.0)	11.0 (6.1)	11.6 (6.4)	24.9 (14.4)	28.7 (16.6)	460 (0.11)	570	700-1000
S40500	405	7.8 (0.28)	200 (29.0)	11.6 (6.4)	12.1 (6.7)	27.0 (15.6)	---	460 (0.11)	600	---
S40900	409	7.8 (0.28)	---	---	---	---	---	---	---	---
S41000	410	7.8 (0.28)	200 (29.0)	11.4 (6.3)	11.6 (6.4)	24.9 (14.4)	28.7 (16.6)	460 (0.11)	570	700-1000
S41400	414	7.8 (0.28)	200 (29.0)	11.0 (6.1)	12.1 (6.7)	24.9 (14.4)	28.7 (16.6)	460 (0.11)	700	---
S41600	416	7.8 (0.28)	200 (29.0)	11.0 (6.1)	11.6 (6.4)	24.9 (14.4)	28.7 (16.6)	460 (0.11)	570	700-1000
S42000	420	7.8 (0.28)	200 (29.0)	10.8 (6.0)	11.7 (6.5)	24.9 (14.4)	---	460 (0.11)	550	---
S42200	422	7.8 (0.28)	---	11.4 (6.3)	11.9 (6.6)	23.9 (13.8)	27.3 (15.8)	460 (0.11)	---	---
S43100	431	7.8 (0.28)	200 (29.0)	12.1 (6.7)	---	20.2 (11.7)	---	460 (0.11)	720	---
S44002	440A	7.8 (0.28)	200 (29.0)	---	---	24.2 (14.0)	---	460 (0.11)	600	---
S44003	440B	7.8 (0.28)	200 (29.0)	---	---	24.2 (14.0)	---	460 (0.11)	600	Ferro-magnetic
S44004	440C	7.8 (0.28)	200 (29.0)	---	---	24.2 (14.0)	---	460 (0.11)	600	---

a. Coefficient of thermal expansion.

Table 3.7 American Standards Cross References – Martensitic and Ferritic Stainless Steels

AMERICAN STANDARDS CROSS REFERENCES

UNS No.	Type	ASTM	SAE	AMS	MIL	FED
S40300	403	A 176 (403), A 276 (403), A 314 (403), A 473 (403), A 479 (403), A 511 (403), A 580 (403)	J405 (51403)	---	---	QQ-S-763 (403)
S40500	405	A 176 (405), A 240 (405), A 268 (405), A 276 (405), A 473 (405), A 479 (405), A 511 (405), A 580 (405)	J405 (51405)	---	---	QQ-S-763 (405)
S40900	409	A 176 (409), A 240 (409), A 268 (409), A 791 (TP 409), A 803 (TP 409)	J405 (51409)	---	---	---
S41000	410	A 176 (410), A 182 (F62, F8a), A 193 (410, B6, B6X), A 194 (410, 6), A 240 (410), A 268 (410), A 276 (410), A 314 (410), A 336 (F6), A 473 (410), A 493 (410), A 511 (410), A 580 (410)	J405 (51410), J412 (51410)	5504, 5505, 5591, 5613, 5776, 7493	---	QQ-S-763 (410), QQ-S-766 (410)
S41400	414	A 276 (414), A 314 (414), A 473 (414), A 479 (414), A 511 (414), A 580 (414)	J405 (51414)	---	---	QQ-S-763 (414)
S41600	416	A 194 (416, 6F), A 314 (416), A 473 (416), A 581 (416), A 582 (416), A 895 (416)	J405 (51416)	5610	---	---
S41623	416Se	A 194 (416 Se, 6F), A 314 (416 Se), A 473 (314 Se), A 511 (416 Se), A 581 (416 Se), A 582 (416 Se), A 895 (416 Se)	J405 (51416 Se)	5610	---	QQ-S-763
S42000	420	A 276 (420), A 314 (420), A 473 (420), A 580 (420)	J405 (51420)	5506, 5621, 7207	---	QQ-S-763 (420), QQ-S-766 (420)
S42020	420F	A 895 (420 F)	J405 (51420F)	5620	---	---
S42200	422	A 565 (616)	J467 (422), J775 (HNV-8)	5655	---	---
S42900	429	A 176 (429), A 182 (429), A 240 (429), A 268 (429), A 276 (429), A 314 (429), A 473 (429), A 493 (429), A 511 (429), A 554 (429), A 815 (WP 429)	J405 (51429)	---	---	QQ-S-763 (429), QQ-S-766 (429)

Table 3.7 American Standards Cross References - Martensitic and Ferritic Stainless Steels (continued)

AMERICAN STANDARDS CROSS REFERENCES

UNS No.	Type	ASTM	SAE	AMS	MIL	FED
S43000	430	A 176 (430), A 182 (430), A 240 (430), A 268 (430), A 276 (430), A 314 (430), A 473 (430), A 479 (430), A 493 (430), A 511 (430), A 554 (430), A 580 (430), A 815 (WP 430)	J405 (51430)	5503, 5627	---	QQ-S-764 (430), QQ-S-766 (430)
S43020	430F	A 314 (430 F), A 473 (430 F), A 581 (430 F), A 582 (430 F), A 895 (430 F)	J405 (51430 F)	---	---	---
S43023	430F Se	A 314 (430 F Se), A 473 (430 F Se), A 581 (430 F Se), A 582 (430 F Se), A 895 (430 F Se)	J405 (51430 F Se)	---	---	---
S43035	439	A 240 (439), A 268 (439), A 479 (439), A 791 (439), A 803 (TP 439)	---	---	MIL-S-18732	---
S43100	431	A 276 (431), A 473 (431), A 493 (431), A 511 (MT 431), A 579 (63), A 580 (431)	J405 (51431)	5628	---	---
S43400	434	---	J405 (51434)	---	---	---
S43600	436	---	J405 (51436)	---	---	---
S44002	440A	A 276 (440 A), A 314 (440 A), A 473 (440 A), A 511 (440 A), A 580 (440 A)	J405 (51440 A)	5631, 5632, 7445	---	QQ-S-763 (440 A)
S44003	440B	A 276 (440 B), A 314 (440 B), A 473 (440 B), A 580 (440 B)	J405 (51440 B)	7445	---	QQ-S-763 (440 B)
S44004	440C	A 276 (440 C), A 314 (440 C), A 473 (440 C), A 493 (440 C), A 580 (440 C)	J405 (51440 C)	5618, 5630, 5880, 7445	---	QQ-S-763 (440 C)
S44200	442	A 176 (442)	J405 (51442)	---	---	---
S44600	446	A 176 (446), A 268 (446-1, 446-2), A 276 (446), A 314 (446), A 473 (446), A 511 (446), A 580 (446), A 815 (WP 446)	J405 (51446)	---	---	QQ-S-763 (446), QQ-S-766 (446)
S50100	501	A 182 (B5, F7), A 193 (501, B5), A 194 (501, 3), A 314 (501), A 387 (5), A 473 (501)	J405 (51501)	---	---	---

CASTI Handbook of Stainless Steels & Nickel Alloys – Second Edition

Table 3.8 Corrosion/Oxidation Resistance and Typical Applications of Martensitic Stainless Steels

UNS No.	Corrosion Resistance	Oxidation Resistance	Typical Applications
S40300	S40300 is resistant to the corrosive action of the atmosphere, fresh water, and various alkalies and mild acids, but is not recommended where resistance to severe corrodants is a prime factor. S40300 does not possess the superior corrosion resistance of the austenitic chromium-nickel stainless steels. It exhibits best corrosion resistance in the hardened and stress relieved condition. Tempering between about 750-1050°F (400-570°C) lowers its resistance in some media.	This grade has good oxidation resistance up to 1250°F (680°C) in continuous service. Scaling becomes excessive above about 1400°F (760°C) in intermittent service.	Turbine blades and highly stressed sections in gas turbines, furnace parts, and burners operating below 1200°F (650°C), valve parts, cutlery, fasteners, hardware, oil refinery equipment, mining machinery, screens, pump parts, fingernail files, sporting goods such as fishing poles, and rifle barrels.
S41000	S41000 is resistant to the corrosive action of the atmosphere, fresh water, and various alkalies and mild acids, but is not recommended where resistance to severe corrodants is a prime factor. S41000 does not possess the superior corrosion resistance of the austenitic chromium-nickel stainless steels. It exhibits best corrosion resistance in the hardened and stress relieved condition. Tempering between about 750-1050°F (400-570°C) lowers its resistance in some media.	This grade has good oxidation resistance up to 1250°F (680°C) in continuous service. Scaling becomes excessive above about 1400°F (760°C) in intermittent service.	Turbine blades and highly stressed sections in gas turbines, furnace parts, and burners operating below 1200°F (650°C), valve parts, cutlery, fasteners, hardware, oil refinery equipment, mining machinery, screens, pump parts, fingernail files, sporting goods such as fishing poles, and rifle barrels.
S41040	S41040 is resistant to the corrosive action of the atmosphere, fresh water, and various alkalies and mild acids, but is not recommended where resistance to severe corrodants is a prime factor. S41040 does not possess the superior corrosion resistance of the austenitic chromium-nickel stainless steels. It exhibits best corrosion resistance in the hardened and stress relieved condition. Tempering between about 750-1050°F (400-570°C) lowers its resistance in some media.	This grade has good oxidation resistance up to 1250°F (680°C) in continuous service. Scaling becomes excessive above about 1400°F (760°C) in intermittent service.	Aircraft and missile components, steam turbine blades, valve parts, and fasteners.
S41400	S41400 is resistant to the corrosive action of the atmosphere, fresh water, and various alkalies and mild acids to an extent slightly better than that of S41000. It exhibits best corrosion resistance in the hardened and tempered condition. Tempering between about 750-1050°F (400-570°C) lowers its resistance in some media.	This grade has good oxidation resistance up to about 1250°F (680°C) in continuous service. Scaling becomes excessive above 1400°F (760°C).	Beater bars, fasteners, gage parts, mild springs, mining equipment, scissors, scraper knives, shafts, spindles, and valve seats.

Table 3.8 Corrosion/Oxidation Resistance and Typical Applications of Martensitic Stainless Steels (continued)

UNS No.	Corrosion Resistance	Oxidation Resistance	Typical Applications
S41500	S41500 is resistant to the corrosive action of the atmosphere, fresh water, and various alkalies and mild acids, but is not recommended where resistance to severe corrodants is a prime factor. S41500 does not possess the superior corrosion resistance of the austenitic chromium-nickel stainless steels. It exhibits best corrosion resistance in the hardened and stress relieved condition. Tempering between about 750-1050°F (400-570°C) lowers its resistance in some media.	This grade has good oxidation resistance up to 1250°F (680°C) in continuous service. Scaling becomes excessive above about 1400°F (760°C) in intermittent service.	Parts with improved weldability over S41000.
S41600	S41600 is resistant to the corrosive action of the atmosphere, fresh water, various alkalies and mild acids. The fine finish obtainable on this grade enhances corrosion resistance.	This grade has good oxidation resistance up to 1200°F (650°C) in continuous service. Scaling becomes excessive above 1400°F (760°C) in intermittent service.	Non-galling and non-seizing corrosion resistant parts machined on automatic screw machines. Various valve parts such as bodies, stems and trim, many types of threaded fasteners, shafts, and pump components are among the applications served by this free machining stainless steel.
S41623	S41623 is resistant to the corrosive action of the atmosphere, fresh water, various alkalies, and mild acids. The fine finish obtainable on this grade enhances corrosion resistance.	This grade has good oxidation resistance up to 1200°F (650°C) in continuous service. Scaling becomes excessive above 1400°F (760°C) in intermittent service.	Improved surface finish over S41600.
S42000	S42000 attains its full corrosion resistance in the hardened and polished condition. Much lower resistance is shown by annealed material and material having its surface contaminated by foreign particles. For this reason, it is advisable to passivate final parts. For most applications, S42000 exhibits corrosion resistance about the same as S41000. S42000 is melted to slightly higher chromium than S41000 in order to offset the detrimental effect of higher carbon.	This grade has good oxidation resistance up to about 1200°F (650°C) in continuous service. Scaling becomes excessive above 1400°F (760°C).	Cutlery, hand tools, dental and surgical instruments, valve trim and parts, shafts, and plastic mold steel.

Table 3.8 Corrosion/Oxidation Resistance and Typical Applications of Martensitic Stainless Steels (continued)

UNS No.	Corrosion Resistance	Oxidation Resistance	Typical Applications
S42010	S42010 attains its full corrosion resistance in the hardened and polished condition. Much lower resistance is shown by annealed material and material having its surface contaminated by foreign particles. For this reason, it is advisable to passivate final parts. For most applications, S42010 exhibits corrosion resistance about the same as S41000. S42010 is melted to slightly higher chromium than S41000 in order to offset the detrimental effect of higher carbon.	This grade has good oxidation resistance up to about 1200°F (650°C) in continuous service. Scaling becomes excessive above 1400°F (760°C).	Fasteners, cutlery, valves, gages, and guides.
S42020	S42020 attains its full corrosion resistance in the hardened and polished condition. Much lower resistance is shown by annealed material and material having its surface contaminated by foreign particles. For this reason, it is advisable to passivate final parts. For most applications, S42020 exhibits corrosion resistance about the same as S41000. S42020 is melted to slightly higher chromium than S41000 in order to offset the detrimental effect of higher carbon.	This grade has good oxidation resistance up to about 1200°F (650°C) in continuous service. Scaling becomes excessive above 1400°F (760°C).	Fasteners, cutlery, valves, gages, and guides.
S42200	S42200 is resistant to the corrosive action of the atmosphere, fresh water and various alkalies, and mild acids, but is not recommended where resistance to severe corrodants is a prime factor. S42200 does not possess the superior corrosion resistance of the austenitic chromium-nickel stainless steels. It exhibits best corrosion resistance in the hardened and stress relieved condition. Tempering between about 750-1050°F (400-570°C) lowers its resistance in some media.	This grade has good oxidation resistance up to 1250°F (680°C) in continuous service. Scaling becomes excessive above about 1400°F (760°C) in intermittent service.	Turbine blades and highly stressed sections in gas turbines, furnace parts, and burners operating below 1200°F (650°C), valve parts, cutlery, fasteners, hardware, oil refinery equipment, mining machinery, screens, pump parts, fingernail files, sporting goods such as fishing poles, and rifle barrels.
S43100	S43100 is resistant to the corrosive action of the atmosphere, various alkalies, and mild acids. It has better resistance to corrosion from marine atmosphere and is considered to have better resistance to stress corrosion than the other martensitic stainless steels.	This grade has good oxidation resistance up to about 1500°F (820°C) in continuous service. Scaling becomes excessive above about 1600°F (870°C).	Aircraft fittings, beater bars, fasteners, conveyor parts, valve parts, pump shafts, and marine hardware.

Table 3.8 Corrosion/Oxidation Resistance and Typical Applications of Martensitic Stainless Steels (continued)

UNS No.	Corrosion Resistance	Oxidation Resistance	Typical Applications
S44002	S44002 attains its full corrosion resistance in the hardened and polished condition. Much lower resistance is shown by annealed material and material having its surface contaminated by foreign particles. For this reason, it is advisable to passivate final parts. The grade is resistant to the corrosive action of the atmosphere, fresh water, perspiration, mild acids, fruit and vegetable juices, foodstuffs, etc.	S44002 is not normally used for elevated temperature service since its resistance to corrosion as well as its hardness and strength are lowered by exposure above about 800°F (430°C). The grade scales appreciably at temperatures above about 1400°F (760°C).	Cutlery, bearings, valves, seaming rolls, and surgical and dental instruments.
S44003	S44003 attains its best corrosion resistance in the hardened and polished condition. Much lower resistance to certain media is shown by annealed material or material having its surface contaminated by foreign particles. For this reason it is advisable to passivate final parts. The grade is resistant to the corrosive action of the atmosphere, fresh water, perspiration, mild acids, fruit and vegetable juices, foodstuffs, etc.	This alloy is not normally used for elevated temperature service since its resistance to corrosion is lowered by exposure above 800°F. The grade scales appreciably at temperatures above about 1400°F (760°C).	Bearings, cutlery, spatula blades, and food processing knives.
S44004	S44004 attains its full corrosion resistance in the hardened and polished condition. Much lower resistance to certain media is shown by annealed material or material having its surface contaminated by foreign particles. For this reason it is advisable to passivate final parts. The grade is resistant to the corrosive action of the atmosphere, fresh water, perspiration, mild acids, fruit and vegetable juices, foodstuffs, etc.	This alloy is not normally used for elevated temperature service, since its resistance to corrosion is lowered by exposure above 800°F. The grade scales appreciably at temperatures above about 1400°F (760°C).	Cutlery, bearings, nozzles, valve parts, pivot pins, balls and seats for oil well pumps.
S44020	S44020 attains its full corrosion resistance in the hardened and polished condition. Much lower resistance to certain media is shown by annealed material or material having its surface contaminated by foreign particles. For this reason it is advisable to passivate final parts. The grade is resistant to the corrosive action of the atmosphere, fresh water, perspiration, mild acids, fruit and vegetable juices, foodstuffs, etc.	This alloy is not normally used for elevated temperature service, since its resistance to corrosion is lowered by exposure above 800°F. The grade scales appreciably at temperatures above about 1400°F (760°C).	Cutlery, bearings, nozzles, valve parts, pivot pins, balls and seats for oil well pumps.

Precipitation-Hardening Stainless Steel

Compositions of most precipitation-hardening or age-hardening stainless steels are carefully balanced to produce hardening by two separate mechanisms:

1. *Transformation*—an allotropic transformation of austenite to martensite produced by thermal austenite-conditioning treatment or by cold working.
2. *Precipitation*—the resulting structure is then given a simple, low-temperature aging treatment that hardens by precipitation of intermetallic compounds and simultaneously tempers the martensite.

Grades that respond to this pattern of hardness development are classed as martensitic or semi-austenitic types. A third category of precipitation-hardening stainless steels does not transform to martensite. These austenitic types retain their austenitic structure at room temperature following solution heat treatment and are hardened by a precipitation treatment. The principal austenitic alloy in general use is S66286 (A-286).

Some true martensitic types include S17400 (Armco 17-4 PH®), S15500 (Armco 15-5 PH®), S13800 (Armco PH 13-8 Mo®), and S45500 (Custom 455®). A martensitic structure forms in these alloys upon cooling from solution-treating or annealing temperatures ranging from 1500 to 1900°F (816 to 1038°C), depending on the alloy. Subsequent aging treatments at 900 to 1150°F (482 to 621°C) develop desired strength and toughness.

Semi-austenitic grades include: S17700 (Armco 17-7 PH®), S15700 (Armco PH 15-7 Mo®), S35000 (AM350®), and S35500 (AM355®). These alloys are principally sheet and strip materials. Compositions, are balanced so that the structure is austenitic in the solution-treated condition (see Table 3.9). This ductile structure permits forming by conventional techniques used for 18-8 stainless steels. Following fabrication, the austenite is transformed to martensite by thermal treatment and subsequent cooling. Maximum strength is then developed by an aging treatment which results in precipitation hardening and tempering of the martensite. For applications requiring maximum strength and only limited formability, the martensite transformation can be accomplished by severe cold working. Strength is then developed, as before, by a simple aging treatment.

Composition and purity of precipitation-hardening stainless steels are critical to achieving precise properties. Many of these proprietary materials are double-vacuum melted or air-melted and vacuum-induction melted to produce predictable fracture toughness behavior and stability at elevated temperatures. The specifications and standards pertinent to these grades are shown in Table 3.10.

Table 3.9 Chemical Composition of Precipitation Hardening Stainless Steels (Weight %)[a]

UNS No.	Alloy	C	Mn	Si	Cr	Ni	Mo	P	S	Other
Martensitic types										
S13800	PH 13-8 Mo	0.05	0.10	0.10	12.25-13.25	7.5-8.5	2.0-2.5	0.01	0.008	0.90-1.35 Al; 0.01 N
S15500	15-5 PH	0.07	1.00	1.00	14.0-15.5	3.5-5.5	---	0.04	0.03	2.5-4.5 Cu; 0.15-0.45 Nb
S17400	17-4 PH	0.07	1.00	1.00	15.0-17.5	3.0-5.0	---	0.04	0.03	3.0-5.0 Cu; 0.15-0.45 Nb
S45000	Custom 450	0.05	1.00	1.00	14.0-16.0	5.0-7.0	0.5-1.0	0.03	0.03	1.25-1.75 Cu; 8 x% C min. Nb
S45500	Custom 455	0.05	0.50	0.50	11.0-12.5	7.5-9.5	0.50	0.04	0.03	1.5-2.5Cu;0.8-1.4Ti;0.1-0.5Nb
Semi-austenitic types										
S15700	PH 15-7 Mo	0.09	1.00	1.00	14.0-16.0	6.50-7.75	2.0-3.0	0.04	0.04	0.75-1.50 Al
S17700	17-7 PH	0.09	1.00	1.00	16.0-18.0	6.50-7.75	---	0.04	0.04	0.75-1.50 Al
S35000	AM-350	0.07-0.11	0.50-1.25	0.50	16.0- 17.0	4.0-5.0	2.50-3.25	0.04	0.03	0.07-0.13 N
S35500	AM-355	0.10-0.15	0.50-1.25	.50	15.0-16.0	4.0-5.0	2.50-3.25	0.04	0.03	0.07-0.13 N
Austenitic types										
S66286	A-286	0.08	2.00	1.00	13.5-16.0	24.0-27.0	1.0-1.5	0.025	0.025	1.90-2.35 Ti; 0.35 max. Al; 0.10-0.50 V; 0.0030-0.0100 B

a. Single values are maximum values unless otherwise indicated.

Table 3.10 Pertinent Specifications and Standards

UNS No.	Specifications
S13800	AISI S13800 AMS 5629; 5864; ASME SA 705 (XM-13) ASTM A 564 (XM-13); A 693 (XM-13); A 705 (XM-13)
S15500	AISI S15500 AMS 5658; 5659; 5826; 5862 ASME SA 705 (XM-12) ASTM A 564 (XM-12); A 693 (XM-12); A 705 (XM-12)
S15700	AMS 5520 ASTM A564 (632); A579 (632); A693 (632); A705 (632) SAE J467 (PH15-7-Mo)
S17400	AISI S17400 AMS 5604; 5622; 5643 ASME SA 564 (630); SA 705 (630) ASTM A 564 (630); A 693 (630); A 705 (630) MIL SPEC MIL-C-24111; MIL-S-81591 SAE J467 (17-4PH)
S17700	AISI S17700 AMS 5528; 5529; 5568; 5644; 5673; 5678 ASME SA 705 (631) ASTM A 313 (631); A 564 (631); A 579 (62); A 693 (631); A 705 (631) MIL SPEC MIL-S-25043 SAE J217; J467 (17-7PH)
S35000	AMS 5546; 5548; 5554; 5745 ASTM A 579 (61); A 693 (633) MIL SPEC MIL-S-8840 SAE J467 (AM-350)
S35500	AMS 5547; 5549; 5743; 5744 ASTM A 564 (634); A 579 (634); A 693 (634); A 705 (634) MIL SPEC MIL-S-8840 SAE J467 (AM-355)
S45000	AMS 5763; 5773; 5859; 5863 ASME SA 564 (XM-25); SA 705 (XM-25) ASTM A 564 (XM-25); A 693 (XM-25); A 705 (XM-25)
S45500	AMS 5578; 5617; 5762; 5860 ASTM A 313 (XM-16); A 564 (XM-16); A 693 (XM-16); A 705 (XM-16) MIL SPEC MIL-S-83311
S66286	AMS 5525; 5726; 5731; 5732; 5734; 5737; 5804; 5805; 5853; 5858; 5895; 7235 ASME SA 638 (660) ASTM A 453 (660); A 638 (660) SAE J467 (A286); J775 (HEV-7)

Even though the changes in precipitation-hardening stainless steels since their introduction over 45 years ago cannot be termed revolutionary, some sharp differences exist between earlier grades and those recently developed. Initial acceptance of these precipitation-hardening (also called age-hardening) materials, characterized by such grades as S17700 and S17400 and S35000, was based on a combination of properties that included ease of fabrication, ability to develop high strength through a low temperature heat treatment, and good corrosion resistance.

The newer precipitation-hardening grades have improved toughness, lower notch sensitivity, greater resistance to stress corrosion cracking, better stability of properties at elevated temperatures, and even higher strength and less sensitivity to directional stresses. However, S17400 bolts and other components for sour well service are acceptable up to a hardness of 33 HRC, according to NACE Standard MR0175, provided they are heat treated in either the H1150M or double aged H1150 condition.

The semi-austenitic grades are basically sheet materials because their austenitic structure before heat treatment offers good formability. Martensitic and austenitic types might be generally classed as bar materials. These distinctions are blurred somewhat because some alloys of all three classes have been produced in both basic forms.

The Cost Factor

Do precipitation-hardening stainless steels cost more than the conventional grades? Yes, if you look only at cost of the material. But material cost is, of course, only a part of the total cost of a part. A ¾ inch diameter bar of S17400 costs about 40% more than one of S30400 (type 304) stainless steel, or about twice as much as S41000 (type 410) bar. Yet the costs of many manufactured parts have been reduced through use of a PH stainless steel–even as a replacement for an alloy steel.

A typical example illustrates this point: high-pressure hydraulic couplings, formerly machined from normalized 4140 bar stock and plated for corrosion protection, are now manufactured from S17400 stainless at a 52% cost reduction. The S17400 alloy permitted simpler and faster machining, elimination of a high temperature heat-treating operation (which caused distortion of thin wall sections in 4140 steel), less costly finishing and deburring, and elimination of a plating operation.

Chemical Composition

Control of chemical composition is extremely important in precipitation-hardening stainless steels (see Table 3.9). In stainless steels such as S13800 (PH 13-8 Mo), slight variations in minor alloying elements can produce major property differences. Phosphorus, sulfur, silicon, and manganese are controlled to extremely low limits–particularly in the vacuum induction-melted alloy–to produce a large improvement in toughness over earlier PH grades. Low carbon content further improves toughness over the earlier grades.

Alloy development has followed the general pattern of lowering carbon content to minimize precipitation of carbides at grain boundaries. At the same time, compositions have been adjusted to eliminate or minimize ferrite content in the fully hardened steel. The newer alloys are characterized by generally lower chromium and carbon content and higher nickel content than earlier types.

Corrosion Resistance

Generally, the precipitation-hardening stainless steels resist corrosive attack better than the hardenable 400 series stainless steels. In most environments, corrosion resistance approaches that of S30400.

S17400 has proven particularly effective in marine environments as a boat-shaft material. Because this alloy resists such severe environments as acidified salt solutions containing hydrogen sulfide, it is also used for valve stems, gates, and seats, pump shafts, and bearing surfaces in chemical and petrochemical applications.

Some newer alloys, with higher nickel and lower carbon contents, are highly resistant to stress corrosion cracking–nearly approaching immunity. But, even with the older bar materials, stress corrosion cracking has not been a significant problem when reasonable precautions (overaging) have been used in applications involving exposure to chloride ions.

General corrosion and stress corrosion are complex and subject to so many variables, that the prospective user of a precipitation hardened stainless is advised to contact the producer regarding specific conditions. Extensive data on corrosion resistance under a variety of conditions as well as "educated predictions" are available.

When to Consider Precipitation-Hardening Steels

Combination of properties is the key feature in considering the use of precipitation-hardening stainless steels. There are stronger steels; there are steels that are more corrosion resistant; and there are steels that are easier to fabricate. But few materials combine all of these features as do the precipitation-hardening types. In application, they have replaced high-alloy steels and other corrosion-resisting stainless steels because they offered advantages not found in either of these types of materials.

Points to consider when evaluating other steels against the utility of precipitation-hardening stainless steels include:

- Does the part require high strength? Precipitation-hardening stainless steels can provide a range of tensile strength levels– depending on heat treatment–from 125,000 to nearly 300,000 psi.
- Will alternative materials require plating for corrosion protection or appearance? Precipitation-hardening steels are generally as corrosion resistant as 18-8 grades.

- Will high temperature heat treatments cause scaling, possible distortion of thin sections, or dimensional changes that may require additional finishing operations? Many of these problems are eliminated or minimized with precipitation-hardening stainless steels because of their ability to be hardened with relatively simple, low temperature heat treatments.
- Would it be desirable to join bar or forged components to sheet sections and then heat-treat the assembly to the desired strength level? This can be done with compatible alloys such as S17400 and S17700.
- Are multiaxial stresses a factor in design? New precipitation-hardening steels are better in this respect than many conventional high strength alloy steels.
- Is toughness after long term exposure to elevated temperatures required? This characteristic is available in S13800.
- Is weldability an important requirement? Most precipitation-hardening stainless steels are readily weldable and do not require preheating as do the hardenable chromium stainless steels and high strength alloy steels. Also, the response of welds to postweld heat treatment is generally similar to that of the parent metal when a precipitation-hardening filler metal is used.

Once it has been determined that a precipitation-hardening stainless steel may be the solution to a design requirement, a specific grade must be selected. First of all, since all grades are not available in all forms, a start on the selection can be made by referring to Table 3.7. Properties, of course, are a major factor in selection. Tables 3.11 and 3.12 list selected room temperature mechanical properties for the alloys under discussion.

Heat Treatment

Heat treatment, too, can be a factor in the selection of an appropriate precipitation-hardening stainless steel. Typical heat treating cycles, listed in Table 3.13, may appear formidable. And the list is by no means complete-special heat treatments have been developed for a number of applications. But the following general comments may serve to group them in usable fashion.

<u>Martensitic alloys</u> have a martensitic structure in all conditions of heat treatment. Solution heat treatment produces a uniform untempered martensite, suitable for machining and other fabrication. Subsequent aging (precipitation hardening) between 900 and 1150°F (482 and 621°C) increases hardness and strength, with typical microstructure shown in Figure 3.3. The stress relieving effect of the age hardening treatment also increases ductility and toughness. Higher aging temperatures and longer times (overaging) improve toughness and ductility further (Note Figure 3.4), at the expense, however, of some strength and hardness, as shown in Table 3.12.

Figure 3.3 Low carbon martensite in PH stainless steels.
Some alloys may also exhibit small stringers of ferrite pools
in the martensitic matrix.

Forgings require both solution and aging treatments to achieve optimum metallurgical structure and mechanical properties. Aging of machined parts to achieve final mechanical properties causes a small predictable contraction and almost no warpage or distortion. Finish-machining operations are not usually required after heat treatment.

Semi-austenitic alloys have austenitic structures in the solution-treated condition (Condition A) as shown in Figure 3.4. This ductile structure permits forming by conventional methods used for 18-8 stainless steels.

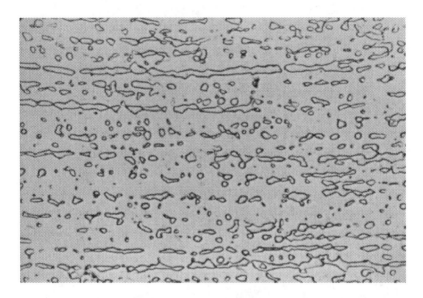

Figure 3.4 Condition A of the semi-austenitic PH grades has an austenitic matrix with ferrite pools up to about 20% which are necessary for the precipitation reaction.

After fabrication, transformation of austenite to martensite is accomplished by a thermal treatment that conditions the austenite for transformation upon cooling. The carbide destabilization mechanism is shown in Figure 3.5. Maximum strength is then developed in the martensite by an aging treatment (like that for the martensitic alloys) which results in precipitation hardening and martensite tempering, as noted in Figure 3.6.

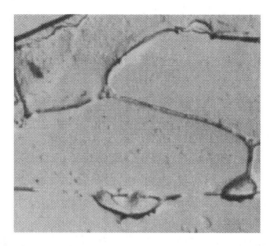

Figure 3.5 Chromium carbide ($Cr_{23}C_6$) precipitation at the ferrite/austenite grain boundary interface to destabilize the structure to induce transformation to martensite. (TEM Mag. 18,000x)

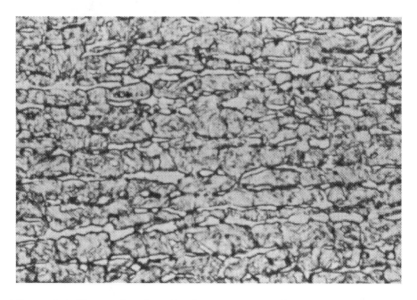

Figure 3.6 Example of aged microstructure of semi-austenitic PH grades showing ferrite pools in tempered martensitic matrix.

Exceptionally high strength-approaching 300,000 psi-can be achieved in cold-rolled (Condition C) sheet and strip because cold working (approximately 60% reduction) produces the austenite-to-martensite transformation. However, such properties are gained at the expense of some toughness and ductility, and the cold-worked product has highly directional properties. Following fabrication, highest strength properties are obtained by a one-step, low temperature aging treatment. For parts requiring only limited formability, this simple heat treatment makes this strong cold-rolled material an economical replacement for alloys that require more extensive heat treatments.

Austenitic alloys have a high nickel content which ensures an austenitic structure at subzero temperatures. Unlike the 300 series austenitic stainless steels, S66286 (alloy A-286) can be strengthened by a precipitation-hardening heat treatment. Solution treating, followed by aging, is recommended for maximum stress rupture strength. Annealing at a slightly lower temperature, followed by aging, provides the best balance between ductility and hardness.

A highly cold-worked and aged modification of S66286 has been developed in bar and wire forms. This alloy-Tensilized A-286-fills the strength and shear requirements of high performance fasteners.

Fabrication

As a class, precipitation-hardening stainless steels offer a wide selection of fabrication alternatives. Generally, the martensitic grades are best for bar, wire, or forging applications. S17400 is available also as sheet and plate for use where extensive cold formability is not required. The semi-austenitic alloys were developed primarily for flat-rolled applications, but several types are available in bar and wire.

Forming - For parts requiring extensive forming, the semi-austenitic alloys are the logical candidate materials. These alloys can be formed in Condition A about the same as S30100 (type 301) stainless steel. They work harden rapidly, however, and intermediate anneals may be required in forming intricate parts or in drawing deep shapes. Springback of these steels is similar to that of S30100.

<u>Forging</u> - Martensitic alloys, particularly S45000 (Custom 450®), S15500 (15-5 PH), and S13800 (PH 13-8 Mo), are generally preferred for forged parts. Because of their high temperature strength, these alloys require about 25% more blows under the hammer than conventional forging steels to produce equivalent deformation. More frequent reheating is also required to restore plasticity. Forging temperatures of about 2150°F (1177°C) are recommended, and the temperature should not be allowed to drop below certain minimums, depending on the alloy.

Two-phase alloys (austenite-ferrite), e.g., S17700 (17-7 PH) and S15700 (PH 15-7 Mo), have pronounced directional properties and, to a lesser extent, so does S17400. These properties increase the possibility of splitting or rupturing, particularly in upsetting or flattening operations. S15500 and S13800–ferrite-free grades–are logical choices for large multiaxially stressed forgings where directionality could be critical.

The austenitic S66286 (A-286) is forged at temperatures from 1800 to 2050°F (982 and 1121°C). Reductions of more than 15% at temperatures below 1800°F (980°C) help prevent excessive grain growth during subsequent reheating.

<u>Machining</u> - In general, precipitation-hardening stainless steels are machined in Condition A, at the low side of the speed range used for conventional stainless steels. Machinability in the hardened condition varies and depends on the final hardness of the part. In general, the precipitation-hardening materials can be machined with good finishes. Every manufacturer has specific recommendations for its materials, which should be followed.

<u>Welding</u> - The metallurgical structure of precipitation-hardening stainless steels and their method of hardening eliminate many of the welding problems usually associated with hardenable stainless grades.

Sound welds are easily produced in the martensitic alloys by any of the arc and resistance-welding processes used for the conventional stainless steels. Even though their structure resembles that of a hardenable grade, they are welded like the chromium-nickel types. Properties comparable to those of the parent metal are obtained in the weld by postweld heat treatment. No preheating is necessary to prevent cracks and excessive hardness. In heavy sections or highly restrained joints, these alloys are often heat treated to a tough, overaged condition before welding. The superior transverse notch-toughness of grades such as S15500 (15-5 PH) and S13800 (PH 13-8 Mo) makes these alloys suitable for heavy, restrained-section weldments in any condition of heat treatment. Regardless of the toughness level of the steel, notch effects that might initiate cracking should be avoided. Weldment design and procedures, as for any high strength steel, should avoid concentration of residual stresses, reaction stress at square corners, unfused notches, and sharp threads.

Semi-austenitic precipitation-hardening stainless steels have similar welding characteristics. The martensitic transformation produced in heat treatment does not impair weldability. They are readily joined by most arc and resistance welding methods and by brazing. However, fusion welding methods using an inert gas should be used to minimize loss of aluminum in the weld metal. Excellent mechanical properties can be developed in weldments without the need for preheating, postweld annealing, or special welding procedures.

The heat of welding austenitizes and solution-treats (anneals) the area adjacent to the weld. Thus, regardless of the base metal condition, the heat-affected zone and the weld metal of these semi-austenitic PH steels have a tough austenitic structure in the as-welded condition. For this reason, these steels can be welded in any condition of heat treatment without reheating and without requiring control of interpass temperature or postweld cooling rate. Strength of fusion welds made with precipitation-hardening stainless welding rods, using recommended procedures, is 90 to 100% of the base metal strength.

Because of the many welding and heat treatment combinations that can be used with the precipitation-hardening stainless steels, detailed information on their effects on weld strength should be obtained from the metal producer before welding specifications are written.

Austenitic S66286 (A-286) can also be welded by both fusion and resistance welding processes. The gas tungsten arc process is the preferred method for fusion welding. Best results are generally obtained by keeping heat input low. This is accomplished by multipass welding, using light stringer beads. Heavy sections should be solution heat treated before welding.

Although S66286 can be resistance welded following the general practices for austenitic stainless steels, allowance should be made for the higher pressures required because of the alloy's high strength at elevated temperatures.

Table 3.11 Mechanical Properties and Specifications of Precipitation-Hardening Stainless Steels

Alloy and Product Form[a]	Condition	ASTM Specification	Min. Tensile Strength, ksi	Min. Yield Strength, ksi	% Elongation, min.	% Reduction of Area, min.	Hardness, HRC[i] Min.	Hardness, HRC[i] Max.
MARTENSITIC								
UNS S13800 (PH 13-8 Mo)								
B, F	H950	A 564, A 705	220	205	10	45; 35[b]	45	---
	H1000	A 564, A 705	205	190	10	50; 40[b]	43	---
	H1025	A 564, A 705	185	175	11	50; 45[b]	41	---
	H1050	A 564, A 705	175	165	12	50; 45[b]	40	---
	H1100	A 564, A 705	150	135	14	50	34	---
	H1150	A 564, A 705	135	90	14	50	30	---
	H1150+H1150	NACE MR0175	135	105	16	50	26	33
	H1150M	A 564, A 705	125	85	16	55	26	---
P, Sh, St	H950	A 693	220	205	6-10[c]	---	45	---
	H1000	A 693	200	190	6-10[c]	---	43	---
UNS S15500 (15-5 PH)								
B, F	H900	A 564, A 705	190	170	10; 6[b]	35; 15[b]	40	---
	H925	A 564, A 705	170	155	10; 7[b]	38; 20[b]	38	---
	H1025	A 564, A 705	155	145	12; 8[b]	45; 27[b]	35	---
	H1075	A 564, A 705	145	125	13; 9[b]	45; 28[b]	32	---
	H1100	A 564, A 705	140	115	14; 10[b]	45; 29[b]	31	---
	H1150	A 564, A 705	135	105	16; 11[b]	50; 30[b]	28	---
	H1150M	A 564, A 705	115	75	18; 14[b]	55; 35[b]	24	---
P, Sh, St	H900	A 693	190	170	5-10[c]	---	40	48
	H1100	A 693	140	115	5-14[c]	---	29	40

Table 3.11 Mechanical Properties and Specifications of Precipitation-Hardening Stainless Steels (continued)

Alloy and Product Form[a]	Condition	ASTM Specification	Min. Tensile Strength, ksi	Min. Yield Strength, ksi	% Elongation, min.	% Reduction of Area, min.	Hardness, HRC[j] Min.	Max.
MARTENSITIC (continued)								
UNS S17400 (17-4 PH)								
B, F	H900[d]	A 564, A 705	190	170	10	40; 35[e]	40	---
	H925[d]	A 564, A 705	170	155	10	44; 38[e]	38	---
	H1025[d]	A 564, A 705	155	145	12	45	35	---
	H1075[d]	A 564, A 705	145	125	13	45	32	---
	H1100[d]	A 564, A 705	140	115	14	45	31	---
	H1150[d]	A 564, A 705	135	105	16	50	28	---
	H1150M[d]	A 564, A 705	115	75	18	55	24	---
P, Sh, St	H900	A 693	190	170	5-10[c]	---	40	48
	H1100	A 693	140	115	5-14[c]	---	29	40
UNS S45000 (Custom 450)								
B, shapes	Annealed	A 564[f]	130[f]	95	10	40	---	32
F, shapes	Annealed	A 705[f]	125[f]	95	10	40	---	33
B, F, shapes	H900	A 564[g], A 705[g]	180[g]	170	6; 10[b]	20; 40[b]	39	---
	H950	A 564[g], A 705[g]	170[g]	160	7; 10[b]	22; 40[b]	37	---
	H1000	A 564[g], A 705	160[g]	150	8; 12[b]	27; 45[b]	36	---
	H1025	A 564[g], A 705	150[g]	140	12	45	34	---
	H1050	A 564[g], A 705	145[g]	135	9; 12[b]	30; 45[b]	34	---
	H1100	A 564[g], A 705	130[g]	105	11; 16[b]	30; 50[b]	30	---
	H1150	A 564[g], A 705	125[g]	75	12-18[h]	33-55[h]	26	---

Table 3.11 Mechanical Properties and Specifications of Precipitation-Hardening Stainless Steels (continued)

Alloy and Product Form[a]	Condition	ASTM Specification	Min. Tensile Strength, ksi	Min. Yield Strength, ksi	% Elongation, min.	% Reduction of Area, min.	Hardness, HRC[i] Min.	Hardness, HRC[i] Max.
MARTENSITIC (continued)								
UNS S45500 (Custom 455)								
P, Sh, St	Annealed	A 693	130-165	90-150	4 min.	---	25	33
	H900	A 693	180	170	3-5[c]	---	40	---
	H1000	A 693	160	150	5-7[c]	---	36	---
	H1150	A 693	125	75	8-10[c]	---	26	---
B, F, Shapes	H900[i]	A 5649, A 705[g]	235	220	8	30	47	---
	H950[i]	A 5649, A 705[g]	220	205	10	40	44	---
	H1000[i]	A 5649, A 705[g]	205	185	10	40	40	---
P, Sh, St	H950	A 693	222	205	≤4	---	44	---
SEMI-AUSTENITIC								
UNS S15700 (PH 15-7 Mo)								
B, F,	RH950	A 564, A 705	200	175	7	25	---	---
	TH1050	A 564, A 705	180	160	8	25	---	---
P, Sh, St	Annealed	A 693	150 max.	65 max.	25 min.	---	---	---
	RH950[d]	A 693	225	200	1-4[c]	---	45-46	---
	TH1050[d]	A 693	190	170	2-5[c]	---	40	---
	Cold rolled Condition C	A 693	200	175	1	---	41	---
	CH900	A 693	240	230	1	---	46	---

Table 3.11 Mechanical Properties and Specifications of Precipitation-Hardening Stainless Steels (continued)

Alloy and Product Form[a]	Condition	ASTM Specification	Min. Tensile Strength, ksi	Min. Yield Strength, ksi	% Elongation, min.	% Reduction of Area, min.	Hardness, HRC[j] Min.	Max.
SEMI-AUSTENITIC (continued)								
UNS S17700 (17-7 PH)								
B, F	RH950[d]	A 564, A 705	185	150	6	10	41	---
	TH1050[d]	A 564, A 705	170	140	6	25	38	---
P, Sh, St	RH950	A 693	210[c]	190[c]	1-6[c]	---	43[c]	44[c]
	TH1050	A 693	180[c]	150[c]	3-7[c]	---	38	---
	Cold rolled Condition C	A 693	200	175	1	---	41	---
	CH900	A 693	240	230	1	---	46	---
W	Cold drawn Condition C	A 313	203-295[c]	---	---	---	---	---
	CH900	A 313	230-365[c]	---	---	---	---	---
UNS S35000 (AM-350)								
P, Sh, St	Annealed	A 693	200 max.	85-90 max.[c]	8-12[c]	---	---	30
	SCT850	A 693	185	150	2-8[c]	---	42	---
	SCT1000	A 693	165	145	2-8[c]	---	36	---
UNS S35500 (AM-355)								
F	SCT1000	A 705	170	155	12	25	37	---
P, Sh, St	SCT850	A 693	190	165	10	---	---	---
	SCT1000	A 693	170	150	12	---	37	---

Table 3.11 Mechanical Properties and Specifications of Precipitation-Hardening Stainless Steels (continued)

Alloy and Product Form[a]	Condition	AMS Specification	Min. Tensile Strength, ksi	Min. Yield Strength, ksi	% Elongation, min.	% Reduction of Area, min.	Hardness, HRC[j] Min.	Hardness, HRC[j] Max.
AUSTENITIC								
UNS S66286 (A-286)								
B, F	ST1650	AMS 5734, 5737	105 max.	---	---	---	---	201 HB
	ST1650A	AMS 5734, 5737	140	95	12	15	277 HB	363 HB
	ST1800	AMS 5731, 5732	105 max.	---	---	---	---	201 HB
	ST1800A	AMS 5731, 5732	130	85	15	20	248 HB	341 HB
P, Sh, St	ST1800	AMS 5525, 5858	105 max.	---	10-25	---	---	90 HRB
	ST1800A	AMS 5525, 5858	125-140	95	4-15	---	24	35

a. B - bar; F - forgings; P - plate; Sh - sheet; St - strip; W - wire.
b. Higher value is longitudinal; lower value is transverse.
c. Values vary with thickness or diameter.
d. Longitudinal properties only.
e. Higher values are for sizes up to and including 75 mm (3 in.); lower values are for sizes over 75 mm (3 in.) up to and including 200 mm (8 in.).
f. Tensile strengths of 860 to 140 MPa (125 to 165 ksi) for sizes up to 13 mm (1/2 in.).
g. Tensile strength only applicable up to sizes of 13 mm (1/2 in.).
h. Varies with section size and test direction.
i. Up to and including 150 mm (6 in.).
j. Unless otherwise noted.

Table 3.12 Typical Mechanical Properties of Precipitation-Hardening Stainless Steels

Alloy and Product Form[a]	Condition[b]	Tensile Strength (ksi)	Yield Strength (ksi)	% Elongation	% Reduction of Area	Hardness, HRC[e]	Impact Charpy V (ft-lb)
MARTENSITIC							
UNS S17400 (17-4 PH)							
B, W	H900	200	185	14	50	44	15
	H1025	170	165	15	56	38	35
	H1075	165	150	16	58	36	40
	H1150	145	125	19	60	33	50
	H1150+H1150	140	110	20	60	31	80
	H1150-M	125	85	22	68	27	100
Sh, St	A	160	145	5	---	35	---
	H900	210	200	7	---	45	---
	H925	200	195	8	---	43	---
	H1025	185	170	8	---	38	---
	H1075	175	165	8	---	37	---
	H1150	160	150	11	---	35	---
	H1150-M	150	130	12	---	33	---
UNS S15500 (15-5 PH)							
B	H900	200	185	14	50	44	15
	H1025	170	165	15	56	38	35
	H1075	165	150	16	58	36	40
	H1150	145	125	19	60	33	50
	H1150-M	125	85	22	68	27	100
UNS S13800 (PH 13-8 Mo)							
Billet, P	H950	225	210	12	50	47	30
B, W	H1000	215	205	13	55	45	40
	H1050	190	180	15	55	43	60
	H1100	170	150	16	60	36	100

Table 3.12 Typical Mechanical Properties of Precipitation-Hardening Stainless Steels (continued)

Alloy and Product Form[a]	Condition[b]	Tensile Strength (ksi)	Yield Strength (ksi)	% Elongation	% Reduction of Area	Hardness, HRC[e]	Impact Charpy V (ft-lb)
UNS S13800 (PH 13-8 Mo) (continued)							
B, W	H1150-M	130	85	22	70	28	120
UNS S45000 (Custom 450)							
B	Annealed 1900	144	117	14	60	28	60
	Aged 900, 4 hr	196	184	14	60	42	35
UNS S45500 (Custom 455)							
1-in. bar	Annealed 1500; WQ	145	115	14	60	31	70
	Annealed 1500; Aged 900, 4 hr	245	235	10	45	49	9
	Annealed 1500; Aged 1000, 4 hr	205	195	16	55	45	20
SEMIAUSTENITIC							
UNS 17700 (17-7 PH)							
Sh, St	A	130	40	35	---	85 HRB	---
	TH1050	200	185	9	---	43	---
	RH950	235	220	6	---	48	---
	C	220	190	5	---	43	---
	CH900	265	260	2	---	49	---
W	C	230-280	---	---	---	---	---
	CH900	282-335	---	---	---	---	---
UNS S15700 (PH 15-7 Mo)							
Sh, St	A	130	55	35	---	88 HRB	---
	TH1050	220	210	7	---	45	---
	RH950	240	225	6	---	48	---
	C	220	190	5	---	45	---
	CH900	265	260	2	---	50	---

Table 3.12 Typical Mechanical Properties of Precipitation-Hardening Stainless Steels (continued)

Alloy and Product Form[a]	Condition[b]	Tensile Strength (ksi)	Yield Strength (ksi)	% Elongation	% Reduction of Area	Hardness, HRC[e]	Impact Charpy V (ft-lb)
UNS S35000 (AM-350)							
Sh, St	H annealed	149	63	39	---	20	---
	SCT 850	200	170	13	---	45	---
	CRT	227-335	214-322	8-11	---	45-50	---
UNS S35500 (AM-355)							
Sh, St	H annealed	187	56	29	---	32	---
	SCT 850	215	180	11	---	40	---
	CRT	210-265	175-260	23-10	---	45-50	---
B	SCT 850	215	169	12	---	---	---
	SCT 1000	174	140	16	---	---	---
AUSTENITIC							
UNS S66286 (A-286)							
B	Annealed 1650	90	35	45	70	83 HRB	---
	A1800, WQ	85	35	48	72	80 HRB	---
	A; aged 1350, 16 hr	150	100	25	40	34	---
	Tensilized[c]	215	200	9.5	30	---	---
W	Annealed 1650	93	35	50	65	80 HRB	---
	A 1800, WQ	91	34	53	68	80 HRB	---
	A; aged 1350, 16 hr	155	100	29	46	34	---
	Tensilized[d]	220	210	8	25	---	---

a. B - bar; F - forging; P - plate; Sh - sheet; St - strip; W - wire.

b. Designation of basic treatment of alloys is as follows (numbers following the letters indicate the temperature of the heat treatment): A=Solution treated; C=Cold rolled; CH=Cold-rolled and precipitation hardened; TH=Austenite conditioned and hardened; RH=Austenite conditioned, cooled to -100°F, and hardened; SRH=Modified RH treatment; H=Precipitation hardened; CRT=Cold-rolled and tempered (or aged); SCT="SCT 850" defines austenite conditioning at 1750°F, followed by subzero cooling for transformation, followed by aging at 850°F. Other abbreviations used are: AC=Air cooled; WQ=Water quenched.   c. Strength in double shear, 123,000 psi.   d. Strength in double shear, 125,000 psi.   e. Unless otherwise noted.

Table 3.13 Typical Heat Treatments for Precipitation-Hardening Steels[a]

UNS No.	Alloy	Condition	Solution Treatment (°F)	Austenite Condition (°F)	Martensite Transformation (°F)	Precipitation Hardening (°F)
S17400	17-4 PH	H900	1900, 30 min.	---	---	900, 1 hr
S15500	15-5 PH	H925-1150	1900, 30 min.	---	---	925-1150, 4 hr
		H1150-M	1900, 30 min.	---	---	1400, 2 hr +1150, 4 hr
S13800	PH 13-8 Mo	H950-1100	1700, 30 min.	---	---	950-1100, 4 hr
		RH 950	1700, 1 hr	1700, 1 hr (optional)	Cool to -100 within 1 hr; hold 8 hr	950, 4 hr
S45500	Custom 455	---	1500-1550	---	---	900-1100, 4 hr
S17700	17-7 PH	TH 1050	1900	1400, 90 min.	Cool to 60 within 1 hr; hold 30 min.	1050, 90 min.
S15700	PH 15-7 Mo	RH950	1900	1750, 10 min.	Cool to -100 within 1 hr; hold 8 hr	950, 1 hr
		CH900-1050	---	---	Cold reduce 60%	900-1500, 1 hr
S35000	AM-350	SCT 850	Supplied cold rolled and aged	1750, 10-60 min.	Cool to -100; hold 3 hr	850, 3 hr
		CRT	Supplied cold rolled and aged	---	---	---
S35500	AM-355	SCT 1000	Supplied cold rolled and aged	1750, 10-60 min.	Cool to -100; hold 3 hr	1000, 3 hr
		CRT	Supplied cold rolled and aged	---	---	---
S66286	A-286	Standard	1800, 1 hr. WQ or annealed, 1650, 2 hr oil quenched	---	---	1300-1400, 16 hr, AC
		Tensilized	Supplied cold worked and aged	---	---	---

a. Designation of heat treatments of alloys is as follows (numbers following the letters indicate the temperature of the heat treatment): CH=Cold-rolled and precipitation hardened; TH=Austenite conditioned and hardened; RH=Austenite conditioned, cooled to -100°F, and hardened; H=Precipitation hardened; CRT=Cold-rolled and tempered (or aged); SCT="SCT 850" defines austenite conditioning at 1750°F, followed by subzero cooling for transformation, followed by aging at 850°F. Other abbreviations used are: AC=Air cooled; WQ=Water quenched.

Table 3.14 Physical Properties of Precipitation-Hardening Stainless Steels

UNS No.	Density g/cm³ (lb/in.³)	Elastic Modulus GPa (10⁶ psi)	Mean CTE[a] from 0°C (32°F) to: 315°C (600°F) μm/m°C (μin./in.°F)	538°C (1000°F) μm/m°C (μin./in.°F)	Thermal Conductivity at 100°C (212°F) W/m°K (Btu/ft h°F)	at 500°C (932°F) W/m°K (Btu/ft h°F)	Specific Heat at 0-100°C (32-212°F) J/kg°K (Btu/lb°F)	Electrical Resistivity ηΩm	Magnetic Permeability
S13800	7.8 (0.28)	203 (29.4)	11.2 (6.2)	11.9 (6.6)	14.0 (8.1)	22.0 (12.7)	460 (0.11)	1020	Ferro-magnetic
S17400	7.8 (0.28)	196 (28.5)	11.6 (6.4)	---	18.3 (10.6)	23.0 (13.1)	460 (0.11)	800	95
S15500	7.8 (0.28)	196 (28.5)	11.4 (6.3)	---	17.8 (10.3)	23.0 (13.1)	420 (0.10)	770	95
S45000	7.8 (0.28)	200 (29.0)	11.2 (6.2)	---	18.0 (10.4)	24.7 (14.3)	460 (0.11)	750	Ferro-magnetic
S45500	7.8 (0.28)	200 (29.0)	11.4 (6.3)	---	18.0 (10.4)	24.7 (14.3)	460 (0.11)	750	Ferro-magnetic
S17700	7.8 (0.28)	204 (29.5)	11.6 (6.4)	---	16.4 (9.5)	21.8 (12.6)	460 (0.11)	830	1.4-3.6 (Condition A), >119 Aged
S15700	7.8 (0.28)	200 (29.0)	11.0 (6.1)	11.9 (6.6)	16.2 (9.4)	21.7 (12.6)	460 (0.11)	820	4.7-5.3 (Condition A), >119 Aged
S35000	7.8 (0.28)	200 (29.0)	12.3 (6.8)	--- (7.2)	15.6 (9.0)	21.5 (12.4)	460 (0.11)	800	14-115
S35500	7.8 (0.28)	200 (29.0)	--- (7.5)	--- (7.7)	15.6 (9.0)	20.9 (12.1)	460 (0.11)	750	87-150
S66286	7.9 (0.29)	200 (29.0)	17.1 (9.5)	17.7 (9.8)	---	21.7 (12.6)	420 (0.10)	910	<1.05

a. Coefficient of thermal expansion.

Ferritic Stainless Steels

Derived from the Latin word *ferrum* meaning iron, ferritic stainless steels are composed of a solid solution of carbon in alpha iron, generally with a low carbon concentration and at least 10.5% chromium. In the annealed condition, ferritic stainless steels look very much like low carbon iron under the microscope as noted in Figure 3.7. The background, or matrix, is made up of grains of chromium iron and black chromium carbides scattered through the grains. Ferrite is a soft and ductile structure and is non-hardenable by heat treatment. Some grades may contain molybdenum, silicon, aluminum, titanium, and niobium to confer particular characteristics. The ferritic alloys have good ductility and formability, but their high temperature strength is relatively poor compared to that of the austenitic grades. Another difference compared to austenitic stainless steels is their ferromagnetic behavior.

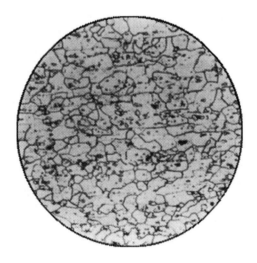

Figure 3.7 Typical microstructure of ferritic stainless steels
which are not hardenable by heat.

The ferritic grades exhibit a transition from ductile to brittle behavior over a narrow temperature range. The brittle fracture can be initiated by intergranular cracking or strain-induced cracking of second phase particles. This ductile to brittle transition can occur at ambient temperatures for chemistries with high carbon, nitrogen, and chromium levels, especially on thick sections. Today, usually with argon-oxygen decarburization (AOD) melting practice, the carbon and nitrogen levels can be controlled further. Progress in melting and processing techniques has helped control the interstitial elements and thereby reduce the carbide and nitride precipitates which are detrimental to toughness. This makes it possible to produce grades with unusually high chromium and molybdenum levels. Based on the level of chromium, the ferritics can be classified into three categories:

1. *Low Cr ferritics* with chromium levels between 10 to 14%. These alloys have carefully controlled chemistries (especially the interstitials and the stabilizers). The prototype of a low Cr ferritic is S40900 (type 409), typically 0.003C-11Cr-0.3Ti.
2. *Intermediate Cr ferritics* with chromium levels between 14% to 19%. Included here are some of the first generation ferritics such as S43000 (type 430) which had high carbon. In terms of usage, the stabilized alloys are preferred for welding applications.
3. *Superferritics* which have Cr levels between 19% to 30%. These are sometimes referred to as the third generation of ferritic stainless steels. These have very low carbon and nitrogen (less then 0.02%).

The chemical compositions of commercially produced 400 series ferritics are shown in Table 3.15. Room temperature mechanical properties for selected ferritic grades are given in Table 3.16. The typical annealed yield strengths range from 35 to 75 ksi (240 to 515 MPa) and the tensile strengths range from 60 to 95 ksi (415 to 655 MPa). Ductility tends to range between 20 and 35%. Austenitic stainless steels have higher coefficients of thermal expansion (typically 60% higher) and lower thermal conductivity as compared to the ferritics. Typical physical properties of selected grades of annealed ferritics are given in Table 3.17.

Table 3.15 Chemical Compositions of Commercially Produced Ferritic Stainless Steels (Weight %)[a]

Alloy	UNS No.	C	Cr	Mo	Ni	Ti	Nb	Other
Low Cr Ferritic Stainless Steels								
405	S40500	0.08	11.5-14.5	---	---	---	---	0.10-0.30 Al
YUS 405Si	---	0.08	12.0	---	---	---	---	0.15 Al, 2 Si
409	S40900	0.08	10.5-11.75	---	0.5	6 x C min.	---	---
409Ni	S40975	---	---	---	---	---	---	---
409Cb[b]	S41045	0.02	12.5	---	0.2	---	0.4	---
410	S41003	0.03	10.50-12.50	---	1.5	---	---	0.030 N
410S	S41008	0.08	11.5-13.5	---	0.6	0.002-0.050	---	1 Si
11CrCb	---	0.01	11.3	---	0.25	0.20	0.35	1.3 Si
12SR	---	0.2	12.0	---	---	0.3	---	1.2 Al
Intermediate Cr Ferritic Stainless Steels								
430	S43000	0.12	16.0-18.0	---	---	---	---	---
430F	S43020	0.12	16.0-18.0	---	---	---	---	0.06 P; 0.15 S
430F Se	S43023	0.12	16.0-18.0	---	---	---	---	0.15 min. Se
434	S43400	0.12	16.0-18.0	0.75-1.25	---	---	---	---
435Mod	---	0.013	19.0-20.0	---	0.29	---	0.6	0.44 Si
436	S43600	0.12	16.0-18.0	0.75-1.25	---	---	0.5-0.6	---
436M2	---	0.02	17.0-17.5	1.2	0.21	---	0.4-0.6	0.5 Si
18CrCb	S44100	0.02	18.0	---	0.3	0.3	0.7	---
AL 433[TMb]	---	0.02	19.0	---	0.3	---	0.4	0.5 Si, 0.4 Cu
AL 466[TMb]	---	0.01	11.5	---	0.2	0.1	0.2	---
AL 468[TMb]	S46800	0.03	20.0	---	0.5	0.07-0.30	0.10-0.60	---
YUS 436S	---	0.01	17.4	1.2	---	0.2	---	---
439	S43035	0.07	17.0-19.0	---	0.5	0.2-1.0	---	---

Table 3.15 Chemical Compositions of Commercially Produced Ferritic Stainless Steels (Weight %)[a] (continued)

Alloy	UNS No.	C	Cr	Mo	Ni	Ti	Nb	Other
Intermediate Cr Ferritic Stainless Steels (continued)								
18SR	--	0.04	18.0	--	--	0.4	--	2.0 Al
406	K91470	0.06	12.0-15.0	--	--	0.6	--	2.75-3.75 Al
Superferritic Stainless Steels								
26-1Ti	S44626	0.02	26.0	1.0	0.25	0.5	--	0.025 N
444	S44400	0.02	18.0	2.0	0.4	0.5	--	0.02 N
SEA-CURE®	S44660	0.02	27.5	3.4	1.7	0.5	--	0.025 N
Nu Monit®	S44635	0.025	25.0	4.0	4.0	0.4	--	0.025 N
AL 29-4C®	S44735	0.030	29.0	4.0	1.0	--	--	0.045 N
E-BRITE 26-1®	S44627	0.002	26.0	1.0	0.1	--	0.1	0.01 N
AL 29-4-2®	S44800	0.005	29.0	4.0	2.0	--	--	0.01 N
SHOMAC 26-4	--	0.003	26.0	4.0	--	--	--	0.005 N
SHOMAC 30-2	--	0.003	30.0	2.0	0.18	--	--	0.007 N
YUS 190L	--	0.004	19.0	2.0	--	--	0.15	0.0085 N
NAR-FC-4	--	0.016	22.0	2.0	--	--	--	--
436LM	--	0.016	22.0	2.0	--	--	--	--

a. Single values are maximums unless otherwise indicated.
b. Typical values.

Table 3.16 Mechanical Properties of Ferritic Stainless Steels[d]

Alloy and Product Form[a]	Condition	Tensile Strength		Yield Strength		% Elongation	Reduction in Area, %	Hardness HRB	ASTM Specification
		MPa	ksi	MPa	ksi				
UNS S30400 (Type 304)									
W	Annealed	515	75	205	30	35	50	---	A 580
P, Sh, St	Annealed	515	75	205	30	40	---	92 max.	A 167
UNS S40500 (Type 405)									
B	Annealed	415	60	170	25	20	45	---	A 479
F	Annealed	415	60	205	30	20	45	---	A 473
W	Annealed	480	70	280	40	20	45	---	A 580
P, Sh, St	Annealed	415	60	170	25	20	---	88 max.	A 176, A 240
UNS S40900 (Type 409)									
P, Sh, St	Annealed	380	55	205	30	20	---	80 max.	A 240
	Annealed	380	55	205	30	22[c]	---	80 max.	A 240
UNS S42900 (Type 429)									
B	Annealed	480	70	275	40	20	45	---	A 276
P, Sh, St	Annealed	450	65	205	30	22[c]	---	88 max.	A 176, A 240
UNS S43000 (Type 430)									
B	Annealed	415	60	205	30	20	45	---	A 276
W	Annealed	480	70	275	40	20	45	---	A 580
P, Sh, St	Annealed	450	65	205	30	22[c]	---	88 max.	A 176, A 240
UNS S43020 (Type 430F)									
F	Annealed	485	70	275	40	20	45	---	A 473
UNS S43035 (Type 439)									
B	Annealed	485	70	275	40	20	45	---	A 479
P, Sh, St	Annealed	450	65	205	30	22	---	88 max.	A 240

Table 3.16 Mechanical Properties of Ferritic Stainless Steels[d] (continued)

Alloy and Product Form[a]	Condition	Tensile Strength		Yield Strength		% Elongation	Reduction in Area, %	Hardness HRB	ASTM Specification
		MPa	ksi	MPa	ksi				
UNS S43036 (Type 430Ti)									
B[b]	Annealed	515	75	310	45	30	65	---	---
UNS S43400 (Type 434)									
W[b]	Annealed	545	79	415	60	33	78	90 max.	---
Sh[b]	Annealed	530	77	365	53	23	---	83 max.	---
UNS S43600 (Type 436)									
Sh, St[b]	Annealed	530	77	365	53	23	---	83 max.	---
UNS S44200 (Type 442)									
B[b]	Annealed	550	80	310	45	20	40	90 max.	---
P, Sh, St	Annealed	515	75	275	40	20	---	95 max.	A 176
UNS S44400 (Type 444)									
P, Sh, St	Annealed	415	60	275	40	20	---	95 max.	A 176
UNS S44600 (Type 446)									
B	Annealed	480	70	275	40	20	45	---	A 276
W	Annealed	480	70	275	40	20	45	---	A 580
	Cold finished	480	70	275	40	16	45	---	A 580
P, Sh, St	Annealed	515	75	275	40	20	---	95 max.	A 176
18SR									
Sh, St[b]	Annealed	620	90	450	65	25	---	90 min.	---
UNS S44627 (E-Brite 26-1)									
B	Annealed	450	65	275	40	20	45	---	A 276
P, Sh, St	Annealed	450	65	275	40	22[c]	---	90 max.	A 176, A 240
UNS S44635 (MONIT)									
P, Sh, St	Annealed	620	90	515	75	20	---	---	A 176, A 240

Table 3.16 Mechanical Properties of Ferritic Stainless Steels[d] (continued)

Alloy and Product Form[a]	Condition	Tensile Strength		Yield Strength		% Elongation	Reduction in Area, %	Hardness HRB	ASTM Specification
		MPa	ksi	MPa	ksi				
UNS S44660 (Sea-Cure/SC-1)									
P , Sh, St	Annealed	585	85	450	65	18	---	100 max.	A 176, A 240
UNS S44735 (29-4C)									
P , Sh, St	Annealed	550	80	415	60	18	---	---	A 176, A 240
UNS S44800 (29-4-2)									
P , Sh, St	Annealed	550	80	415	60	20	---	98 max.	A 176, A 240
B	Cold finished	520	75	415	60	15	30	---	A 276
	Annealed	480	70	380	55	20	40	---	A 479

a. B - bar; F - forging; W - wire; P - plate; Sh - sheet; St - strip
b. Typical values.
c. 20% reduction for 1.3mm (0.050 in.) and under in thickness.
d. Single values are maximums unless otherwise indicated.

Table 3.17 Typical Physical Properties of Wrought Stainless Steels in the Annealed Condition

UNS No.	Type	Density g/cm³	Elastic Modulus GPa	Mean CTE[a] from 0°C to:			Thermal Conductivity		Specific Heat[b] J/kg°K	Electrical Resistivity nΩm	Magnetic Permeability[c]	Melting Range (°C)
				100°C mm/m°C	315°C mm/°C	538°C mm/°C	at 100°C W/m°K	at 500°C W/m°K				
S30400	304	8.0	193	17.2	17.8	18.4	16.2	21.5	500	720	1.02	1400-1450
S40500	405	7.8	200	10.8	11.6	12.1	27.0	---	460	600	---	1480-1530
S40900	409	7.8	---	11.7	---	---	---	---	---	---	---	1480-1530
S42900	429	7.8	200	10.3	---	---	25.6	---	460	590	---	1450-1510
S43000	430	7.8	200	10.4	11.0	11.4	26.1	26.3	460	600	600-1100	1425-1510
S43020	430F	7.8	200	10.4	11.0	11.4	26.1	26.3	460	600	---	1425-1510
S43100	431	7.8	200	10.2	12.1	---	20.2	---	460	720	---	---
S43400	434	7.8	200	10.4	11.0	11.4	---	26.3	460	600	600-1100	1425-1510
S43600	436	7.8	200	9.3	---	---	23.9	26.0	460	600	600-1100	1425-1510
S43035	439	7.8	200	10.4	11.0	11.4	24.2	---	460	630	---	---
S44400	444	7.8	200	10.0	10.6	11.4	26.8	---	420	620	---	---
S44600	446	7.5	200	10.4	10.8	11.2	20.9	24.4	500	670	400-700	1425-1510
S44627	26-1	7.6	200	10.6	---	---	---	---	---	---	---	---
S44800	29-4-2	7.6	200	9.4	---	---	---	---	---	---	---	---

a. Coefficient of thermal expansion.
b. At 0 to 100°C.
c. Approximate values.

Applications

Among the various ferritic stainless steels, S40900 (type 409) is the most widely used (more than 300,000 tons/year in the U.S.). The total amount of ferritics produced in the U.S. is more than 500,000 tons/year with, the automotive market being the largest user.

To avoid the embrittlement problems, ferritic stainless steels are mostly used in relatively thin gages, especially in alloys that are high in chromium, such as the superferritics. Figure 3.8 shows the ductile-to-brittle transition temperatures (DBTT) for ferritic stainless steels increase with section thickness. Ferritic stainless steels are subject to another embrittlement phenomenon when exposed to temperatures of 700-950°F (371-510°C) over an extended period of time. This phenomenon is called 885°F embrittlement because that is the temperature at which embrittlement is most pronounced. This embrittlement results in low ductility, poor impact strength, and increased hardness and tensile strength at room temperature which can cause the alloy to fracture, 885°F embrittlement can be removed by heat treating at 1100°F (593°C) or above, followed by air cooling.

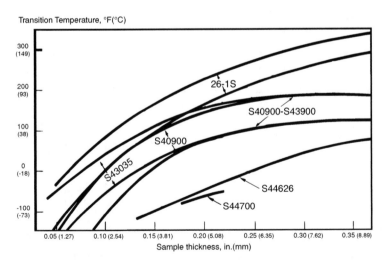

Figure 3.8 Ductile-to-brittle transition temperatures (DBTT)
for ferritic stainless steels increase with section thickness.
Bands for 26-1S, S40900 (type 409), and
S43035 (type 439) indicate data scatter.

Ferritic stainless steels are also subject to sigma-phase embrittlement when exposed to temperatures of 1000-1600°F (538-871°C) over extended periods of time. This results in a loss of ductility at room temperature. Sigma phase embrittlement can be removed by heat treatment at 1850°F (1010°C) followed by air cooling.

Ferritic stainless steels can give moderate to excellent corrosion resistance compared to their austenitic counterparts at a lower cost (since a lower amount of alloying elements is needed in conditions which lead to localized corrosion). In addition, these grades are highly resistant to chloride stress corrosion cracking. The use of ferritics in various applications is listed in Tables 18a and 18b. The low chromium (11%) alloys have reasonable corrosion and oxidation resistance and fabricability at low cost. Therefore, use of S40900 and its modifications will continue to increase in the future since it can replace coated carbon steel and brass. The largest market for these alloys is in automotive exhaust systems. S40500 (type 405) and S40900 have been used in petroleum refinery equipment as fractionation column trays, vessel cladding, catalyst screens, heat exchanger tubing, valve trim, pump shafts, pump cases, impellers, and wear rings.

The intermediate chromium alloys, such as S43000 (type 430), resist mild oxidizing acids and organic acids and are used in food handling equipment, sinks, etc. S43000 is successfully used as tubing in crude distillation, hydrodesulfurizing and hydrocracking services. The intermediate chromium alloys are also used as automotive trim and cooking utensils. These alloys are not as easily fabricated as the lower chromium alloys. However, newer stabilized alloys such as S43035 (type 439), 18CrCb, and 436M2 make fabrication easier. These grades are suitable for equipment exposed to aqueous chloride environments, heat transfer applications, and condenser tubing for fresh water power plants. Stainless steels of S43000, S43400 (type 434), 436M2, and 436LM are also used for auto trim and architectural applications. In particular, the thermal expansion of these ferritics is less than S30400 (type 304), which is advantageous in terms of design flexibility. It could therefore replace S30400 for roofs.

Superferritics are resistant to chloride stress corrosion cracking and are preferred over austenitics for this reason. However, it has to be recognized that the fabricability of these alloys, especially welding, is not as good as austenitic stainless steels. Therefore, they are generally used in applications requiring a high level of corrosion and oxidation resistance such as in heat exchangers and piping systems for chloride-bearing aqueous solutions and seawater. S44400 (type 444) is used in wineries, breweries, meat processing plants, and institutional cooking equipment. The superferritics are also used in the chemical and petrochemical industries, synthesis of urea, and the pulp and paper industry. S44626 (alloy 26-1) has been successfully used in crude distillation, hydrogen separation, crude oil refining, sour water stripping, and in handling hot caustic solutions. Some of the superferritics have been used with reducing acids. One example is the use of S44800 (29-4-2) alloy in the acid hydrolysis process for making alcohol from starch. The same alloy has also been used as flue gas scrubbers and reheater tubing where acid chlorides are present. Superferritics have also been used for architectural applications. Table 3.18b shows the major industrial markets and applications for ferritic stainless steel.

Corrosion

Selection of a suitable stainless steel for a specific environment requires consideration of several criteria. Corrosion tests have been designed to single out one particular mechanism; therefore, a particular application of stainless steel will usually require consideration of more than one type of test. In this section the corrosion resistance properties of ferritic stainless steels will be described.

Table 3.18a Applications and Major Properties
of Ferritic Stainless Steels

Alloy	Characteristics	Applications
Low Cr Ferritic Stainless Steel		
S40500 (405)	Good weldability	Annealing boxes, quenching racks, vessel linings, and precision rolled profiles for steam turbine parts
YUS 405Si	Oxidation resistance	Oil stoves, furnace materials
S40900 (409)	High workability	Automotive exhaust system parts
S40975 (409Ni)	Formability	Flanges for exhaust system
11CrCb	Strength and Oxidation Resistance	Gas turbine silencer housings
Intermediate Cr Ferritic Stainless Steel		
S43000 (430)	Good corrosion resistance	Appliances, decorative trim, and chemical handling
S43020 (430F) S43023 (430F Se)	Good machinability	Equipment, oil burner parts, television cones, and kitchens
S43400 (434)	Good pitting corrosion resistance	Appliances, appliance molding, and plumbing supplies
S43600 (436)	Good corrosion and heat resistance	Solenoid valves
Uniloy446	Good scaling resistance	Annealing boxes, heaters, salt bath electrodes, and oil burner parts
YUS 436S	High workability	Automotive mufflers, automotive exhaust system parts
S43900 (439)	Corrosion resistance equal to 304	Automotive exhaust components, and hot water tanks
12SR	Good oxidation resistance	---
18SR	High oxidation resistance	Muffler wraps

Table 3.18b Usage of Ferritic Stainless Steels
in Various Industrial Sectors

Application	Comments
Corrosion in Pulp and Paper Plant Environment	Superferritics such as 29-4-2 and 29-4 should be considered.
Food and Beverage Industry	Type 444 is used.
Oil and Gas Industry	S43000 (type 430) and type 444 are used for exchanger tubing.
Power Industry	Ferritics are used for exchangers and condensers.
Transportation Industry - Automotive Applications	S43000 (type 430), S43400 (type 434) used for trim, wheel covers, fasteners, windshield wiper arms; type 409, type 436S used in exhaust systems.
- Railroad Cars, Trucks and Buses	S40975 (type 409Ni) is used as a structural component in buses. S43000 (type 430) used for exposed functional parts.

Atmospheric and General Corrosion

Atmospheric corrosion resistance is an essential property for exterior applications. Ferritic stainless steels can give moderate to outstanding atmospheric corrosion resistance, due to the passive films that are mainly composed of chromium oxy-hydroxides. Breakdown of passivity leads to formation of small pits, which tend to appear as stain (red rust) on the surface.

Tables 3.19 compares some of the ferritic stainless steels exposed to marine atmospheres. All alloys had a bright annealed (BA) finish before exposure. The panels in this case were exposed at Kure Beach in North Carolina. The average pit depths for most of the alloys reported in the table was less than 1 mil (25 microns) after one year of exposure. The atmospheric corrosion resistance is improved by increasing chromium and molybdenum content in the alloy.

Table 3.19 Corrosion of Ferritic Stainless Steels
in a Marine Atmosphere at Kure Beach 25m from Ocean

Alloy	UNS No.	Cr (wt %)	Duration	Appearance
430	S43000	17%	4weeks	Rust staining on 39% of surface
434	S43400	17%	4weeks	Rust staining on 24% of surface
436LM	---	21%	4weeks	Rust staining on 13% of surface
436M2	---	17%	4weeks	Rust staining on 12% of surface
444	S44400	18%	4weeks	Rust staining on 9% of surface
304	S30400	18%	4weeks	Rust staining on 25% of surface
304	S30400	18%	15years	Spotted with slight rust stain on 15% of surface[a]

a. Kure Beach 250m from ocean

Other factors influencing atmospheric corrosion resistance are the level of residuals (i.e., carbon, nitrogen, titanium, niobium, etc.) and the surface finish. In Table 3.20 the excellent corrosion resistance of type 436M2 is attributed to the lower level of residuals. In Japan, superferritics (30 Cr-2 Mo and 22 Cr-2 Mo) have been used for roofing applications.

Ferritic steels have good corrosion resistance to strongly oxidizing acids, such as nitric acid, and have excellent resistance to most organic acids. The general corrosion characteristics of ferritic stainless steel in various acids are summarized in Table 3.20. The general corrosion behavior of the superferritics is equivalent to austenitics in oxidizing acids. Stainless steels exhibit stable passivity above 93% H_2SO_4 concentration. In dilute concentrations (such as 2% H_2SO_4), which are reducing in nature, the ferritics are attacked severely.

Table 3.20 General Corrosion of Ferritic Stainless Steels in Acids

UNS (Type)	Corrosion rate	
	mm/year	Mils/year
Boiling 40% HNO$_3$		
S40900 (409)	9.65	380
S43000 (430)	0.72	28.3
S43400 (434)	0.76	29.9
(436M2)	0.14	5.5
S44400 (444)	0.036	1.4
S30400 (304)	0.036	1.4
Boiling 80% Acetic acid		
S40900 (409)	9.13	359
S43000 (430)	0.65	25.6
S43400 (434)	0.29	11.4
(436M2)	0.14	5.5
S44400 (444)	0.073	2.87
S30400 (304)	0.073	2.87
2% Sulfuric acid at 86°F (30°C)		
(444)	10.2	400
S30400 (304)	0.03	1.3
60% NaOH at 212°F (100°C)		
(444)	0.61	24
S30400 (304)	0.07	3.0

Pitting and Crevice Corrosion

The evolution of corrosion pits on stainless steels has been studied extensively. The chromium, molybdenum, and general cleanliness of the steel, along with the presence of stabilizers to tie up the interstitials, all contribute to the pitting and crevice corrosion resistance of ferritics. The surface finish can also have a significant effect on pitting corrosion. The tendency to pitting can be measured by the pitting potentials. The higher the pitting potential, the less tendency there is for pitting (see Table 3.21).

Table 3.21 Pitting Potential of Stainless Steels
in 1M NaCl at 86°F (30°C)

UNS (Type)	Potential vs. Standard Calomel Electrode, mV
S30400 (304)	220
S43035 (439)	280
S44400 (444)	430

Localized corrosion can also be measured using weight loss measurements after exposure to either ferric chloride solution (as per ASTM G 48) or sodium chloride solution. The ferric chloride immersion test is very aggressive for low and intermediate chromium ferritics. Testing in NaCl solution is preferred for such alloys. The pitting behavior for several ferritic grades in a variety of oxygen-saturated NaCl solutions at 194°F (90°C) for 30 days is shown in the Table 3.22.

Table 3.22 Pitting Behavior of Stainless Steels
in Oxygen-Saturated NaCl Solutions

Solution	Weight Loss, mm/year			
	S43000	S43400	444	S30400
3% NaCl, 18200 ppm Cl⁻	0.21	---	0.0	0.0
1200 ppm Cl⁻, 400 ppm Cu^{++}	4.21	1.87	0.86	0.87
180 ppm Cl⁻, 60 ppm Cu^{++}	0.18	0.06	0.004	0.02
600 ppm Cl⁻, 1 ppm Cu^{++}, 5 ppm Zn^{++}	---	0.02	0.002	0.03
600 ppm Cl⁻, 5 ppm Cu^{++}	---	0.11	0.015	0.06
600 ppm Cl⁻, 20 ppm Cu^{++}	---	0.38	0.03	0.39

The pitting behavior in ferric chloride solution is reported in Table 3.23. Note the high weight loss in this test.

Table 3.23 Pitting Corrosion Behavior in Ferric Chloride Solution
(per ASTM G 48) at 72°F (22°C) for 24 Hours

UNS (Type)	Weight Loss	
	mils/year	mm/year
S43000 (430)	2146	54.5
S43400 (434)	2992	76
(436M2)	638	16.2
S44400 (444)	1185	30.1
(436LM)	197	5.0
S40900 (409)	3493	88.7
S30400 (304)	1670	42.4

Intergranular Corrosion Resistance (IGC)

Susceptibility of the ferritics to intergranular corrosion (IGC) is due to chromium depletion caused by precipitation of chromium carbides and nitrides at grain boundaries. In conventional ferritic alloys, sensitization is caused by heating above 1700°F (925°C). Sensitization of ferritics is commonly found in welds and their heat affected zone.

IGC can be mitigated by either reducing carbon to very low levels or adding titanium or niobium to tie up the carbon and nitrogen. Recently the ASTM A 240 specification has been updated to include three different compositions (S40910, S40920, and S40930) under the S40900 designation. This was done in recognition of the fact that the previous stabilization requirements (Ti/C ≥6) for avoiding sensitization of the S40900 alloy did not consider the detrimental effects of nitrogen. Hence, product could have been produced in compliance with the ASTM A 240 specification that was potentially susceptible to intergranular attack at welds and heat affected zones.

To avoid this shortcoming three different compositions have been approved which provide for different levels of stabilization. Each of the new stabilization requirements do take into account the detrimental effects of nitrogen. The S40930 composition also allows the use of dual stabilization with additions Ti + Cb. The degree of stabilization increases from S40910 to the S40930 composition. The stabilization requirements for the S40910, S40920 and S40930 designations are %Ti/(%C + %N) $\geq$ 6, %Ti/(%C + %N) $\geq$ 8, and (%Ti + %Cb) $\geq$ 0.08 +8(%C + %N) respectively. This revision now allows the user to specify a composition that best matches the stabilization needs of the application.

Postweld heat treatment is another way to eliminate IGC in weldments. Altering the chemistry is the preferred method, since post fabrication processes can increase the costs substantially.

For S44400 (18Cr-2Mo) alloys to be immune to IGC a maximum level of carbon plus nitrogen (C+N) is 60 to 80 ppm. This level rises to around 150 ppm for a superferritic such as S44626 (26Cr-1Mo) alloy.

Susceptibility to intergranular corrosion by the ferritic stainless steels can be evaluated using the standard test method ASTM A 763, practices X, Y and Z (see Table 3.24). The tests are run with and without sensitizing treatment or with a weld.

Table 3.24 ASTM Standard Tests for Susceptibility to
Intergranular Corrosion in Ferritic Stainless Steels

ASTM	Test Medium & Duration	Alloys	Phases detected
A 763, practice X	$Fe_2(SO_4)3-H_2SO_4$; 24-120hr	29-1, S44700 (29-4), S44800 (29-4-2)	Chromium carbide and nitride
A 763, practice Y	$CuSO_4-50\%H_2SO_4$; 96-120hr	S44700 (29-4), S44800 (29-4-2)	Chromium carbide and nitride
A 763, practice Z	$CuSO_4-16\%H_2SO_4$; 24hr	S43000, S43400, S43600, S43035 (439), S44400	Chromium carbide and nitride

Stress Corrosion Cracking (SCC)

Ferritic stainless steels are highly resistant to stress corrosion cracking (SCC). Table 3.25 shows the SCC resistance of ferritics in comparison to austenitics. Factors that have been identified as detrimental to the chloride SCC resistance of ferritic stainless steels include the presence of certain alloying elements, sensitization (induced by heat treatment or welding), cold work, and high temperature. The alloying elements that have been identified as detrimental to the chloride SCC resistance include copper, nickel, cobalt (in the presence of molybdenum), ruthenium, and carbon. sulfur, either in the alloy or as sulfur-containing gas, can initiate SCC in chloride environments.

Table 3.25   SCC Resistance of Ferritic Stainless Steels

UNS No. (Alloy)	SCC Test (U-bend tests, stressed beyond yielding)		
	Boiling 42% MgCl$_2$	Wick test	Boiling 25% NaCl
S40900 (type 409)	P^b	P	P
S43035 (type 439)	P	P	P
S44400 (alloy 444)	P (>1700hr)	P (>1440hr)	P (>1440hr)
S44626 (alloy 26-1)	P	P	P
S44660 (Sea-Cure)	F	P	P
S44635 (Monit)	F	P	P
S44700 (alloy 29-4)	P	P	P
S44800 (alloy 29-4-2)	F	P	P
S44735 (alloy 29-4C)	P	P	P
S30400 (type 304)	F^a (<2hr)	F (<120hr)	F (<16hr)
S31600 (type 316)	F (<16hr)	F (<360hr)	F (<240hr)

a.   Fails, cracking observed.
b.   Passes, no cracking observed.

ASTM G 36 test method determines the effects of composition, heat treatment, surface finish, microstructure, and stress on the susceptibility of these materials to chloride SCC. The test is run in boiling 42% MgCl$_2$ for 1000 hours. The wick test is used to evaluate the chloride cracking characteristics of thermal insulation for applications in the chemical process industry. ASTM C 692 covers the methodology and apparatus used to conduct this procedure.

Corrosion in Pulp and Paper Plant Environments

Ferritic stainless steels have not been considered for use in bleach plant environments because of their poor corrosion resistance, particularly after welding. However, the superferritics (with the low residuals) can have performance from moderate to very high corrosion resistance depending on the alloy content. Thin-sections can find applications as corrugated deck or tube. Metal samples with polytetrafluoroethylene spacers were exposed to bleaching environments in the washer vat below the C (chlorination), D (chlorine dioxide), and H (hypochlorite) stages of several paper mills. The results of their tests were reported by the Metals Subcommittee of the Technical Association of the Pulp and Paper Industry (TAPPI). The results are reported in Table 3.26.

Table 3.26 Corrosion in Pulp Bleach Plant Environment

UNS No. (Trade Name)	Pitting	Crevice
S44735 (AL 29-4®)	0.5 mm	1.7 mm
S44660 (SEA-CURE)	0.9 mm	1.0 mm
(Shomac 30-2)	1.0 mm	1.0 mm
S44635 (MONIT)	2.1 mm	3.7 mm
S44800 (AL 29-4-2)	2.6 mm	2.1 mm
Nitronic 50	17.0 mm	7.5 mm
S44627 (26-1)	19.2 mm	8.9 mm
S31703 (type 317L)	28 mm	10 mm

Data Source: ASM Specialty Handbook - Stainless Steels

Superferritics such as S44700 (alloy 29-4) have shown usefulness in handling alkaline cooking liquors used in Kraft paper pulp processes.

Corrosion in Caustic Applications

Superferritics such as S44700 (alloy 29-4) and S44800 (alloy 29-4-2) show excellent resistance to general corrosion by hot caustic solutions. Laboratory tests show corrosion rates of less than 0.025mm/year in 50% concentration of boiling caustic solution. S44700 tubing has therefore become a standard material of construction for caustic evaporator steam chests and other associated heat exchangers. S44700 has, in fact, excellent corrosion resistance in the presence of chlorate in caustic.

Corrosion in Automotive Applications

Ferritic stainless steels are used extensively in the automotive industry, especially in the exhaust systems. They are also used as trim materials. As automobile prices continue to increase, durability and reduced operating costs continue to be major concerns for consumers and automakers alike. Ferritic stainless steels often meet both these requirements for exhaust systems. In this section, the corrosion performance of ferritics in the auto exhaust environment is described.

In the 1950s and 60s, the hot end (closest to the engine) of a car's exhaust system was normally a heavy cast iron manifold bolted to a straight heavy wall carbon steel downpipe which snaked its way back to carbon steel or coated steel muffler. Exhaust replacements came every 2 to 3 years and often included the entire exhaust system, except for the manifold.

Generally, big V-8 engines ran on leaded fuel and got 10 miles to the gallon (4.25 km/L). The advent of pollution control separated the exhaust system into a hot end (Figure 3.9a, downpipe/converter) and a cold section (Figure 3.9b, I-pipe-muffler-tail stub). The EPA has now mandated a new hot end warranty of 8 years/80,000 miles, with lower levels of permissible HC, CO, and NO_x emissions targeted.

Hot End

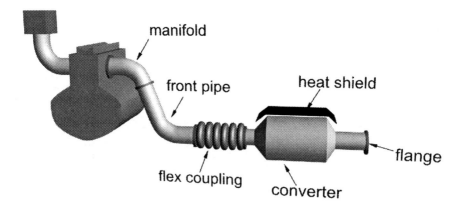

Figure 3.9a Schematic of a typical exhaust system
showing the hot end, downpipe/converter.

Cold End

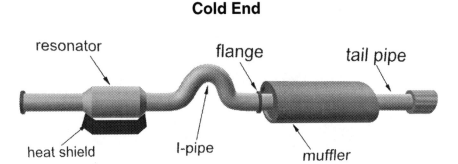

Figure 3.9b Schematic of a typical exhaust system
showing the cold end, I-pipe-muffler-tail stub.

The primary material requirements for an exhaust system include oxidation resistance, high temperature strength, formability, weldability, and condensate corrosion resistance. The exhaust system, especially the hot end, is subject to high temperatures and engine vibrations. Table 3.27a lists the various alloys used in the exhaust system with their nominal chemical compositions, while Table 3.27b lists international cross-references of ferritic stainless steels used for this application.

Table 3.27a Ferritic Exhaust Chemical Compositions

UNS No.	Type/Alloy	Nominal Compositions (Weight %)					
		C	Cr	Ni	Ti	Cb	Other
S40900	409	0.01	11.2	---	0.20	---	---
S40975	409Ni	0.01	11.0	0.85	0.20	---	0.75 Mn
S43035 (S43932)	439 (439LT)	0.015	17.3	---	0.30 0.28	--- 0.06	---
S44100	18CrCb	0.015	18	---	0.25	0.55	---
S44500	430M	0.015	19.2	---	---	0.4	0.4 Cu
---	436S	0.015	18.0	---	0.35	---	1.0 Mo
S44400	444	0.015	17.7	---	0.25	0.15	2.0 Mo
---	18SR	0.015	17.3	---	0.25	---	1.7 Al
---	11CrCb	0.010	11.3	---	0.20	0.35	1.3 Si

Table 3.27b Ferritic Exhaust Alloy International Cross References

USA	Europe		Japan	
Alloy	DIN	Trade Names	SUS	Trade Names
409	1.4512 X6CrTi12	Polarit 853 UGINOX F12T	SUH409	409D YUS409D
---	1.4003 X2Cr11	3Cr12/409Ni UGINOX F12N	---	YUS410WH (no Ti)
439	1.4510 X6CrTi17	ACX515 UGINOX F17T	430LX	430D 430LX
441	1.4509 X6CrTiNb18	441 UGINOX FX	---	YUS 180M (no Ti)
433 435M	---	---	430MT 430EM	YUS 180 NSS442
434Ti	1.4513 X2CrMoTi16-1	---	436MT	YUS 436S NAS436LS
444	1.4521 X2CrMoTi18-2	ACX555 UGINOXF18MT	444	YUS 190 NSSEM2
---	---	INOXIUM 180AL	---	R18-3SR (3%Al)
---	---	---	425T1 (14.5Cr)	R409SR (No Cb)

Quantitative strength at temperature comparisons and estimated upper temperature oxidation limits for typical exhaust alloys are shown in Table 3.28.

Table 3.28 High Temperature Properties
of Typical Exhaust Materials[a]

Material	Upper Oxidation Temp. Limit	Yield Strength		Tensile Strength		Fatigue Strength[b]	Stress Rupture @1500°F (816°C)
		70°F (21°C)	1000°F (538°C)	70°F (21°C)	1000°F (538°C)	1500°F (816°C)	1000hrs
S40920	1450°F	33.9	59.0	17.4	34.9	1	0.9
S43035	1625°F	41.6	66.0	21.5	37.8	1.4	1.0
S43932	1625°F	44.6	68.2	22.7	41.3	2	1.1
11CrCb	1625°F	51.0	74.0	27.3	51.7	3	1.4
18CrCb	1625°F	50.0	74.0	25.4	54.3	3	1.8
436S	1650°F	42.0	68.0				

a. All units are in ksi.
b. Stress for 10^7 cycles

Oxidation resistance of alloy 18SR is the best among all ferritics. This is attributed to the addition of 2% Al and 1% Si to the alloy. Figures 3.10, 3.11 and 3.12 compare typical exhaust alloys in different types of accelerated corrosion tests.

Figure 3.10 shows results of a test conducted in a typical condensate solution containing 3500 ppm of chloride with samples half immersed in the solution at 27°F (80°C). This test is used to evaluate materials for the cold end of the exhaust system. And as the chromium increases along with the addition of molybdenum, the pit growth rate decreases substantially.

In this test the samples are exposed to 128°F (260°C) for 1 hour once every week, with the remainder of the time undergoing 6 hour cycles which include immersion in salt solution for 15 minutes followed by drying for 20 minutes at 4°F (40°C) and 35%RH followed by 5 hours and 25 minutes of humidity exposure at 16°F (60°C) and 85%RH.

Note that S43932, 436S and aluminized 409 all perform similarly until 588 hours, at which point the pit depths measure for aluminized 409 increase. Aluminized 409 is typically used in the cold end of the exhaust system to provide cosmetic corrosion resistance. Pre-painted stainless steels are also being used for enhanced cosmetic corrosion resistance (BLACKCOAT™). The BLACKCOAT system uses a high temperature resistant silicone resin on type 409 stainless steel. The system is black in color, and therefore hides the muffler. On-vehicle testing has shown no visible red rust after 35,000 miles (56,000 km).

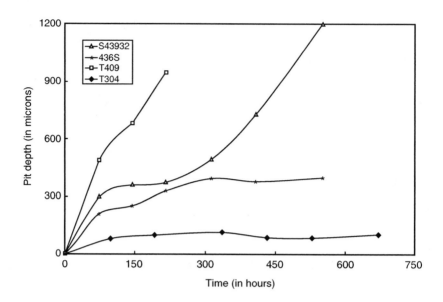

Figure 3.10 Pit depth vs. time for 409, S43932, 436S and T304 in the continuous condensate with 3500ppm of chloride.

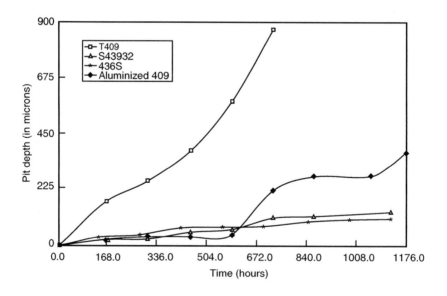

Figure 3.11 Pit depth as a function of time for T409, S43932, 436S and aluminized 409 in the cyclical cold-end corrosion test.

The hot end of the exhaust system is evaluated in a different accelerated test with exposures to much higher temperatures. Figure 3.12 shows results of panels in a cyclical test designed to evaluate materials for the hot end. Here samples are exposed to 1000°F (538°C) for 1 hour for 5 days per week, with the remainder of the time undergoing 6 hour cycles which include immersion in salt solution for 15 minutes followed by drying for 20 minutes at 104°F (40°C) and 35%RH followed by 5 hours and 25 minutes of humidity exposure at 140°F (60°C) and 85%RH.

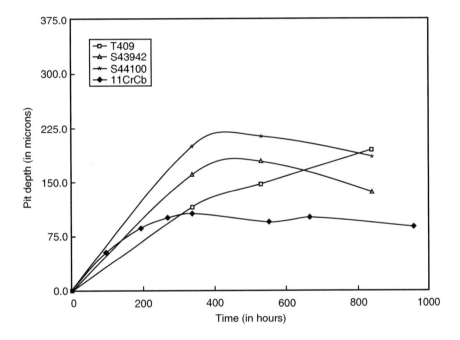

Figure 3.12 Pit depth as a function for time for 409, S43932, S44100, and 11CrCb in hot end cyclical corrosion test.

Pitting corrosion can also be evaluated in a relatively short time using electrochemical techniques. The pitting potential can be used to rank materials although the ranking is probabilistic. Figure 3.13 describes the probability of formation of a stable pit as a function of the potential for typical auto exhaust materials. The higher the pitting potential, the better the pitting corrosion resistance. The pitting potential at 80% probability is shown in Table 3.29.

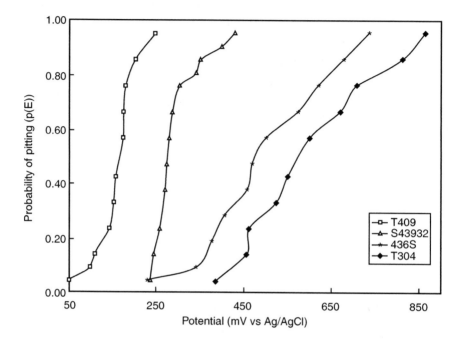

Figure 3.13 Plot of the pitting probability with critical pitting potential measured with respect to Ag/AgCl reference electrodes

Table 3.29 Pitting Potentials of Typical Exhaust Alloys
in 3.5% NaCl Solution

Alloy	Potential in mV vs. Ag/AgCl @ p(E) = 0.8
S40900	180
S43932	280
436S	600
S30400	750

The overall ranking of the various auto exhaust materials based on various corrosion tests for the hot and cold end of the exhaust system is summarized in Table 3.30.

Table 3.30 Corrosion Properties of Typical Exhaust Materials

Alloy	Hot End[a] Hot Salt Resistance	Cold End[b] Corrosion Durability	Cosmetic Resistance
S40900	D	5-7 years	0.5 year
11CrCb	B+	5-7 years	0.5 year
S43932	B-	10 years	1-3 years
18CrCb	B+	10+ years	1-3 years
S30400	–	10+ years	2-4 years

a. A= Most Desirable, D= Least Desirable but not necessarily bad
b. Expected life until the onset of significant corrosion

Summary

The martensitic, ferritic, and precipitation-hardening stainless steels offer a wide variety of mechanical and corrosion properties. They range from the barely stainless S40900 (type 409) to superferritics like S44800 (alloy 29-4-2) and from 25 ksi yield strength (S40500; type 405) to 275 ksi (S44004; type 440C). The optimum selection of an alloy is often a compromise of several factors such as: environmental corrosion and load-carrying properties, available forms, forming, machining, welding properties, surface finish, and cost. Often, the initial higher cost of stainless is far offset by less maintenance, improved durability, and reduced downtime to achieve lower life cycle costs.

References

1. *Stainless Steels,* Universal-Cyclops, Pittsburgh, PA, 1977.

2. *Development of the Stainless Steels,* Armco Inc., Middletown, OH, 1983.

3. Davis, J. R., *ASM Specialty Handbook-Stainless Steels,* Materials Park, OH, 1994.

4. White, K. L., "Precipitation-Hardening Stainless Steels," *Machine Design,* Jan. 1969.

5. Lula, R. A., "Source Book on the Ferritic Stainless Steels," ASM, 1982.

6. "Corrosion in the Pulp and Paper Industry," Volume 13, *ASM Handbook,* 1996.

7. "High Temperature Characteristics of Stainless Steel", *Designer Handbook Series*, AISI, April 1979.

8. Sabata, A, Brossia C.S., and Behling, M., "Localized Corrosion Resistance of Automotive Exhaust Alloys" NACE International Corrosion 98, Paper #549.

9. Baroux, B., "Corrosion Mechanisms in Theory and Practice," edited by P. Marcus and J. Oudar, Marcel Dekker, Inc., 1995.

Chapter

4

AUSTENITIC STAINLESS STEELS

C.W. Kovach and J.D. Redmond

Technical Marketing Resources, Inc.

Pittsburgh, Pennsylvania

Introduction

The austenitic stainless steels possess an outstanding combination of corrosion resistance, mechanical properties, and fabricability properties. For this reason they are used in a wide variety of applications ranging from consumer goods, where aesthetics are of primary importance, to applications in many industrial, power, aerospace settings and in many kinds of equipment where service performance is paramount.

As a result of this great versatility, these steels are among the most important commercial alloys. They are essential to the functioning of modern society, and they constitute a major part of the metal producing and metal working industries around the world.

The rate of growth of austenitic stainless steel production and consumption in recent years has exceeded that of most other classes of engineering metals. This trend is expected to continue as a result of the increasing technical complexity of our society, strong market demand in energy and environmental segments of the world economy, and anticipated improvements in cost relative to other materials.

General Characteristics and Applications

The basic austenitic stainless is UNS S30400 (type 304) sometimes called "18-8 stainless." It is an iron-based alloy containing nominally 18% chromium and 8.5% nickel, minor amounts of carbon, nitrogen, manganese, and silicon. This alloy had its origins in the second decade of this century when it was simultaneously discovered in England and Germany that an addition of about 12% chromium to iron would prevent the formation of rust that normally would develop in moist air.

This effect, illustrated in Figure 4.1, was subsequently found to hold in many environments and provides the basis for the outstanding corrosion resistance characteristic of all modern stainless steels. Shortly after, it was discovered that an addition of nickel would stabilize the high temperature allotrope of iron, austenite, at room temperature. Austenite, with a face-centered cubic crystal structure, provides highly desirable mechanical properties in terms of strength, ductility, and toughness. Thus, through a fortuitous combination of discovery and thermodynamics, this family of steels was born.

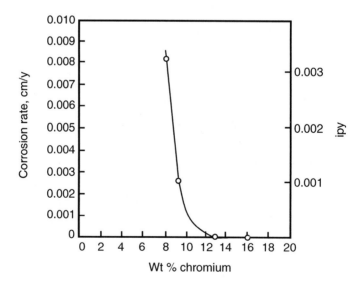

Figure 4.1 Corrosion rates of chromium-iron alloys in intermittent water spray, room temperature.

From this beginning a family of dozens of variations on the original "18-8" alloy have evolved. These are known as the "standard austenitic stainless steels" and a representative list is shown in Table 4.1.

Many modifications of these standard alloys have been developed for special applications, and their evolution continues as a result of technological progress and the demands of the marketplace. Technological progress has been based on an improved understanding of factors which control mechanical properties, corrosion resistance, fabricability, and cost.

Mechanical property development has largely been based on defining optimum combinations of chromium, nickel, carbon, and nitrogen in relation to the work hardening and toughness characteristics of austenite. Corrosion resistance development has used additions of molybdenum and nitrogen for better performance in aqueous environments, and silicon and rare earth elements for resistance at high temperatures. Improvements in fabricability and cost have come from new steel refining techniques that cost effectively eliminate impurities such as carbon and sulfur, and from the use of continuous casting and other advanced production techniques. Thus a continuing specialization of these alloys exists, and they are being used in an ever wider range of applications.

Major market segments for the austenitic stainless steels include:

- consumer products,
- transportation,
- architecture,
- food and beverage,
- chemical and petrochemical,
- paper,
- pharmaceutical and biotech,
- semiconductor,
- energy,
- environmental, and
- aerospace.

Table 4.1 Chemical Compositions of Austenitic Stainless Steels (Weight %)[a]

UNS No.	Grade	EN	DIN	C	Cr	Ni	Mo	Cu	N	Other
Corrosion Resistant										
S20100	201	1.4372	---	0.15	16.0-18.0	3.5-5.5	---[b]	---	0.25	5.5-7.5 Mn
S20200	202	---	---	0.15	17.0-19.0	4.0-6.0	---	---	0.25	7.5-10.0 Mn
S20400	Nitronic 30[c]	---	---	0.03	15.0-17.0	1.50-3.00	---	---	0.15-0.30	7.0-9.0 Mn
S30100	301	1.4310	1.4310	0.15	16.0-18.0	6.0-8.0	---	---	0.10	---
S30200	302	---	---	0.15	17.0-19.0	8.0-10.0	---	---	0.10	---
S30430	302HQ	---	---	0.03	17.0-19.0	8.0-10.0	---	3.00-4.00	---	---
S30400	304	1.4301	1.4301	0.08	18.0-20.0	8.0-10.5	---	---	0.10	---
S30403	304L	1.4307	---	0.030	18.0-20.0	8.0-12.0	---	---	0.10	---
S30451	304N	1.6907	1.6907	0.08	18.0-20.0	8.0-10.5	---	---	0.10-0.16	---
S30500	305	1.4303	1.4303	0.12	17.0-19.0	10.5-13.0	---	---	---	---
S31600	316	1.4401	1.4401	0.08	16.0-18.0	10.0-14.0	2.00-3.00	---	0.10	---
S31603	316L	1.4404	1.4404	0.030	16.0-18.0	10.0-14.0	2.00-3.00	---	0.10	---
S31703	317L	1.4438	1.4438	0.030	18.0-20.0	11.0-15.0	3.00-4.00	---	0.10	---
S31726	317LMN	1.4439	1.4439	0.030	17.0-20.0	13.5-17.5	4.0-5.0	---	0.10-0.20	---
S32100	321	1.4541	1.4541	0.080	17.0-19.0	9.0-12.0	---	---	0.10	5 x (C+N) Ti min. 0.70 max.
S21800	Nitronic 60[c]	---	---	0.10	16.0-18.0	8.0-9.0	---	---	0.08-0.18	7.0-9.0 Mn; 3.5-4.5 Si
S20161	Gall-Tough[e]	---	---	0.15	15.0-18.0	4.0-6.0	---	---	0.08-0.20	4.0-6.0 Mn; 3.00-4.00 Si
S30300	303	1.4304	1.4304	0.15	17.0-19.0	8.0-10.0	---	---	---	0.15 S min.
S30323	303Se	---	---	0.15	17.0-19.0	8.0-10.0	---	---	---	0.15 Se min.
S30800	308	---	---	0.08	19.0-21.0	10.0-12.0	---	---	---	---
S30883	308L	---	---	0.03	19.5-22.0	9.0-11.0	0.05	0.75	---	0.30-0.65 Si

Table 4.1 Chemical Compositions of Austenitic Stainless Steels (Weight %)[a] (continued)

UNS No.	Grade	EN	DIN	C	Cr	Ni	Mo	Cu	N	Other
Heat Resistant										
S30409	304H	1.4948	1.4948	0.04-0.10	18.0-20.0	8.0-10.5	---	---	---	---
S34700	347	1.4450	1.4450	0.080	17.0-19.0	9.0-13.0	---	---	---	10 x C Cb min. 1.00 max.
S31609	316H	---	---	0.04-0.10	16.0-18.0	10.0-14.0	2.00-3.00	---	0.10	---
S30815	253MA[d]	1.4835	---	0.05-0.10	20.0-22.0	10.0-12.0	---	---	0.14-0.20	1.00-2.00 Si; 0.03-0.08 Ce
S30900	309	---	---	0.20	22.0-24.0	12.0-15.0	---	---	---	---
S30908	309S	1.4833	1.4833	0.08	22.0-24.0	12.0-15.0	---	---	---	---
S30909	309H	---	---	0.04-0.10	22.0-24.0	12.0-15.0	---	---	---	---
S31000	310	---	---	0.25	24.0-26.0	19.0-22.0	---	---	---	---
S31008	310S	1.4845	1.4845	0.08	24.0-26.0	19.0-22.0	---	---	---	---
S31009	310H	---	---	0.04-0.10	24.0-26.0	19.0-22.0	---	---	---	---

a. Maximum unless range or minimum is indicated.
b. None required in the specification.
c. Nitronic is a trademark of Armco Inc.
d. 253MA is a trademark of Avesta Sheffield AB.
e. Gall-Tough is a trademark of Carpenter Technology.

The consumer market is very broad and includes items such as appliances, sinks, and pots and pans. Transportation consists of automotive trim and emission control applications, truck components, and siding and structural applications in rail and rapid transit cars. Architectural applications include roofing, curtain walls, elevators, and interior and exterior hardware items. The food segment extends from food processing equipment to restaurant counters and cooking stoves and ovens.

Most chemical plants and refineries use tanks, pressure vessels, piping and heat exchangers constructed of austenitic stainless steels, and this is also true of the paper industry. Much of the equipment used to produce pharmaceuticals benefits from the hygienic and cleanability properties of these steels. The semiconductor industry relies on vacuum equipment constructed of stainless steel. In the energy sector, stainless steels are used in piping and heat exchanger applications.

Growing environmental applications range from equipment used in sewage treatment plants to scrubbers to remove polluting gases and particulates from flue gases. Aerospace applications range from aircraft hydraulic lines to pressure vessels handling gases in missiles.

From these applications, it seems clear that growth in the use of austenitic stainless steels will keep pace with the growing world population and especially the expanding economy as third world countries improve their standard of living. New technological developments related to the steels themselves will also contribute to growth. These include developing a continuous steelmaking process from melting to finished coil, which will reduce production costs; thin strip casting, which is near to commercialization, represents further potential cost savings.

Also, technical developments in other areas such as fuel cells and advanced power plants will also further stimulate demand. Fortunately, ample supplies of chromium and nickel exist, and capital markets seem ever ready to provide expanded production capacity.

There appears to be no limit to the bright future of the austenitic stainless steels.

Mechanical and Physical Properties and Specifications

The wrought austenitic stainless steels were originally assigned "type" numbers by the American Iron and Steel Institute (AISI). Stainless steels are still known by this designation, e.g., type 304. Recently the American Society for Testing Materials (ASTM) has begun assigning similar common names to stainless steels which are widely used, not trademarked, and not associated with a single producer.

In addition, all stainless steels carry a Unified Numbering System (UNS) designation. Grades which were originally assigned AISI type numbers have UNS designations which reflect that history. For example, type 304 is UNS S30400.

The ASTM specifications usually identify stainless steels by their UNS number and include their common name or type number where applicable. The materials specifications applicable to stainless steels are the ASTM "A" specifications for ferrous alloys, stainless, and heat resisting steels.

All of the austenitic stainless steels are covered in one or more of these specifications for virtually all product forms. These steels are also included in ASTM testing specifications, such as A 262 for detecting susceptibility to intergranular corrosion. In addition these steels are covered in many government, military, industrial, and other international specifications.

Tables 4.2a and 4.2b are partial listings of applicable ASTM specifications for wrought austenitic stainless steels arranged by product form and grade.

For example, A 240 is for plate, sheet, and strip, while A 358 covers pipe welded with filler metal. The general requirements specifications shown are used in conjunction with the product form specifications.

The general requirements for specifications, e.g., A 480 and A 530, cover items such as ordering information, finishes, heat treatment, test methods, flatness, and size tolerances, packaging, and marking. The product specifications define specific grade chemical composition and mechanical property requirements.

In Table 4.1 the heat resisting grades have been grouped separately from the corrosion-resistant grades. The heat-resistant grades generally carry more restrictive annealing and grain size requirements because of the effect of these parameters on high temperature strength.

Some of the corrosion-resistant grades appear in a limited number of specifications. Typically, these grades are used in a limited number of applications because they are specialized in regard to one of their properties.

Those listed in many specifications are more versatile and, therefore, are more widely used and more readily available. Examples include S30400 (type 304) and S31600 (type 316), and their low carbon "L" modifications, S30403 (type 304L)and S31603 (type 316L), which are by far the most extensively used austenitic stainless steels.

Table 4.2a ASTM Specifications for Corrosion Resistant Austenitic Stainless Steels

Corrosion-Resistant Austenitic Stainless Steels

Designation		201	202	Nitronic® 30	301	302	304	304L	304N	302HQ	305	316	316L	317L	317LMN
UNS No.		S20100	S20200	S20400	S30100	S30200	S30400	S30403	S30451	S30430	S30500	S31600	S31603	S31703	S31726
ASTM Specification - Flat-Rolled															
A 480	General requirements	•	•	•							•	•	•	•	•
A 240	Plate, sheet	•	•	•	•		•	•	•		•	•	•	•	•
A 793	Floor plate						•	•				•	•		
ASTM Specification - Long Products															
A 484	General requirements		•			•	•	•	•		•	•	•	•	•
A 276	Bar, shapes	•	•			•	•	•	•		•	•	•	•	•
A 479	Bar, shapes					•	•	•	•			•	•	•	•
A 493	Wire, wire rods					•	•	•		•		•	•		
A 580	Wire					•		•			•	•	•		
A 581	Free-machining wire and rod														
A 582	Free-machining bar														
A 182	Fittings						•	•	•			•	•	•	•
ASTM Specification – Long Products															
A 555	Wire general requirements	•					•	•			•	•	•		
ASTM Specification – Tubing															
A 450	General requirements	•	•				•	•			•	•	•	•	•
A 249	Welded tube	•	•				•	•			•	•	•	•	•

CASTI Handbook of Stainless Steels & Nickel Alloys – Second Edition

Table 4.2a ASTM Specifications for Corrosion Resistant Austenitic Stainless Steels (continued)

Corrosion-Resistant Austenitic Stainless Steels														
Designation	201	202	Nitronic 30	301	302	304	304L	304N	302HQ	305	316	316L	317L	317LMN
UNS No.	S20100	S20200	S20400	S30100	S30200	S30400	S30403	S30451	S30430	S30500	S31600	S31603	S31703	S31726
ASTM Specification - Tubing (continued)														
A 269 Welded tube						•	•				•	•		•
A 270 Sanitary tube						•	•				•	•		
ASTM Specification – Pipe														
A 999 General requirements						•	•	•			•	•	•	•
A 312 Pipe, seamless & welded w/o filler						•	•	•			•	•	•	•
A 358 Welded w/ filler						•	•	•			•	•	•	•
A 778 Welded, unanneal							•					•		
ASTM Specification - Testing Specifications														
A 262 Carbides/ austenitic stainless steels	•	•	•	•	•	•	•	•	•	•	•	•	•	
E 112 Grain size	•	•	•	•	•	•	•	•	•	•	•	•	•	•

• - Listed in specification.
w/o – without
w - with

Table 4.2b ASTM Specifications for Heat Resistant Austenitic Stainless Steels (continued)

Heat Resisting Austenitic Stainless Steels												
Designation		304H	321	347	316H	253 MA	309	309S	309H	310	310S	310H
UNS No.		S30409	S32100	S34700	S31609	S30815	S30900	S30908	S30909	S31000	S31008	S31009
ASTM Specification - Flat-Rolled												
A 480	General requirements	•	•	•	•	•		•	•		•	•
A 240	Plate, sheet	•	•	•	•	•		•	•		•	•
A 793	Floor plate											
ASTM Specification – Long Products												
A 484	General req.	•	•	•	•	•		•	•		•	•
A 276	Bar, shapes	•	•	•	•	•	•	•	•	•	•	•
ASTM Specification – Long Products Continued												
A 479	Bar, shapes	•	•	•	•	•		•			•	
A 493	Wire, wire rods											
A 580	Wire		•	•				•			•	
A 581	Free-machining wire and rod											
A 582	Free-machining bar											
A 182	Fittings	•	•	•	•	•						
A 555	Wire general requirements		•	•				•			•	
ASTM Specification – Tubing												
A 450	General req.	•	•	•	•	•		•	•		•	•
A 249	Welded tube	•	•	•	•	•		•	•		•	•
A 269	Welded tube		•	•								
A 270	Sanitary tube											

Table 4.2b ASTM Specifications for Heat Resistant Austenitic Stainless Steels (continued)

Heat Resisting Austenitic Stainless Steels											
Designation	304H	321	347	316H	253 MA	309	309S	309H	310	310S	310H
UNS No.	S30409	S32100	S34700	S31609	S30815	S30900	S30908	S30909	S31000	S31008	S31009
ASTM Specification – Pipe											
A 530 General requirements	•	•	•	•	•		•	•		•	•
A 312 Pipe, seamless & welded w/o filler	•	•	•	•	•		•	•		•	•
A 358 Welded w/filler	•	•	•	•	•		•			•	
A 778 Welded, unanneal		•	•								
ASTM Specification - Testing Specifications											
A 262 Carbides/austenitic stainless steels	•	•	•				•			•	
E 112 Grain size	•	•	•	•	•		•	•		•	•

• - Listed in specification.

w/o - without

w - with

There are many similar cast grades to the wrought austenitic stainless steels (see Table 4.3). The similar cast grades have their own UNS numbers and ASTM specifications.

Table 4.3 Wrought Austenitic Stainless Steels
and Similar Cast Grades

Wrought Grade	UNS No. (Wrought)	Nominal Composition	Similar Cast Grade	UNS No. (Cast)	ASTM Casting Specification
304	S30400	18Cr-8Ni	CF-8	J92600	A 351, A 743, A 744
304L	S30403	18Cr-8Ni	CF-3	J92500	A 351, A 743, A 744
316	S31600	16Cr-12Ni-2Mo	CF-8M	J92900	A 351, A 743, A 744
316L	S31603	16Cr-12Ni-2Mo	CF-3M	J92800	A 351, A 743, A 744
317L	S31703	18Cr-13Ni-3Mo	CG-3M	J92999	A 743, A 744
Nitronic 60[a]	S21800	18Cr-8Ni-4Si-N	CF-10SMnN	J92972	A 351, A 743
303	S30300	18Cr-8Ni	CF-16Fa	---	A 743
303Se	S30323	18Cr-8Ni	CF-16F	J92701	A 743
304H	S30409	18Cr-8Ni	CF-8	J92590	A 351, A 452
321	S32100	18Cr-8Ni-Ti	---	J92630	MIL-S-81591
347	S34700	18Cr-9Ni-Nb	CF-8C	J92710	MIL-S-81591, A 452
316H	S31609	16Cr-12Ni-2Mo	CF-8M	J92920	A 452
309	S30900	23Cr-12Ni	CH-20	J93402	A 351, A 743
310	S31000	25Cr-20Ni	CK-20	J94202	A 351, A 743

a. Nitronic is a trademark of Armco Inc.

Table 4.4 summarizes the ASTM minimum room temperature annealed mechanical property requirements for each grade for wrought product specifications.

For most grades the minimum yield strength is 30 ksi (205 MPa). Exceptions are S30403 (type 304L) and S31603 (type 316L) with 25 ksi (170 MPa) minimum yield strength because of their lower carbon content. Grades such as S30451 (type 304N) and S31726 (317LMN), have a small nitrogen addition which raises the yield strength to 35 ksi (240 MPa), and the high nitrogen grades which have substantially higher strength.

Minimum tensile strength values generally follow the yield strength values.

The ASTM specifications list minimum properties for annealed material. However, because of their high work hardening rates, the austenitic grades, as-delivered, may have much higher yield strength as a result of:

- mill straightening and flattening operations,
- intentional mill processing to produce higher mechanical properties by cold working, and
- cold fabrication operations by the end user.

Many of the austenitic stainless steels are approved for high temperature service by the ASME Boiler and Pressure Vessel Code. This code provides for maximum allowable design stresses based on either short-term tensile properties at lower temperatures, or on creep properties at higher temperatures.

The maximum design stress values as a function of temperature for wrought plate and castings, as they appear in various sections of the ASME Boiler and Pressure Vessel Code, are given in Table 4.5.

It should be noted that other product forms may have slightly different design stress values, and product forms containing a weldment generally carry a stress value reduced by a multiplying factor of 0.85. For more details on this subject, see the *CASTI Guidebooks* to the ASME Boiler and Pressure Vessel Code, Volumes 1 to 4, published by *CASTI* Publishing.

The physical properties of the austenitic stainless steels are different than those of iron and steel in that the elastic moduli are slightly lower and the thermal conductivity is substantially lower, while the thermal expansivity and electrical resistivity are significantly higher.

The difference in thermal expansivity between carbon steel and these stainless steels is great enough to require that this be taken into account when designing structures containing both metals for elevated temperature service.

The austenitic stainless steels are also nonmagnetic in the annealed condition if they contain no delta ferrite. Small amounts of delta ferrite that result from welding, or martensite produced by cold work, will produce a small amount of ferromagnetism.

Some typical physical property values are listed in Table 4.6.

Table 4.4 ASTM Mechanical Properties (A 240, A 276, A 479, A 493, A 582)

UNS No.	Alloy	Condition	Tensile Strength, min.		Yield Strength, min.		% Elongation	Hardness, max.	
			ksi	MPa	ksi	MPa		HB	HRB
S20100	201	Annealed	95	655	38	260	40.0	---	95
S20200	202	Annealed	90	620	38	260	40.0	241	---
S20400	Nitronic 30	Annealed	95	655	48	330	35.0	241	100
S30100	301	Annealed	75	515	30	205	40.0	217	95
S30200	302	Annealed	75	515	30	205	40.0	201	92
S30430	302HQ	Annealed	88	605	---	---	---	---	---
S30400	304	Annealed	75	515	30	205	40.0	201	92
S30403	304L	Annealed	70	485	25	170	40.0	201	92
S30451	304N	Annealed	80	550	35	240	30.0	201	92
S30500	305	Annealed	75	515	30	205	40.0	183	88
S31600	316	Annealed	75	515	30	205	40.0	217	95
S31603	316L	Annealed	70	485	25	170	40.0	217	95
S31703	317L	Annealed	75	515	30	205	40.0	217	95
S31726	317LMN	Annealed	80	550	35	240	40.0	223	96
S21800	Nitronic 60	Annealed	95	655	50	345	35.0	241	---
S20161	Gall-Tough	Annealed	125	860	50	345	40.0	255	---
S30300	303	Annealed	---	---	---	---	---	262	---
S30323	303Se	Annealed	---	---	---	---	---	262	---
S30409	304H	Annealed	75	515	30	205	40.0	201	92
S32100	321	Annealed	75	515	30	205	40.0	217	95
S34700	347	Annealed	75	515	30	205	40.0	201	92
S31609	316H	Annealed	75	515	30	205	40.0	217	95
S30815	253MA	Annealed	87	600	45	310	40.0	217	95
S30900	309	Annealed	75	515	30	205	40.0	217	95
S30908	309S	Annealed	75	515	30	205	40.0	217	95

Table 4.4 ASTM Mechanical Properties (A 240, A 276, A 479, A 493, A 582) (continued)

UNS No.	Alloy	Condition	Tensile Strength, min.		Yield Strength, min.		% Elongation	Hardness, max.	
			ksi	MPa	ksi	MPa		HB	HRB
S30909	309H	Annealed	75	515	30	205	40.0	217	95
S31000	310	Annealed	75	515	30	205	40.0	217	95
S31008	310S	Annealed	75	515	30	205	40.0	217	95
S31009	310H	Annealed	75	515	30	205	40.0	217	95

Table 4.5 Maximum Allowable Stress Values (ksi), Plate as a Function of Temperature (°F)

ASME Section I; Section III, Class 2 and 3; Section VIII, Division 1															
UNS No.	Alloy	-20 to 100	200	300	400	500	600	650	700	750	800	850	900	950
S20100	201-1	25.3	19.3	16.6	---a	---	---	---	---	---	---	---	---	---
S20400	Nitronic 30	27.1	23.6	20.3	17.9	16.5	15.8	15.6	15.5	15.3	15.1	14.8	14.3	---
S30400	304	20.0	16.7	15.0	13.8	12.9	12.3	12.0	11.7	11.5	11.2	11.0	10.8	10.6
J92600	CF-8	20.0	16.7	15.0	13.8	12.9	12.3	12.0	11.7	11.5	11.2	11.0	10.8	10.6
S30403	304L	16.7	14.3	12.8	11.7	10.9	10.4	10.2	10.0	9.8	9.7	---	---	---
J92500	CF-3	20.0	16.7	15.0	13.8	12.9	12.3	12.0	11.7	11.5	11.2	---	---	---
S30451	304N	22.9	19.1	16.7	15.1	14.0	13.3	13.0	12.8	12.5	12.3	12.1	11.8	11.6
S31600	316	20.0	17.3	15.6	14.3	13.3	12.6	12.3	12.1	11.9	11.8	11.6	11.5	11.4
J92900	CF-8M	20.0	17.2	15.5	14.2	13.3	12.6	12.3	12.1	11.9	11.7	11.6	11.5	11.4
S31603	316L	16.7	14.2	12.7	11.7	10.9	10.4	10.2	10.0	9.8	9.6	9.4	---	---
J92800	CF-3M	20.0	17.2	15.5	14.2	13.3	12.6	12.3	12.1	11.9	11.7	11.6	---	---
S31703	317L	20.0	17.0	15.2	14.0	13.1	12.5	12.2	12.0	11.7	11.5	11.3	---	---
S30409	304H	20.0	16.7	15.0	13.8	12.9	12.3	12.0	11.7	11.5	11.2	11.0	10.8	10.6
S32100	321	20.0	18.0	16.5	15.3	14.3	13.5	13.2	13.0	12.7	12.6	12.4	12.3	12.1
S34700	347	20.0	18.4	17.1	16.0	15.0	14.3	14.0	13.8	13.7	13.6	13.5	13.4	13.4
S31609	316H	20.0	17.3	15.6	14.3	13.3	12.6	12.3	12.1	11.9	11.8	11.6	11.5	11.4
S30815	253MA	24.9	24.7	22.0	19.9	18.5	17.7	17.4	17.2	17.0	16.8	16.6	16.4	16.2
S30908	309S	20.0	17.5	16.1	15.1	14.4	13.9	13.7	13.5	13.3	13.1	12.9	12.7	12.5
S30909	309H	20.0	17.5	16.1	15.1	14.4	13.9	13.7	13.5	13.3	13.1	12.9	12.7	12.5
S31008	310S	20.0	17.6	16.1	15.1	14.3	13.7	13.5	13.3	13.1	12.9	12.7	12.5	12.3
S31009	310H	20.0	17.6	16.1	15.1	14.3	13.7	13.5	13.3	13.1	12.9	12.7	12.5	12.3

Table 4.5 Maximum Allowable Stress Values (ksi), Plate as a Function of Temperature (°F) (continued)

UNS No.	Alloy	1000	1050	1100	1150	1200	1250	1300	1350	1400	1450	1500	1550	1600	1650	1700
S20100	201	---	---	---	---	---	---	---	---	---	---	---	---	---	---	---
S20400	Nitronic30	---	---	---	---	---	---	---	---	---	---	---	---	---	---	---
S30400	304	10.4	10.1	9.8	7.7	6.1	4.7	3.7	2.9	2.3	1.8	1.4	---	---	---	---
J92600	CF-8	10.4	9.5	7.5	6.0	4.8	3.9	3.3	2.7	2.3	2.0	1.7	---	---	---	---
S30403	304L	---	---	---	---	---	---	---	---	---	---	---	---	---	---	---
J92500	CF-3	---	---	---	---	---	---	---	---	---	---	---	---	---	---	---
S30451	304N	11.3	11.0	9.8	7.7	6.1	---	---	---	---	---	---	---	---	---	---
S31600	316	11.3	11.2	11.1	9.8	7.4	5.5	4.1	3.1	2.3	1.7	1.3	---	---	---	---
J92900	CF-8M	11.3	11.2	8.9	6.9	5.4	4.3	3.4	2.8	2.3	1.9	1.6	---	---	---	---
S31603	316L	---	---	---	---	---	---	---	---	---	---	---	---	---	---	---
J92800	CF-3M	---	---	---	---	---	---	---	---	---	---	---	---	---	---	---
S31703	317L	---	---	---	---	---	---	---	---	---	---	---	---	---	---	---
S30409	304H	10.4	10.1	9.8	7.7	6.1	4.7	3.7	2.9	2.3	1.8	1.4	---	---	---	---
S32100	321	12.0	9.6	6.9	5.0	3.6	2.6	1.7	1.1	0.80	0.50	0.30	---	---	---	---
S34700	347	13.4	12.1	9.1	6.1	4.4	3.3	2.2	1.5	1.2	0.90	0.80	---	---	---	---
S31609	316H	11.3	11.2	11.1	9.8	7.4	5.5	4.1	3.1	2.3	1.7	1.3	---	---	---	---
S30815	253MA	14.9	11.6	9.0	6.9	5.2	4.0	3.1	2.4	1.9	1.6	1.3	1.0	0.86	0.71	---
S30908	309S	9.9	7.1	5.0	3.6	2.5	1.5	0.80	0.50	0.40	0.30	0.20	---	---	---	---
S30909	309H	12.3	10.3	7.6	5.5	4.0	3.0	2.2	1.7	1.3	1.0	0.75	---	---	---	---
S31008	310S	9.9	7.1	5.0	3.6	2.5	1.5	0.80	0.50	0.40	0.30	0.20	---	---	---	---
S31009	310H	12.1	10.3	7.6	5.5	4.0	3.0	2.2	1.7	1.3	0.97	0.75	---	---	---	---

a. No allowable stress values at this temperature; material not recommended for service at this temperature.

Table 4.6 Physical Properties[a]

UNS No.	Type or Alloy	Density lb/in.³	Modulus of Elasticity 10⁶ psi	Linear Expansion 68-212°F x 10⁻⁶/°F	Thermal Conductivity Btu/ft•hr•°F	Heat Capacity Btu/lb•°F	Electrical Resistivity μΩ•in.
S20100	201	0.28	29	9.2	9.4	0.12	27.2
S20200	202	0.28	29	9.4	9.4	0.12	27.2
S20400	Nitronic 30	0.28	28	9.4	---	---	---
S30100	301	0.29	29	9.4	9.4	0.12	28.3
S30200	302	0.29	29	9.4	9.4	0.12	28.3
S30430	302HQ	0.29	28	9.4	9.4	0.12	28.3
S30400	304	0.29	29	9.4	8.7	0.12	27.6
S30403	304L	0.29	29	9.4	8.7	0.12	27.6
S30451	304N	0.29	29	9.4	8.7	0.12	27.6
S30500	305	0.29	29	9.4	8.7	0.12	29.8
S31600	316	0.29	29	9.2	7.8	0.12	29.5
S31603	316L	0.29	29	9.2	7.8	0.12	29.5
S31703	317L	0.29	29	8.9	7.8	0.12	29.5
S31726	317LMN	0.29	29	8.9	7.8	0.12	33.5
S21800	Nitronic 60	0.28	26.2	8.8	---	0.12	---
S20161	Gall-Tough	0.28	24.8	9.6	7.1	0.12	28.2
S30300	303	0.28	---	10.4	---	0.12	28.3
S30323	303Se	0.28	---	10.4	---	0.12	28.3
S30800	308	0.29	29	9.6	8.8	0.12	28.3
S30883	308L	0.29	29	9.6	8.8	0.12	28.3
S30409	304H	0.29	29	9.4	8.7	0.12	27.6
S32100	321	0.29	29	9.4	8.7	0.12	27.6
S34700	347	0.29	28	10.4	---	0.12	28.3
S31609	316H	0.29	29	9.2	7.8	0.12	29.5
S30815	253MA	0.28	29	9.4	8.7	0.12	33.5

Table 4.6 Physical Properties[a] (continued)

UNS No.	Type or Alloy	Density lb/in.³	Modulus of Elasticity 10⁶ psi	Linear Expansion 68-212°F x 10⁻⁶/°F	Thermal Conductivity Btu/ft•hr•°F	Heat Capacity Btu/lb•°F	Electrical Resistivity μΩ•in.
S30908	309S	0.29	29	10	---	0.12	30.7
S30909	309H	0.29	29	10	---	0.12	30.7
S31008	310S	0.29	29	8.6	7.0	0.12	31.5
S31009	310H	0.29	29	8.6	7.0	0.12	31.5

a. Various industry sources. Values shown for room temperature unless a temperature range is specified.

Corrosion Properties

Stainless steels rely on a "passive film" of oxygen and chromium to provide their corrosion-resistant properties. This film will form on a fresh surface in all natural environments and in many chemical environments as long as the alloy contains at least about 12% chromium at its surface. It is this protective film which confers corrosion resistance. An understanding of this property and its implications is important to the successful application of stainless steels in corrosive environments. Some important principles related to this property are:

- passivity is an interrelated material–environment property,
- environmental variables such as temperature and minor contaminates can effect passivity,
- material variations such as improper heat treatment and surface contamination can effect passivity,
- if passivity is lost, either locally or generally, very high localized or general corrosion rates can occur.

In most stainless steel applications, it is rare that a generalized breakdown of passivity will occur unless there has been a gross grade misapplication or the environment deviates widely from what was anticipated. The more usual case is that of the occurrence of a localized breakdown in passivity which will then result in localized corrosion. This will usually take the form of pitting, crevice corrosion, or stress corrosion cracking. For this reason, the concept of a "corrosion allowance" has no meaning for stainless steels. The localized passivity breakdown is usually caused by some local disturbance in the environment, or sometimes by some local disturbance on the metal surface. Some examples are:

- solution concentration changes associated with localized boiling,
- the use of gaskets or bolted fasteners which create crevices on the stainless surface,
- deposits which create crevices,
- chromium-depleted surface layers which may form beneath heat tints associated with weld structures.

Unfortunately, once passivity has been lost it will not restore itself until the conditions which caused it are removed. In terms of localized corrosion, the "Achilles heel" of stainless steels is the chloride ion, and to a lesser extent the other halogen ions which are less frequently encountered. Therefore, when considering stainless steels for any application, it is important to consider the presence of chloride, as well as those factors which might promote localized corrosion. When selecting austenitic stainless steels for corrosive service, the design, method of operation, cleaning methods and frequency, etc. are just as important as the grade and environment, because they will have a profound effect on the potential for localized corrosion. Fortunately, when sound corrosion engineering practices are followed, the austenitic stainless will give good service under many demanding situations.

Atmospheric Environments

Atmospheric environments are generally classified as: rural, industrial/urban, or marine/coastal. Rural environments are very mild in terms of corrosivity to stainless steels, and virtually any grade of austenitic stainless steel will withstand many years of exposure with essentially no discoloration, rust staining, or pitting. S30400 (type 304) stainless is normally used in most architectural and other outdoor applications in rural environments.

Urban/industrial environments are quite variable with respect to corrosivity, and the performance of stainless in these environments will depend on the specifics of the situation. Those environments which contain sulfur dioxide or chloride-containing hydrocarbons can be very corrosive, especially in locations with little rainfall but high humidity. In northern urban areas where deicing salts are employed, the environment near streets and sidewalks carries chlorides which will be corrosive. In these situations S31600 (type 316) stainless is often chosen over S30400 because of its improved resistance to acid-chloride attack. The nature of the exposure is also very important to the degree of resistance exhibited by either S30400 or S31600. If it involves exposure to prevailing wind and rain or at heights well above the street level, the natural cleaning action of the weather will often allow the use of S30400 in major urban centers. If little natural or intentional cleaning is possible, S31600 is usually warranted. If the surface must remain free of any discoloration, then sometimes S31703 (type 317L) or an even more resistant grade might be required. Both S30400 and S31600 are widely used in architectural applications such as roofing and flashing, curtain walls, handrails and hardware, rapid transit cars, and in many decorative items. In those applications where appearance is important, some degree of maintenance cleaning is helpful.

Marine/coastal environments are defined as those close enough to the ocean where the local prevailing winds can deposit sea salt on exposed metal surfaces. The distance over which this effect occurs varies greatly with the location but five miles (eight kilometers) may be considered a conservative estimate of the limit of salt deposition. S31600 is the alloy of choice within the coastal zone, if some cleaning takes place, or if some discoloration or rust staining can be tolerated. Otherwise the use of S31703 may be warranted.

Natural and Purified Waters

All natural waters contain more than enough dissolved oxygen to maintain passivity on a stainless steel surface. Thus, the austenitic stainless steels are ideally suited for applications involved in the handling of waters of all kinds. One advantage that the stainless steels have in these applications is that they do not depend on a corrosion product film to provide resistance, and so they are essentially non-contaminating in food, pharmaceutical, and similar applications where maintenance of product purity is a necessity.

Applications involving water usually consist of water handling or control, water used in heat exchangers, or water used in processes ranging from boiler feedwater to a high purity feedstock in a biotech application. Water handling and heat exchanger applications may involve water that has been clarified but which still retains most dissolved solids and other impurities. The most frequently encountered dissolved solid in water which can be corrosive to stainless is the chloride ion. Because of the high corrosivity of chloride, the resistance of austenitic stainless steels is strongly dependent on water chloride content. With the exception of coastal sites, the chloride content of most natural waters rarely exceeds about 200 ppm which is about the service limit of S30400 at ambient temperature. Thus S30400 is suitable for many applications involving natural waters[1]. As temperature and other corrosivity factors increase, grades containing increased levels of chromium, molybdenum, and nitrogen are required. The effect of temperature and chloride on grade selection for water handling and heat exchanger service is shown in Figure 4.2. This graph shows advantages gained by selecting S31600 or S31726 for severe chloride service, but also illustrates that even these grades are not suitable for brackish or seawater service except in special circumstances. Some of these exceptions include cases where the stainless is cathodically protected, as in fasteners used for steel piling, boat hardware that is routinely cleaned, and evaporators where the water is deaerated.

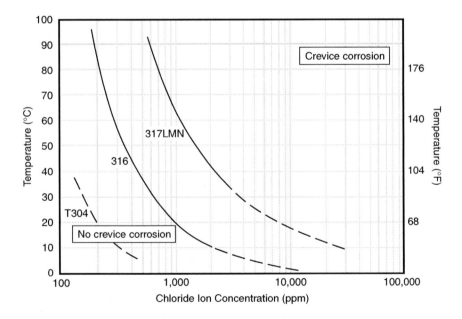

Figure 4.2 Effect of temperature and chloride on the initiation of crevice corrosion on austenitic alloys in aerated synthetic sea salt solutions at pH 2.[2]

Most contaminates in natural waters have little effect on their corrosivity to austenitic stainless grades. Some of the more commonly encountered contaminates falling into this category are trace amounts of dissolved acidifying gases such as hydrogen sulfide and carbon dioxide. Stainless steels can handle chlorine at the levels necessary for biofouling control provided certain limits of temperature, free chlorine in solution, and dissolved chlorine ion are not exceeded. There are two not infrequently encountered circumstances involving contaminates in natural waters, however, where serious localized corrosion can be encountered even with the more highly alloyed S31600 grade. The first of these involves the presence of certain microbes which can lead to what is termed "microbiologically influenced corrosion" or MIC[3]. Certain sulfur metabolizing microbes can produce tubercules, or deposits, on the metal surface which are acidified within by hydrogen sulfide created by the microbes. These regions are also low in oxygen compared to the surrounding water environment. Thus a specialized form of acidified oxygen concentration cell can occur leading to rapid localized attack. These conditions are most likely to occur in stagnant or low flowing water conditions when the water contains the required microbial species. The other circumstance involves waters that contain high dissolved manganese[4]. Manganese in itself is not detrimental in its normal valance state of +2 when present in a deposit as manganese oxide. However, if chlorination is being used, or if manganese oxidizing microbes are present in the presence of the chloride ion, the manganese can be oxidized to the +7 valence state which produces very aggressive pitting conditions. Either of these conditions can lead to very rapid pitting in S30400, and, unfortunately, S31600 and S31700 offer little improved resistance to this form of corrosion. The usual approach in handling waters of this kind is through proper water treatment or use of one of the 6% molybdenum austenitic stainless steels which are very resistant to MIC and permanganate associated attack.

High purity waters are those which contain very low dissolved solids and gases as the result of some special treatment. From a corrosion standpoint, these waters will usually be very low in chloride ion but will still contain enough oxygen to allow passivity of the stainless surface. Consequently, all austenitic stainless steels have very good resistance and are widely used in handling waters of this kind. In fact their range of usefulness with respect to temperature is greatly extended because of the low chloride ion concentration usually present. For example, S30400 is widely used to handle boiler feedwater and for piping in nuclear power plants at temperatures approaching 600°F (315°C).

Chemical Environments

Because the austenitic stainless steels can maintain stable passive films in a wide variety of chemical environments, they are widely used in many chemical handling and process applications. However, the point at which passivity can no longer be maintained must be considered. In general, the oxidizing chemicals which promote passive film stability, such a nitric acids or caustics, can be handled over a wide range of temperatures and concentrations. However, under certain conditions, these passive films can be damaged with resultant corrosion attack. For instance, reducing acids, such as hydrochloric acid, can only be handled under a limited rage of conditions. Oxidizing chlorides, such as ferric chloride and other chemical species, must be considered in evaluating the suitability of austenitic stainless steels for the intended service.

<u>Strong Reducing Acids, Hydrochloric and Hydrofluoric</u> The standard austenitic stainless steels have limited usefulness and can handle only very dilute solutions of these acids at ambient temperatures. The passive film is not stable under the strongly reducing conditions present in these solutions. At elevated temperatures corrosion rates can be very rapid and produce hydrogen evolution. The effect of temperature on corrosion rates in hydrochloric acid solutions is illustrated in Figure 4.3. Even in very dilute solutions pitting can occur, especially if surfaces that have contacted the acid are allowed to dry and then are exposed to a moist environment. Any stainless steel which has been exposed to a hydrochloric or hydrofluoric acid environment should always be thoroughly rinsed afterwards.

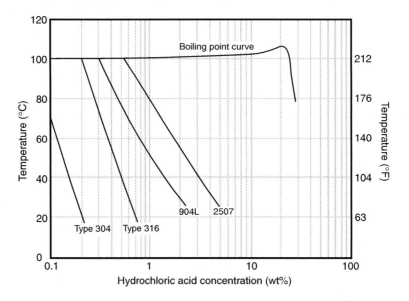

Figure 4.3 Corrosion in pure hydrochloric acid solutions.
Iso-corrosion curves at 0.1 mm/yr (4 mpy).[2]

<u>Sulfuric Acid Solutions</u> Sulfuric acid solutions have varied effects on
the passive behavior of austenitic stainless steels. Solutions at both
the low and high ends of the acid concentration range produce passive
behavior and allow useful service. Mid-acid concentration ranges and
elevated temperatures produce mildly reducing conditions and high
corrosion rates. The effect of acid concentration and temperature on
S31600 in pure acid solutions is illustrated in Figure 4.4. The acid
concentrations at which passivity is retained or lost are highly
dependent on other constituents that may be in the solutions, and the
iso-corrosion lines in Figure 4.4 may shift considerably one way or the
other in impure acids.

Aeration and oxidizing ions such as ferric, cupric, and chromate
increase the oxidizing potential of dilute solutions and allow
austenitic stainless steels to maintain passive behavior over a
broader range of acid concentration and temperature. On the other
hand, the presence of chloride may lead to pitting when an alloy
would otherwise be expected to display passive behavior.

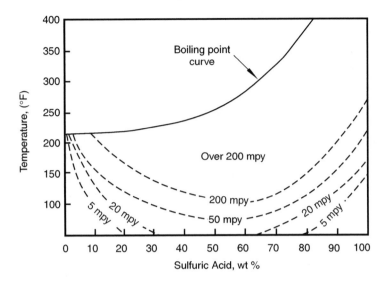

Figure 4.4 Corrosion of S31600 stainless steel by sulfuric acid
as a function of temperature.[5]

Nickel, copper and molybdenum are very effective in conferring
passivity to stainless steels in sulfuric acid solutions. For this reason
S31600 is used for a variety of storage, transfer, and moderate
temperature process applications involving mixed acids or solutions
containing oxidizing species. S31600 would normally be chosen over
S30400 for these applications because it is better able to withstand
unanticipated temperature or concentration variables. However, for
mid-range acid concentrations and high temperatures, a high
performance stainless steel or nickel-based alloy containing high
nickel and molybdenum should be considered.

Nitric Acid Solutions The strong oxidizing power of nitric acid means
that the austenitic stainless steels will have good resistance to this
acid. Very low corrosion rates occur over all temperatures up to the
boiling point for all concentrations except very concentrated solutions
as shown in Figure 4.5. Even above the boiling point, good resistance
is obtained at the lower acid concentrations with S30403. Because
high chromium improves resistance to this acid, higher chromium
grades such as S30908 or S31000 stabilized with columbium are

preferred for mid-range concentrations above the boiling point. Molybdenum and nickel are not helpful in this acid so S31600 offers no advantage except in dilute solutions containing chloride. Because high chromium is necessary to maintain passivity, any grain boundary chromium depletion associated with carbide precipitation or sensitization is highly detrimental to resistance in this and any strongly oxidizing acid. Therefore, the low carbon version, S30403, or the columbium stabilized version, S34700, are usually specified for nitric acid service. The titanium stabilized grade, S32100, is not suitable because nitric acid attacks titanium carbonitride. S30403 and the other austenitic grades are widely used in nitric acid manufacture. They are also used in many chemical processes based on nitric acid solutions because nitric acid will reduce the corrosivity of other acids such as sulfuric, acetic, and phosphoric.

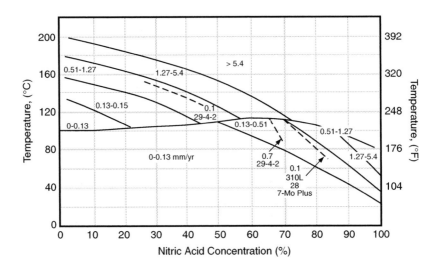

Figure 4.5 Corrosion of S30400 stainless in pure nitric acid solutions compared to some high performance stainless steels. Iso-corrosion curves in mm/yr.[2]

Phosphoric Acid Solutions   Pure phosphoric acid solutions are less aggressive than sulfuric acid solutions but are similar in the sense that, while classified as oxidizing acids, they have low oxidizing power. Therefore, austenitic stainless steel corrosion rates can be high in the higher acid concentrations at high temperatures, and they are also sensitive to ions that affect oxidizing potential or that may initiate pitting. S30400 and S31600 have good resistance to all concentrations of the pure acid at ambient and moderate temperatures and are widely used in the production and handling of this acid. S31600 extends the range of usefulness to near the boiling point in solutions containing up to 30% acid. Oxidizers such as nitrate and ferric ions will reduce corrosion rates while chloride and fluoride will increase the corrosivity. Grades which contain high chromium and molybdenum maintain passivity over a broader range of solution conditions.

Sulfurous Acid   Sulfurous acid is a relatively weak acid often encountered as condensate in flue gases. When chloride is present, it will cause pitting of S30400. As the acidity, chloride, and temperature of sulfurous acid condensates increase, grades containing higher chromium, molybdenum, and nitrogen are required. The order of progression in terms of increased performance is S30400–S31600–S31700. When flue gas condensates contain sulfuric acid accompanied by high concentrations of halogen ions, a 6% molybdenum stainless steel or nickel-based alloy is required.

Carbonic Acid   The dissociation constant of carbonic acid is very low but sufficient to cause corrosion of carbon steels in many applications involving gas and steam condensates. The austenitic stainless steels are unaffected by carbonic acid and are extensively used in applications involving carbon dioxide in air, gas, and steam condensates.

Organic Acids   The austenitic stainless steels are very resistant to most organic acids and compounds. Even the strongest organic acids, formic and acetic, can be handled by S30400 at most concentrations and temperatures below their boiling points. Exceptions are formic acid solutions in the 30% to 90% concentration range near the boiling point, and hot concentrated acetic acid solutions. With these solutions, improved resistance is obtained if oxidants such as air are present, and the molybdenum-containing grades such as S31600 and S31703 will perform better than S30400. However, there are many lower temperature process applications involving cellulose, acetates, and foods where S30400 and S31600 give good service and are widely used. Halogenated organic compounds in certain circumstances, however, can present a corrosive threat. This is in circumstances where the halogenated compound can decompose to form halogen salt which in the presence of water can hydrolyze to produce the corresponding acid. Also, some high temperature processes involving fatty acids will become corrosive to S30400. In this case S31600 and S31703 will provide improved corrosion performance. However, in aggressive conditions such as in tall oil distillation, the 6% molybdenum austenitic grades have become the standard material of construction.

Alkalies   The austenitic stainless steels will resist strong alkalies such as soda ash (sodium hydroxide) and caustic potash (potassium hydroxide) at ambient and moderate temperatures. The weaker alkalies such as sodium carbonate are not corrosive to these stainless steels up to solution boiling temperatures. When high temperatures and strong alkalies are involved, the usefulness of austenitic stainless steels is limited by caustic cracking in the mid-concentration range near the boiling point, and by general corrosion at higher caustic concentrations as shown in Figure 4.6. S30400 and S31600 can be used at temperatures at which carbon steels are limited by caustic cracking, but not in the high temperature ends of caustic evaporators.

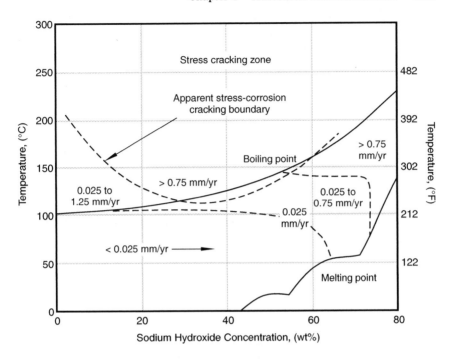

Figure 4.6 Corrosion of S30400 and S31600 in sodium hydroxide solutions. Iso-corrosion curves in mm/yr.[6]

<u>Non-Halogen Salts</u>   The austenitic stainless steels are highly resistant to a broad range of neutral and alkaline salts. This is also true for acid salts which do not contain halogens. Corrosion rates are very low to nil over a broad range of concentrations and temperatures. S30400 is suitable for most applications, but for the more acid salts at higher temperatures S31600 provides better resistance. Because they resist such a broad range of chemicals, both grades are suitable for general chemical storage and transportation applications although S31600 is preferred because of its greater versatility. They are also used in a variety of crystallizer and evaporator applications involving these salts.

<u>Halogen Salts</u>   Halogen salts represent a special case for stainless steels because of the tendency of the halogen ion to locally break down the passive film and produce pitting, crevice, or stress

corrosion. However, with proper attention the specifics of the application, even S30400 can give good service in certain circumstances. General considerations involving the handling of halogen salts include:

- For a given salt concentration, there is usually a critical temperature below which localized corrosion will not occur. This temperature decreases with increasing acidity (pH),
- High oxygen increases this critical temperature,
- Deposits or other crevice formers are very detrimental, especially in the presence of oxygen,
- Resistance to halogen salts increases significantly with increasing steel chromium and molybdenum content. Therefore, types S31600 or S31703 are usually chosen for applications involving these salts.

Corrosion by Foods

The demonstrated high resistance of the austenitic stainless steels to acids and salts makes them suitable for food processing, cooking, and handling applications. Corrosion rates are so low that they are of no consequence in these applications. Many tests have also been conducted showing that the introduction of metal species into food by food processing in stainless steels is negligible and of no health consequence.[7, 8, 9, 10] A factor of utmost importance in food handling is the tendency of the surface contacting the food to harbor harmful bacteria. The austenitic stainless steels are outstanding in their ability to resist bacteria retention and to be cleaned and sterilized. This advantage applies to pharmaceutical and biotech applications as well.

High Temperature Corrosion

A fortuitous feature of the metallurgy of chromium and iron is that the 12% chromium required to confer passivity in many aqueous environments also produces a marked improvement in high temperature oxidation resistance. This is not through a passivating effect but rather because chromium stabilizes the high temperature chromite and chromic oxides which are protective at high temperature. The nickel in austenitic stainless steels also improves high temperature strength and carburization resistance. Therefore, the austenitic steels have outstanding high temperature properties and are widely used in heat resisting applications.

As with aqueous corrosion, the beneficial effect of chromium on heat resistance continues with increasing chromium levels and is temperature dependent as illustrated in Figure 4.7. The resistance of any austenitic stainless steel will depend on the anticipated temperature of application as well as the chromium content. As a result, a family of austenitic stainless heat-resistant grades exists to meet a range of requirements. This range in resistance is illustrated in Figure 4.8 for several austenitic grades exposed to air under cycling conditions at 1800°F (980°C). At this temperature S30400, with only 18% chromium, has little useful resistance under cycling conditions. On the other hand S31000 (type 310), with 25% chromium, is highly resistant at this temperature. Similar curves could be constructed at different temperatures for the same alloys and they would show a characteristic upper useful temperature limit for each grade in cyclic, oxidizing exposure. Limits have been established for air exposure in both continuous and cycling service and are summarized in Table 4.7. The maximum temperature limit ranges from a low of 1500°F (815°C) for S20100 (type 201) to a high of 2100°F (1150°C) for S31000. These temperatures are well above the approximately 1100°F (595°C) limit for carbon steel and 1500°F (790°C) limit for most chromium ferritic stainless steels.

Other alloying elements are also useful in various ways to improve the heat resistance of these steels. Silicon improves resistance in oxidizing and carburizing atmospheres and is used in S30815 (253MA®) and in S31008 (type 310S). Titanium, columbium, and nitrogen, as well as nickel improve strength and they are used in S32100, S34700, and S30815 respectively. Because nickel improves austenite stability with respect to both ferrite and sigma phase and improves carburization resistance, it is used at high levels in several of these stainless steels. The "H" grades (0.04 to 0.10% C) also use high carbon and large grain size to increase high temperature strength.

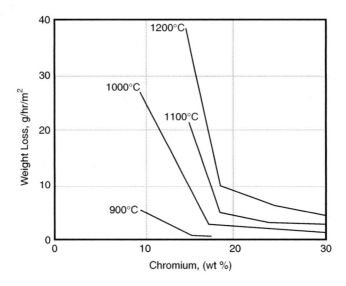

Figure 4.7 Effect of alloyed chromium on oxidation of steels
containing 0.5% C, 220 hr.[11]

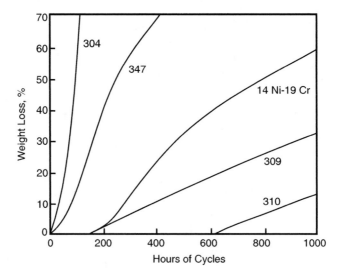

Figure 4.8 Scaling resistance of some iron-chromium-nickel alloys in
cycling-temperature conditions at 1800°F (982°C). Cycle consisted of
15 minutes in the furnace and 5 minutes in air. Sheet specimens
0.031 in. (0.737 mm) thick were exposed on both sides.[12]

Table 4.7. Suggested Maximum Service Temperatures in Air[13]

Grade	Intermittent Service		Continuous Service	
	°F	°C	°F	°C
201	1500	815	1550	845
202	1500	815	1550	845
301	1550	840	1650	900
302	1600	870	1700	925
304	1600	870	1700	925
308	1700	925	1800	980
309	1800	980	2000	1095
310	1900	1035	2100	1150
316	1600	870	1700	925
317	1600	870	1700	925
321	1600	870	1700	925
347	1600	870	1700	925

The chemical composition of a high temperature atmosphere has a strong influence on the performance of stainless steels. Best performance usually occurs in a dry air atmosphere, regardless of grade. The presence of moisture, typically found in combustion gases, will often lead to premature failure of the protective oxide and require a reduced operating temperature limit. If the atmosphere is oxidizing and contains sulfur oxides, all of the austenitic stainless steels will still perform fairly well. In a reducing environment, sulfidation can lead to premature failure of the high nickel grades such as S31000. Reducing atmospheres rich in carbon produce carburization in all of these steels, but S31000 is reasonably resistant because of its high nickel content. Halogen contaminates increase corrosivity under most conditions, especially when conditions are reducing.

Fabrication

The austenitic stainless steels can be readily fabricated by all common methods employed for steels and most other metals. The main considerations involved are those that may effect the corrosion resistance the specific parameters necessary for successful hot and cold working, annealing, and machining.

Hot Working

The upper hot working limit of austenitic stainless steels is controlled either by the initiation of excessive scaling which occurs above about 2100°F (1140°C), or by a rapid loss in hot ductility above 2250°F (1230°C). Hot ductility is very good at all lower temperatures but strength increases rapidly below 1700°F (955°C). Thus the normal hot working range for these steels is between 1700°F and 2100°F. Heating atmospheres will effect the degree of scaling and a slightly oxidizing atmosphere is preferred to minimize scaling and avoid carburization. Contaminates such as sulfur, zinc, copper, and lead should not be present on the surface prior to heating nor in the furnace atmosphere because they could lead to hot cracking.

Annealing

Annealing may be required to take carbides, sigma phase, or ferrite back into solution after hot working, or to recrystallize the material after cold forming. Typical annealing temperatures are in the range of 1800 to 2000°F (980 to 1095°C). Some special annealing treatments are employed for grain size control in the heat-resistant grades or to insure proper stabilization for S32100 and S34700 which contain carbon stabilizing elements. For all but the heat-resistant grades, rapid cooling from the annealing temperature is necessary to avoid carbide precipitation and its detrimental effect on corrosion resistance. Carbon content has a strong affect on the cooling rate required to avoid sensitization, low carbon content being beneficial as shown in Figure 4.9. The low carbon "L" grades are useful for heavy sections or other circumstances where the required cooling rates for the normal carbon grades cannot be achieved. Surface cleanliness and atmosphere control are important from the standpoint of providing a corrosion-resistant surface. All parts should be thoroughly degreased and cleaned before annealing to avoid surface carburization, and furnace atmospheres should be designed to avoid carburization and nitriding.

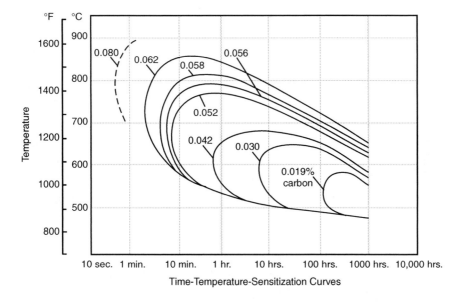

Figure 4.9 Time required for formation of carbide precipitation in stainless steels with various carbon contents. Carbide precipitation forms in the areas to the right of the various carbon-curves. Within time-periods applicable to welding, chromium-nickel stainless steels with 0.05% carbon would be quite free from grain boundary precipitation.[14]

Because the austenitic stainless steels are essentially single phase at all temperatures, they cannot be hardened by heat treatment. The only heat treat operation that might be employed is stress relieving to reduce cold working, welding, or machining stresses. When this is contemplated, grade selection should take into consideration the cooling rate that can be obtained because sensitization can occur with slow cooling rates. If slow cooling rates are anticipated, selection of one of the low carbon grades might be warranted. Stress relief temperatures should either be below 900°F (480°C) or above 1600°F (900°C) to avoid sensitization from carbide precipitation. Low temperature stress relieving provides little benefit in terms of stress relief while the high temperatures will introduce distortion problems. Therefore, stress relieving is seldom a satisfactory solution to residual stress problems in stainless steels.

Cold Working

The high work hardening rate and ductility of the austenitic stainless steels will have an effect on forming equipment, loading, and power, clearance, and lubrication requirements compared with equivalent operations on carbon steels. For equal thicknesses, power requirements are higher for stainless because of the high strength developed by work hardening. Strong rigid equipment is necessary when shearing or forming equivalent thicknesses, or thickness must be reduced by about one half compared to carbon steels. Shearing and blanking clearances are larger than for carbon steel; a clearance of about 0.001 to 0.002 inches (0.03 to 0.06 mm) is suitable for thin gage sheet and the clearance should increase as the material thickness increases. To avoid possible contamination of the stainless surface, tooling should be sharp and it should not have been used previously to cut carbon steel or copper alloys. Forming operations require high pressure lubricants and allowance for greater springback compared with carbon steels.

Welding

The austenitic stainless steels lend themselves to all of the common welding practices used in metal fabrication. Weldability is very good in joints made to other stainless steels, carbon steels, and nickel-based alloys. Filler metals are available to provide properties which match the corrosion resistance and strength properties of the base metal. These filler metals are designed to provide a small amount of ferrite in the weld metal to minimize any tendency for hot cracking. An important consideration in welding is to avoid conditions which will produce sensitization in the weld heat-affected zone. This is caused by carbide precipitation which produces a loss in corrosion resistance. Sensitization is most likely to occur when welding heavy sections and where high heat inputs are involved. For these applications the low carbon grades such as S30403 or S31603 are selected, or stabilized grades such as S32100 may be used. It is important to use inert gas protection and to properly clean the joints prior to welding to prevent contamination which could subsequently reduce the as-welded corrosion resistance.

Post-Fabrication Cleaning

Maintaining good surface cleanliness is very important during all stages of fabrication to avoid a number of possible surface degradation effects related to contaminates. In addition to possible carburization or hot cracking from copper contamination, the most commonly encountered contamination problems are those related to iron deposits left by metalworking tools, and weld oxide and weld splatter. Many cleaning procedures are available to meet specific cleaning needs. These fall into three categories:

- detergent or solvent cleaning to removes dirt and oils,
- passivation to remove iron contamination,
- acid and/or mechanical cleaning to remove surface chromium depletion associated with heat tints and weld oxides.

In all cases it is important to avoid cleaners which might leave deposits of chloride or acid on the surface, and to ensure the surface is neutralized and thoroughly rinsed with clean water after cleaning. When mechanically cleaning with wire brushes, the brush should be made of stainless and not previously used on carbon steel. Similarly, grinding wheels and abrasive belts should not have been used previously on carbon steel, and care should be taken to avoid heating the surface and forming new oxides.

Basis for Alloy Selection

Corrosion Resistance

The first consideration in selecting a stainless steel for a given application usually involves the environment to which the material will be exposed, the required service life, or surface appearance required over the anticipated service life. All grades will withstand exposure to normal atmospheric or mildly corrosive environments. However, some grades may develop slight staining, or tarnishing, or corrosion in mild environments. Because surface appearance is a very important consideration in many consumer applications, grade selection becomes an important matter even though environments in these applications may be mild. The grades are most likely to show some corrosion in mild environments are those which contain the lower amounts of chromium, molybdenum, and nitrogen, and those with silicon and high carbon. Some examples are S20100 and S30100 which have 16% minimum chromium, S30409 (type 304H) which may have as much as 0.10% carbon, and the free machining grade, type S30300, which contains a large sulfur addition. Normally S30400 is adequate for most applications involving atmospheric environments where surface appearance is a consideration. However, in coastal locations or harsh industrial atmospheres S31600 is preferred.

The basic S30400 stainless will resist a wide variety of organic and inorganic chemicals, acids, and bases at ambient temperatures. For this reason S30400 and S31600 are often selected for chemical storage and transportation. An exception is the case of strong reducing acids such as hydrochloric acid or mid-concentration range sulfuric acid solutions which can be very aggressive even at ambient temperature. The basic S30400 and S31600 grades will also withstand many chemicals at somewhat elevated temperatures, and improved resistance is usually obtained by selecting those grades with the higher chromium, nickel, and molybdenum contents such as S31603 or S31703.

In aggressive environments resistance to localized corrosion or severe general attack becomes of primary importance. The austenitic stainless steels are susceptible to localized pitting or crevice corrosion and to stress corrosion in aggressive aqueous environments containing the chloride ion. They also will not withstand strong reducing acids, especially at elevated temperatures. Many environments contain the chloride ion, and it is often necessary to consider grades with increased chromium, molybdenum, and nitrogen to provide improved corrosion performance. This includes S31703 which contains 3% molybdenum or S31726 with 4.3% molybdenum and a 0.15% nitrogen addition. These grades will withstand pitting and crevice corrosion in coastal architectural applications and many process applications involving nearly neutral but high chloride waters. The higher nickel also present in these alloys will provide some improvement in stress corrosion resistance compared to S30400, but will not offer immunity. Stress corrosion is not likely with these grades at ambient temperatures. However, temperatures above 120°F (50°C), chlorides, and high tensile residual stresses easily lead to susceptibility in all standard austenitic stainless steels. Practical engineering solutions to chloride stress corrosion cracking are provided by the 6% molybdenum austenitic stainless steels and duplex stainless steels such as alloy 2205 (S31803, S32205).

The austenitic stainless steels also have very good resistance to high temperature oxidizing environments and are often used for heat resisting applications. S30400 is the most commonly used grade for heat resistance; it will withstand oxidizing and mildly reducing environments at temperatures up to about 1400°F (760°C). For higher temperatures and carburizing environments, the higher nickel grade, S31000, is often required. For elevated temperatures in an oxidizing or sulfidizing environment, S30900 which has higher chromium and lower nickel has given good service.

Mechanical Properties

When grade selection is based on mechanical properties, the usual considerations involve strength, low temperature toughness, or ductility as it relates to formability. The mechanical properties of all of the austenitic stainless steels are similar in the sense they do not exhibit a defined yield point and have a relatively low yield strength as evaluated by the 0.2% offset method. However, they all work harden considerably during plastic deformation and will develop quite high strength while maintaining good ductility. However, when high yield strength is needed, those grades alloyed with nitrogen can deliver annealed yield strengths in the order of 50 ksi (345 MPa) compared to 25–30 ksi (170–205 MPa) for the standard grades. If high tensile strength is needed, a grade with a high rate of work hardening can give tensile strengths in excess of 150 ksi (1035 MPa), and much higher strengths can be obtained in the cold worked condition. S30100 (type 301) is widely used in spring applications where a combination of corrosion resistance and good spring properties is required. Compared to carbon steel and many nonferrous alloys, strength is retained at elevated temperatures, and so the high carbon and "H" grades are used in structural applications involving temperatures up to about 1400°F (760°C).

All of these grades deliver very good toughness down to temperatures well below ambient and are suitable for all structural applications involving normal atmospheric conditions. Most of them retain their toughness to cryogenic temperatures and are widely used in liquefied natural gas (LNG) service. However, the grades which contain high carbon or nitrogen, or are metastable, begin to lose toughness at the extremely low temperatures of liquid hydrogen.

Good formability is often the most important property required for austenitic stainless steels because of their wide application in tanks, vessels, containers, etc. Fortunately, the metastable nature of the austenitic structure of these alloys (tendency to transform to martensite upon cold deformation) has been used to create a number of alloys with a wide range of formability for any kind of forming operation. S30100, which work hardens rapidly, is well suited to severe stretch forming operations. S30400 can be modified to meet a range of stretch versus draw requirements to optimize forming of a wide variety of complex parts. Where severe drawing or cold heading is required, steels with high nickel and very stable austenite, such as S30500, are available.

Machining is more difficult with austenitic stainless steels than it is with carbon steels because of their high work hardening characteristics. When extensive machining is required, a resulfurized version of S30400, S30300, is available. This grade provides very high productivity on screw machines and other high productivity machining centers. However, S30300 has reduced corrosion resistance and weldability compared with the standard, low sulfur grades.

Design and Application

Austenitic stainless steels offer the designer and engineer practical and economic application opportunities. While many traditional applications depend entirely on corrosion resistance, many other opportunities can be realized by considering the combination of properties which can be delivered by a specific design for a specific application. Life-cycle cost analyses often justify the use of stainless steels because of their long-term durability. Cost savings with stainless steels include eliminating a corrosion allowance in structural applications and eliminating periodic painting to provide corrosion protection. Other examples of common applications are provided in Table 4.8. Sophisticated and complex applications which use a combination of stainless steels' unique characteristics include:

- Architecture, where unique finishes are used to combine pleasing appearance and color with durability.
- Transportation, where high strength structural members in light sections without a corrosion allowance make possible light weight rail and transit cars.
- Toxic gas handling in semiconductor manufacturing which relies on specially melted stainless steel for tubing that is electropolished on the inside to insure gas purity.
- Strong, tough, nonmagnetic structural supports for magnets operating at cryogenic temperatures.

The list of applications is almost endless and will continue to grow as long as designers and engineers continue to explore the almost unlimited possibilities for these versatile steels.

Table 4.8 Typical Applications

UNS No.	Grade	Typical Applications
S20100	201	Hose clamps, cookware
S20200	202	Hose clamps, cookware
S20400	Nitronic 30	Bulk solids handling equipment, truck and car frames
S30100	301	Food equipment, hose clamps, wheel covers
S30200	302	Springs
S30430	302HQ	Cold headed parts, fasteners
S30400	304	General purpose grade, kitchen equipment, architecture components, chemical industry tanks, vessels, heat exchangers
S30403	304L	Welded applications of 304
S30451	304N	Higher strength applications of 304
S30500	305	Deep drawn components
S31600	316	Chemical industry equipment, pharmaceutical production, coastal architectural applications
S31603	316L	Welded applications of 316
S31703	317L	Paper mill equipment
S31726	317LMN	Flue gas desulfurization scrubbers
S32100	321	Oil refinery, petrochemical equipment

Table 4.8 Typical Applications (Continued)

UNS No.	Grade	Typical Applications
S21800	Nitronic 60	Fasteners
S20161	Gall-Tough	Fasteners
S30300	303	Machined parts
S30323	303Se	Machined parts
S30800	308	Welding filler metal for type 304
S30883	308L	Welding filler metal for type 304L
S30409	304H	High temperature applications of 304
S34700	347	Oil refinery, petrochemical equipment
S31609	316H	High temperature applications of 316
S30815	253MA	Furnace components, thermal oxidizers
S30908	309S	Furnace components, thermal oxidizers
S30909	309H	Furnace components, thermal oxidizers
S31008	310S	Furnace components, thermal oxidizers
S31009	310H	Furnace components, thermal oxidizers

References

1. "Nickel Stainless Steels for Marine Environments Natural Waters and Brines", Nickel Development Institute Reference Book Series No. 11 003, Nickel Development Institute, Toronto, 1987.

2. "High Performance Stainless Steels", Nickel Development Institute Reference Book Series No. 11 003, Nickel Development Institute, Toronto, 2000.

3. G. Kobrin, "Corrosion by Microbiological Organisms in Natural Waters", Materials Performance", July 1976, p. 38, NACE International, Houston, 1976.

4. C.W. Kovach, "Types 304 and 316 Stainless Steels can Experience Permanganate Pitting in Water-Handling Systems", Materials Performance, Vol 38 #9, September 1999.

5. Fontana, M. G., "Corrosion Series," Ind. Eng. Chem., Vol. 44, March 1952, p. 86

6. Nelson, T. K., Alkalies and Hypochlorates, "Process Industries Corrosion," Moniz, B. J., Pollock, W. I., National Association of Corrosion Engineers, Houston, Texas, 1986, p. 257

7. G.N. Flint, D.K. Worn, "Hygiene and Other Health and Safety Aspects of Stainless Steel in Food-handling and Processing Plant", Proceedings of the Conference Processes & Materials Innovation Stainless Steel, Vol. 1, p.1.43, Associazione Italiana di Metallurgia, Milano, Italy, 1993.

8. K.W. Tupholme, D. Dulieu, J. Wilkenson, N.B. Ward, "Stainless Steel for the Food Industries", Proceedings of the Conference Processes & Materials Innovation Stainless Steel, Vol. 1, p.1.49, Associazione Italiana di Metallurgia, Milano, Italy, 1993.

9.   P. Haudrechy, H. Brüning-Pfaue, I. Lopez de Ahumada, M.J. Guio, H.J. Grabke, P.J. Cunat, "Innocuousness of Stainless Steels in Contact with Food or Skin", Proceedings of the Conference Stainless Steels '96, p. 229, Verein Deutscher Eisenhüttenleute, Düsseldorf, 1996.

10.   M.O. Lewus, D. Dulieu, K.W. Tupholme, S.C. Hobson, "A Study of the Potential for the Migration of Metals from Stainless Steel Systems into Chloride and Hypochlorite Bearing Waters", Proceedings of the Conference Stainlees Steels '96, p.237, Verein Deutscher Eisenhüttenleute, Düsseldorf, 1996.

11.   E. Houdremont, Handbuch der Sondersuhlkunde, Vol. I, Springer, 1956, p. 815.

12.   Industry data reported in "High Temperature Characteristics of Stainless Steels," No. 9004, Nickel Development Institute, originally published by American Iron and Steel Institute.

13.   "Design Guidelines for the Selection and Use of Stainless Steels", Specialty Steel Industry of North America, Washington, DC.

14.   Svetsaren, English Edition 1-2, 1969, p. 5.

DUPLEX STAINLESS STEELS

Gary Coates
Nickel Development Institute
Toronto, Ontario, Canada

Introduction

Duplex stainless steels (DSS) are not new, having been developed in the 1930s (see Figure 5.1). They have received significant attention for industrial applications in the last two decades, offering a combination of high strength and corrosion resistance, good fabricability, and reasonable cost, which make them the most cost effective materials for many applications.

The term *duplex* refers to the alloy's two phase microstructure, containing both austenite and ferrite. An alloy can be considered a duplex alloy if it nominally contains a minimum of 25-30% of any of those two phases. (Some austenitic stainless castings and austenitic weld metals may contain as much as 25% ferrite in their austenitic microstructure, and are sometimes described as having duplex microstructures, but will not be addressed in this chapter.)

Most commercial wrought duplex stainless steels have 50-55% austenite and 45-50% ferrite. This somewhat more complex microstructure necessitates both controlled processing by the manufacturer and extra consideration when fabricating and welding. Care in selection of qualified fabricators is advised, similar to what would be advised for the higher alloyed austenitic stainless steels.

Figure 5.1 First known duplex stainless steel application in the pulp
and paper industry, a Brobeck cooler for a sulfite mill.
Picture taken in 1932.

Also, the possibilities of upset conditions, especially regarding
temperature excursions, need to be fully explored prior to selecting a
DSS.

There are three basic categories of DSS–low alloyed, intermediate
alloyed, and high alloyed ("superduplex") DSS–grouped according to
their PREN (Pitting Resistance Equivalent Number) and roughly

corresponding to the range of austenitic stainless steels. Table 5.1a lists the more common wrought duplex grades, and 5.1b lists some of the more standardized cast alloys.

The "2205" alloy is by far the most common DSS grade and is the most readily available. It is classified under two UNS numbers, the older S31803 as well as the more restrictive S32205[*].

An example of a modern lower alloyed DSS would be S32304 (alloy 2304). There are a plethora of different superduplex grades, nominally with 25% chromium and at least 3% molybdenum.

Figure 5.2 An oxygen delignification vessel built of
S31803 (2205) plate for a pulp mill
being prepared for delivery.

---

[*] In this chapter, where the designation S31803 (2205) is used, the S32205 grade is
not only directly applicable, but is preferred.

Alloys/Properties/Specifications

Table 5.1a Typical Chemical Compositions of the More Common Wrought Duplex Stainless Steels

UNS No.	Common Name	Grouping[a]	Weight %						Nominal PREN[a]
			C max.	Cr	Ni	Mo	N	Others	
S32900	AISI 329	LD	0.08	25	4.5	1.5	0.05	---	30.8
S32001	19D	LD	0.025	21.3	1.2	---	0.14	5% Mn	23.6
S31500	3RE60	LD	0.030	18.5	4.7	2.7	0.08	1.7% Si	28.7
S32404	UR 50	LD	0.04	21	7	2.5	0.05	1.5% Cu	30.1
S32304	2304	LD	0.030	23	4.5	0.1	0.12	---	25.3
S31803	2205	MD	0.030	22	5.5	3.0	0.14	---	34.1
S32205	2205 +	MD	0.030	22.5	5.5	3.2	0.17	---	35.8
S32950	7-Mo Plus	MD	0.030	26.5	4.8	1.5	0.20	---	34.7
S31200	44LN	MD	0.030	25	6	1.7	0.15	---	33.0
S31260	DP-3	SD	0.030	25	7	3.0	0.15	0.5% Cu, 0.3% W	37.3
S39274	DP-3W	SD	0.030	25	7	3.0	0.27	0.5% Cu, 2 %W	39.2
S32550	255	SD	0.040	25	6	3.3	0.20	2% Cu	38.1
S32520	UR 52N	SD	0.030	25	6.5	4.0	0.25	1.5% Cu	42.2
S32750	2507	SD	0.030	25	7	4.0	0.28	---	42.7
S32760	Zeron 100	SD	0.03	25	7	3.5	0.25	0.7% Cu, 0.7% W	40.6

a. Duplex alloys grouped in terms of PREN (Pitting Resistance Equivalent Number), where PREN = %Cr + 3.3 x %Mo + 16 x %N. Tungsten (W) not included in calculations, but may improve pitting resistance.

LD = Low Alloy Duplex, PREN under 32
MD = Moderately Alloyed Duplex, PREN between 32 and 38
SD = Superduplex, PREN over 38

Table 5.1b Typical Chemical Compositions of the More Common Cast Duplex Stainless Steels

UNS No.	ACI Name	Other Names	C	Cr	Ni	Mo	N	Others
J93370	CD-4MCu	1A	0.04	25.5	5.5	2.0	---	3% Cu
J93372	CD-4MCuN	1B	0.04	25.5	5.5	2.0	0.17	3% Cu
J92205	CD-3MN	2205, 4A	0.03	22.5	5.5	3.0	0.16	---
J93345	CE8MN	45D, 2A	0.08	25	9	3.7	0.18	---
J93371	CD6MN	3A	0.06	25.5	5	2	0.20	---
J93380	CD-3MWCuN	Z-100, 6A	0.03	25	7.5	3.5	0.25	0.7% Cu, W
J93404	CE3MN	A958, 5A	0.03	25	7	4.5	0.20	---

History

It is very instructive to examine the history of the development of DSS, with their advantages and their limitations.[1] The first recorded intentional production and use of duplex stainless steels was in the early 1930s. The refining technology of the day resulted in relatively high carbon levels (0.08% was considered low) in all stainless steels, and the duplex grades were found to offer superior intergranular corrosion resistance compared to the austenitic grades. For this reason, duplex alloys were used, for example, in sulfite liquors (pulp mills) and in the manufacture of explosives using nitric acid. Improvement in refining techniques in the electric arc furnace in the 1960s made low carbon stainless steels relatively inexpensive. Intergranular corrosion was no longer such an issue, with the result that the low carbon 300 series stainless steels came to predominate.

Duplex alloys were also somewhat easier to cast, and the largest usage in their early history was as cast products. Additionally, DSS offered superior erosion corrosion and abrasive wear properties, making them ideal for pump casings and impellers, a property well used today for both cast and wrought DSS.

The early duplex grades (e.g., AISI type 329) typically had higher chromium contents than the standard austenitic grades, resulting in a higher general corrosion resistance in many environments. They also offered significantly improved resistance to chloride stress corrosion cracking (SCC), to which austenitic stainless steels are quite susceptible.

S31500 (alloy 3RE60) was one of the first duplex grades developed specifically with low carbon. It was aimed at applications where S30403 (type 304L) and S31603 (type 316L) suffered from chloride SCC, and for the most part performed up to expectations. However, especially during welding of heavier sections, it was possible to end up with too high a ferrite content in the heat affected zone (HAZ) of the base metal and also in the weld metal itself. This resulted in premature corrosion and/or mechanical failures from low ductility. All of the early duplex alloys had suffered from this problem, but to a much greater degree than does S31500 (alloy 3RE60).

By adding nitrogen to these early alloys, a more balanced structure was achieved, especially after welding. The original alloy 2205, developed in the mid-1970s and assigned UNS S31803 (Table 5.2), permitted a wide range of nitrogen, 0.08%-0.20%. However, at the lower end of the nitrogen range, significant problems have occurred with this alloy due to high ferrite levels in the HAZ.

A modified version, called UNS S32205, has a chemistry completely within the S31803 range, but in the upper end of the range for chromium, molybdenum, and nitrogen. This high nitrogen (0.14% min.) has ensured proper austenite/ferrite balance in the HAZ in all but the most extreme cases. For the superduplex alloys, even higher nitrogen contents are necessary. Nitrogen alloying also provides other benefits, which will be discussed later.

The duplex microstructure results in a material with significantly higher yield strength, making it suitable for shafts, rotating components, and structural applications. Table 5.2 shows typical mechanical properties of several duplex stainless steel plate grades. The minimum strength depends on the chemical composition, but yield strengths are typically 100-200% greater than those of S30403 (type 304L) or S31603 (type 316L). The increase in tensile strength is less.

This higher strength has its advantages in the use of DSS in pressure vessels built to various national codes, including the ASME Boiler and Pressure Vessel Code, as well as in numerous non-code applications. Similarly, the improved corrosion fatigue properties of the DSS are well documented[2], especially for suction press rolls in the pulp and paper industry.

Table 5.2 Mechanical Properties of Some Duplex Stainless Steels (from ASTM A 240/A 240M)

UNS No.	Common Name	Yield Strength ksi (MPa) min.	Tensile Strength ksi (MPa) min.	% Elongation min.	Hardness Brinell (HB max.)
S32900	AISI 329	70 (485)	90 (620)	15	269
S31500[a]	3RE60	64 (440)	92 (630)	30	290
S32304	2304	58 (400)	87 (600)	25	290
S31803	2205	65 (450)	90 (620)	25	293
S32950	7-Mo Plus	70 (485)	100 (690)	15	293
S31200	44LN	65 (450)	100 (690)	25	293
S32550	Alloy 255	80 (550)	110 (760)	15	302
S32750	2507	80 (550)	116 (795)	15	310
S32760	Z-100	80 (550)	108 (750)	25	270
S31260	DP-3	70 (485)	100 (690)	20	290

a. UNS S31500 not in ASTM A 240/A 240M; mechanical values from ASTM A 790/A 790M.

Mechanical and physical properties of DSS and austenitic stainless steels are compared in Table 5.3.

Table 5.3 Property Comparisons Between Duplex and
Austenitic Stainless Steels

Property	Comparison
Yield strength	About twice that of S31603 (316L) or S31703 (317L). This can be used to advantage in applications where strength can be used to reduce wall thickness. S31803 (2205) has greater spring-back than annealed austenitic stainless steel which may necessitate greater compensation during cold forming applications (e.g., spinning heads).
Tensile strength	About 35% stronger than S31603 (316L) and 25% stronger than S31703 (317L) at 68°F (20°C), higher at slightly elevated temperatures. ASME code calculations (allowable stresses) based primarily on these figures.
Low temperature ductility	S31803 (2205) wrought material is suitable down to at least -40°F (-40°C), if properly heat treated, and perhaps slightly lower. Welds are usually the limiting factor. (Note: Read ASME Section VIII Div. 1, UHA 51 for impact test requirements, which may be required at any minimum design metal temperature.) S31600 (316) and S31700 (317) wrought grades are suitable down to -320°F (-196°C) without UHA 51 requirement for impact testing, although welds require testing below -150°F (-101°C).
High temperature	S31803 (2205) is normally used up to only 570°F (300°C). Above that temperature, embrittlement and significant loss of corrosion resistance can occur. S31703 (317L) can be used up to about 1200°F (650°C), before sigma phase precipitation becomes an issue. S31603 (316L) is often used up to its oxidation limit, about 1570°F (850°C).
Thermal expansion	S31803 (2205) expands at a rate closer to mild steels, with S31603 (316L) and S31703 (317L) about 25-50% higher than S31803 (2205), e.g., in the interval 68-212°F (20-100°C), S31803 (2205) is 13 x 10^{-6}/°C, whereas for S31603 (316L) and S31703 (317L), the thermal expansion is 16.5 x 10^{-6}/°C and 16 x 10^{-6}/°C, respectively.
Thermal conductivity	No major differences. At 68°F (20°C), S31803 (2205) is approximately 14.0 W/m°C, S31603 (316L) and S31703 (317L) is approximately 13.5 W/m°C.
Electrical resistivity	S31803 (2205) is slightly more electrically resistant. At 68°F (20°C), S31803 (2205) is approximately 850 nΩm versus approximately 750 nΩm for S31603 (316L) and S31703 (317L).

Metallurgy as it Relates to Fabrication

Austenite/Ferrite Ratio and Welding

Although a full discussion of phase transformations is outside the scope of this practical handbook, there are a few very significant points that should be understood when welding and heat treating DSS.

By heating a DSS, e.g., S31803 (2205) which typically has about 55% austenite and 45% ferrite at room temperature, to just below the melting point at around 2550°F (1400°C), the resulting microstructure would be totally ferritic. All of the austenite phase would have transformed into ferrite. As the temperature is then slowly lowered, some of the ferrite transforms back to austenite, so that when the duplex alloy is held at 1830°F (1000°C) for sufficient time, a normal (55/45) structure results again. Very little ferrite-to-austenite transformation takes place below 1830°F (1000°C).

Figure 5.3 shows the ferrite/austenite ratio for a DSS as a function of temperature. If rapidly cooled from the near 100% ferrite temperature range, there is insufficient time for equilibrium conditions to be reached, and the ferrite is not allowed to transform completely. The result is a structure with too high ferrite, which results in poor ductility and poor corrosion resistance.

However, by heating the base material back into the 1830°F (1000°C) temperature range or higher (annealing), and holding for sufficient time, the proper ferrite/austenite ratio can be re-obtained.

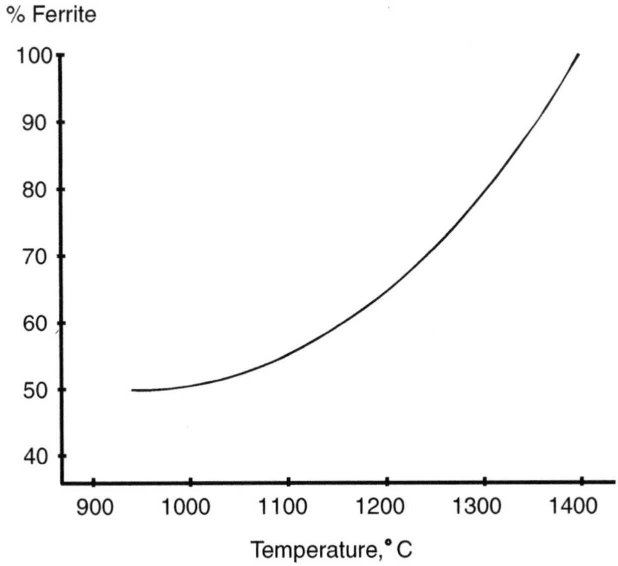

Figure 5.3 Equilibrium ferrite content
as a function of temperature.

Intermetallic and Other Phases

Originally, some duplex stainless steels were used for high
temperature applications, primarily because their slightly higher
chromium content provided better oxidation resistance. However, this
type of application is rare today, in part because there are many more
suitable austenitic alloys.

Generally, duplex stainless steels are not used at temperatures above
570°F (300°C). Above this temperature and up to about 1830°F
(1000°C), several detrimental microstructural changes occur. First,
the ferrite phase is susceptible to a transformation to alpha prime
(α'), which is a hard, brittle phase, although within a ductile
austenite. This is often called 885°F (475°C) embrittlement, based on
the temperature at which the embrittling happens quickest, i.e., the
nose of the curve (see Figure 5.4). The net result is that the ductility
of the alloy is lowered, often catastrophically. Embrittlement can also

occur at temperatures as low as 500-535°F (260-280°C), depending on the alloy, so great care should be exercised when dealing with applications involving long term exposure at or above this temperature.

The rate of embrittlement at any temperature is also composition specific. The highest temperature for alpha prime embrittlement for S31803 (2205) is about 975°F (525°C).

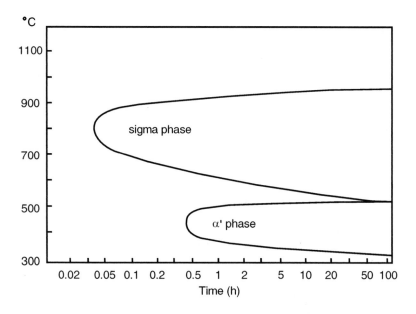

Figure 5.4 TTT curve for S31803 (2205) showing both sigma phase and alpha prime.[3]

In the temperature range of 840-1470°F (450-800°C), chromium carbide precipitation may occur, as it may in the austenitic grades. However, DSS are less sensitive to this phenomenon and this is rarely observed to be a problem, especially since all wrought DSS today are low carbon grades. Chromium nitrides may also form at around 1650°F (900°C) in nitrogen alloyed DSS, but these are rarely seen except in situations where the ferrite is too high or the nitrogen is too high. These situations arise occasionally at welds.

There are many different intermetallic phases, with complex chemistries, which can form in DSS. For ease of discussion, they are often grouped together and referred to as sigma phase, the most predominant of the phases. These may form in the temperature range of 1100-1830°F (600-1000°C), with the higher chromium and molybdenum alloyed stainless steels being most affected. The time to formation of these phases is composition specific, but can be as short as several seconds. These phases can negatively affect the corrosion resistance and the ductility of the material, and are the most serious threats to the successful application of these alloys. In small amounts they are not necessarily detrimental and may be impossible to avoid in the superduplex grades.

Fabrication Characteristics

Intermetallic phases can develop in the production of the base material during welding and during other fabrication steps that involve heating. The intermetallic phases that form as a result of the casting and hot working portions of the steelmaker's manufacturing cycle are normally removed by proper heat treating (annealing). Any carbides, nitrides, and alpha prime are similarly removed. The annealing temperature is grade specific but is normally 1920-2010°F (1050-1100°C). Time and temperature are the factors involved in completely dissolving the detrimental phases and precipitates. Rapid cooling from the annealing temperature to about 570°F (300°C) is necessary to avoid reformation of intermetallic phases. Heavier sections should be water quenched, while thinner sections may be rapidly quenched in air or by other means.

Fabrication procedures that involve elevated temperatures include hot forming, stress relieving, etc. Heads, for example, are often hot formed and often need to be re-annealed after forming. Occasionally, components being cold formed are "helped along" with the aid of locally applied heating, many times without consideration of possible affects or monitoring of the temperatures.

In the case of S30403 (type 304L) and S31603 (type 316L) there are no detrimental affects when short time localized heating is applied during cold forming, but duplex alloys are less forgiving, so all thermal exposures should be evaluated. Duplex stainless steels have relatively high strength and more springback.

Stress relieving treatments must also be carefully considered and thoroughly evaluated before specifying and applying to DSS. In general, they are not necessary and should be avoided, unlike the austenitic grades which may require stress relieving to minimize the possibility of stress corrosion cracking. When absolutely necessary, stress relieving of DSS should be done at or just below the annealing temperature, depending on the grade, followed by a rapid cool. Stress relieving at around 1100°F (600°C) may be possible, as chromium carbide sensitization is the only major concern. In a construction containing DSS where welded carbon or low alloyed steel must be stress relieved, this latter temperature is usually chosen as being suitable, i.e., least detrimental. The ASME Section VIII Division 1 code does not require stress relieving of welded duplex stainless steel components.

Figure 5.5 shows an example of welding DSS where both the weld metal and the base metal near the weld go through critical temperature ranges. Consequently, if the weldment cools very quickly, so that there is insufficient time for transformation of ferrite to austenite, both the weld metal and the HAZ may have too high ferrite content. Nitrogen, mentioned earlier, promotes the rapid formation of austenite during cooling from high temperatures. (Nitrogen in stainless steels is of the mono-atomic form, i.e., it is N not N_2 gas). So the problem of too high ferrite in the HAZ is rare today on modern duplex stainless steels with their higher nitrogen content. However, it is still possible to have this problem with all duplex stainless steels if the cooling rate is extreme. An example of this could be a small, low heat input weld (e.g., a GTAW "dress" or blend pass) performed on a heavy section. It may also occur in a DSS with a lean nitrogen composition.

Figure 5.5 Field installation of a continuous digester
built of S31803 (2205) plate, being prepared
for start up at a pulp mill.

To maintain the proper phase balance in weld metal, most but not all
duplex filler metals are overalloyed with nickel. Nickel, like nitrogen,
promotes the ferrite-to-austenite transformation. For welding S31803
(2205), the standard filler metal is AWS E2209 or AWS ER2209,
which contains 3-4% more nickel than the base material. Nitrogen
levels in the filler metals are also important and should not be at the

low end of the range. Weld metal will have a much broader range in ferrite content because of the great variability in cooling rates. Often, a range of 30%-65% ferrite is specified as being acceptable. Although having lower than 30% ferrite may create some concern, especially regarding stress corrosion cracking resistance, the detrimental effects of ferrite levels higher than 65% are normally of much greater concern. AWS/ASME specifications for DSS filler metals are listed in Table 5.4.

Table 5.4 AWS/ASME Specifications for Duplex Filler Metals

AWS	ASME	Same spec. as for 300 series?	Product Forms
A5.4	SFA 5.4	Yes	Coated electrodes
A5.9	SFA 5.9	Yes	Bare wire
A5.22	SFA 5.22	Yes	Flux core electrodes

Note: Only a few duplex filler metals are currently listed in these specifications.

For welds that will be annealed afterward, filler metals overalloyed in nickel may not be needed. Some examples are weld repair of castings where the final operation is a full anneal and longitudinal welds in pipe production that will be later annealed.

ASTM A 923 addresses the issue of testing for detrimental intermetallic phases in wrought duplex stainless steels. It ensures that the base metal meets a certain minimum criteria to permit welding. Currently, only the S31803 is covered, but other alloys are being added. Figures 5.6a and 5.6b show photomicrographs from ASTM A 923 that illustrate unaffected and affected structures in S31803.

ASTM A 923 is used in combination with the applicable product standard, e.g., ASTM A 240 or ASME SA 240 for plate, sheet, and strip. (See Tables 5.5a, 5.5b, and 5.5c for applicable standards.) Specifying ASTM A 923 prevents the situation in which an otherwise acceptable base material would contain sufficient sigma phase to adverse affect the HAZ after normal welding. Tests other than specified in ASTM A 923 may be required in more critical applications and are common for offshore oil and gas platforms.

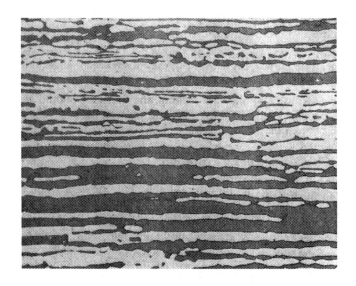

Figure 5.6a ASTM A 923 unaffected structure in S31803
(longitudinal section).

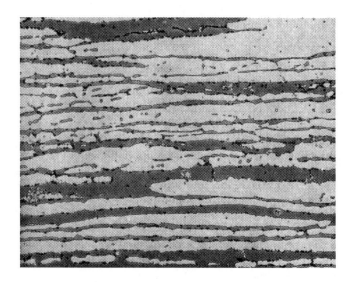

Figure 5.6b ASTM A 923 affected structure in S31803
(longitudinal section).

Table 5.5a ASTM/ASME Specifications for Wrought and Cast Products

ASTM	ASME	Same spec. as for 300 series?	Product Forms
A 182	SA 182	Y	Forged components - flanges, fittings, forged parts
A 240	SA 240	Y	Flat rolled - plate sheet, strip
A 276	---	Y	Bars and shapes
A 314	---	Y	Billets and bars for forging
A 473	---	Y	Forgings
A 479	SA 479	Y	Bars and shapes
A 789	SA 789	N	Tubing - welded and seamless
A 790	SA 790	N	Pipe - welded and seamless
A 815	SA 815	N	Butt weld fittings
A 890	---	N	Castings
A 923	---	N	Specification for freedom from detrimental intermetallic phases in mill produced product
A 928	pending	N	Welded pipe with filler metal

Note: Not all duplex grades are listed in these specifications. New DSS are being added and older ones are being deleted continually. Consult the latest version for current information.

Table 5.5b Grades and Product Forms Listed in ASME Section II Part D (1998 Ed.) for Section VIII Div. 1 Usage

UNS No.	ASME Nominal Composition	Product Forms Listed
S32900	26Cr-4Ni-Mo	Plate, seamless and welded tube, seamless and welded pipe
S32950	26Cr-4Ni-Mo-N	Plate, seamless and welded tube, seamless and welded pipe
S31803	22Cr-5Ni-3Mo-N	Forgings, plate, seamless and welded tube, seamless and welded pipe, seamless and welded fittings
S32304	23Cr-4Ni-Mo-Cu-N	Plate, seamless and welded tube, seamless and welded pipe
J93345	24Cr-10Ni-4Mo-N	Castings
J93370	25Cr-5Ni-3Mo-2Cu	Castings
S32250	25Cr-5Ni-3Mo-2Cu	Plate, bar, seamless and welded tube, seamless and welded pipe
S31200	25Cr-6Ni-Mo-N	Plate
S31260	25Cr-6.5Ni-3Mo-N	Plate, seamless and welded tube, seamless and welded pipe
S32750	25Cr-7Ni-4Mo-N	Seamless and welded tube, seamless and welded pipe
S32760	Code Case 2245	Forgings, plate, seamless and welded tube, seamless and welded pipe, seamless and welded fittings
J93380	Code Case 2244	Castings

Table 5.5c Grades and Product Forms Listed in ASME Section II Part D (1998 Ed.) for Section VIII Div. 2 Usage

UNS No.	ASME Nominal Composition	Product Forms Listed
S32304	23Cr-4Ni-Mo-Cu-N	Plate
S31200	25Cr-6Ni-Mo-N	Plate
S31500	18Cr-5Ni-3Mo	Seamless and welded tube, seamless and welded pipe
S32550	Code Case 2068-2	Plate
S31803	Code Case 2067-2	Forgings, plate, seamless and welded tube, seamless and welded pipe

Alloy Selection/Limitations

The total amount of DSS currently in use is estimated to be just over 100,000 MT[4], i.e., less than 1% of the total stainless steel market of 13,000,000 MT (1995). Among certain stainless steel producers, it is as high as 3-4%. Many stainless steel mills do not currently produce duplex grades or produce only minimal quantities.

Duplex stainless steels are known as "problem-solver" alloys. Hence, they are often used to replace the standard S30403 (type 304L) and S31603 (type 316L) alloys where they have failed for one reason or another, or where total life cycle costing concepts are applied and the long term cost of the duplex alloys is lower.

Each application needs to be fully evaluated to determine which duplex alloy is most suitable as a replacement alloy. In addition, even though DSS may offer increased SCC resistance fatigue strength, corrosion fatigue properties, and wear resistance to the standard austenitic stainless steel, all the duplex alloys have their limitations.

Where S30403 (type 304L) and S31603 (type 316L) are successfully used, there are a few applications where the total installed cost of a duplex alloy is still lower, mainly where the high strength can be used to maximum advantage.

There are also times when a particular duplex property, e.g., the low thermal expansion rate, can weigh the decision in favor of a duplex alloy. When a minimum of S31703 (type 317L) is required for corrosion resistance, duplex alloy S31803 (2205) should always be considered.

Major Markets/Applications

Rather than list applications by individual alloys, they are listed by group in this section. See Table 5.1a for this division, where the groups are based on their PREN (Pitting Resistance Equivalent Number).

The PREN can be calculated by the following equation.

$$PREN = \%Cr + 3.3 \text{ x } \%Mo + 16 \text{ x } \%N^{\dagger}$$

The groups are Low Alloy Duplex (LD), with a PREN < 32, Moderately Alloyed Duplex (MD), with a PREN between 32 and 37, and Super Duplex (SD), with a PREN > 37. This is a somewhat arbitrary division, and an alloy's pitting resistance often will have nothing to do with its actual corrosion resistance in a application. However, these categories (if not the names) are generally recognized. Tables 5.6a through 5.6e list the applications of DSS and the typical alloy they would replace within various industries; figure 5.7 illustrates one of these examples.

Figure 5.7 Final stages of fabrication of a sulfate digester,
13.1 ft (4 m) diameter by 57 ft (17.4 m) long,
made with 55 tons of S31803 (2205) plate, for a pulp mill.

---

† Many PREN formulas have been proposed. The "%C + 3.3 x %Mo" portion is generally accepted. The nitrogen coefficient, however, is more contentious, with values of 22 and 30 often widely cited.

Table 5.6a Chemical Process/Petrochemical Industry

Application	Duplex Alloys Used	Replacing UNS No.
PVC stripper columns, reactors	MD	S31603
Alcohol production	MD	S31603
Production of organic acids and intermediates	LD, MD, SD	S31603, S31703, N08904
Formaldehyde evaporators, columns	MD	S31603
Brackish water heat exchangers	LD, MD, SD	S30403, S31603
Seawater heat exchangers	SD	6% Mo austenitic stainless steel
Production and storage of phosphoric acid (rakes, tanks, pipes, pumps, etc.)	MD, SD	S31603, S31703, N08904
Nitric acid equipment	23% Cr LD	S30403, S31002
Piping e.g., carry hot fluids where exterior is possibly subject to SCC	MD	S30403, S31603
Air coolers	MD	Carbon steel, S30403, S31603
Various heat exchangers, heating coils, condensers, etc. for process side conditions	LD, MD, SD	S30403, S31603, S31703

Table 5.6b Offshore Industry

Application	Duplex Alloys Used	Replacing UNS No.
Offshore platforms - pressure vessels, separators, risers, flow-lines, sub-sea manifolds and umbilicals, pumps, valves	MD, SD	S31603, S31603 clad, carbon and low alloyed steels
Offshore platforms - seawater systems	SD	Option to 6%Mo austenitic stainless steel
Gas processing plants, pipelines	MD	S31603, S31603 clad, carbon and low alloyed steels
Explosion and blast walls	LD	S31603

Table 5.6c Pulp and Paper Industry

Application	Duplex Alloys Used	Replacing
Kraft digesters, blow lines, blow tanks	LD, MD, SD clad	Carbon Steel, S30403, S31603
Digester target plate	MD, SD	AR plate
Liquor heater tubes	LD, MD, SD	S30403, S31603
Sulfite digesters and tanks	MD	S31603, S31703
Oxygen delignification vessels	MD	S31603
Brown stock washers	LD, MD	CS, S31603, S31703
Bleach plant washers (D-stage)	MD	S31603, S31703
Disc filters	LD, SD	S31603
Pumps	LD, MD, SD	Various
Wash presses	MD	S31603, S31703
White liquor oxidizers	MD, SD	S31603
Screw conveyors	LD, MD	S30403, S31603
TMP Steaming vessels	MD, SD	S30403, S31603, S31703
Steam nozzles	MD	S31603
Paper machine suction rolls	LD, MD	S31603, bronze, martensitic stainless steel
Lining of carbon steel for various applications, e.g., liquor storage tanks, hog fuel dryers, blow tanks, ducting, precipitator walls, etc.	LD, MD, SD	S30403, S31603, S31703
Screen plates	MD	S30403, S31603
Pressure feeders	MD	S30403, 400 series stainless steel
Pump shafts	MD, SD	S30403, S31603, S31703

Table 5.6d Mining/Hydrometallurgy

Application	Duplex Alloys Used	Replacing
Autoclaves and piping	SD	Nickel alloys, N08904, N08020
Agitators, thickener rakes, leach tanks, reactors, pipes, pumps, valves, etc.	LD, MD, SD	S31603, S31703, N08904
Abrasive/corrosive applications	LD, MD, SD	AR plate, S31603

Table 5.6e Other Industries

Application	Duplex Alloys Used	Replacing
Transport of chemicals (trucks, boats, RR cars)	MD	S31603, S31653, S31753
FGD (flue gas desulfurization) - agitators, absorber internals, dampers, inlet ducting, outlet ducting	MD, SD	S31603, S31703, S31726
Fans and blowers, compressors	MD, SD	S31603, S31703
Food industry - production of vegetable fats, catsup, pectin; boiling pasta kettles, brewery hot water piping, mussel boiler heating coils, vibrating	MD, SD	S31603, S31703
Building industry - rebar, fasteners	MD	Carbon steel, S30400
Boats - propeller blades and shafts	MD, SD	S30400, S31603
Domestic hot water heaters	LD	Ferritic stainless steel, S30400

Corrosion Properties

As a family of alloys, from low alloyed to very highly alloyed, DSS show a very wide range of corrosion resistance. The emphasis in this section will be on environments where testing has shown them to be of significant interest, and there have been some applications. It should be emphasized that testing should be performed to ensure that any alloy is suitable in any particular environment. It appears that comparable austenitic alloys with nickel contents higher than their DSS counterparts are usually more forgiving to upset conditions, where a material switches temporarily to an active from a passive state. In some environments, this phenomenon is particularly important, e.g., sulfuric acid. When using laboratory tests to evaluate suitability for service, it is wise to electrochemically activate the metal coupons during testing to monitor their ability to repassivate.

Acids

Sulfuric Acid

DSS can show excellent corrosion resistance in weak (<20%) sulfuric acid solutions (Figure 5.8)[5] and especially in the addition of oxidants, such as metallic ions or oxygen. S31803 and S32205 should be superior to S31603 (type 316L) and S31703 (type 317L) in most weak acid solutions and can sometimes be better than even higher alloyed grades such as N08904 (904L) and N08020 (alloy 20). The 25% Cr superduplex stainless steel provide even greater corrosion resistance in weak sulfuric acid solutions. In the case of hydrometallurgical autoclaves, an extremely severe application with sulfuric acid concentrations usually less than 1% and temperatures well in excess of 100°C, these 25% Cr superduplex stainless steel often are the only ferrous alloys that perform well. Even nickel-based alloys have high corrosion rates in this type of environment. Weak sulfuric acid contaminated with chlorides (and also often oxidants) is another situation where DSS are currently used. Figure 5.9 shows iso-corrosion curves for several alloys in sulfuric acid with 2000 ppm chloride.[5]

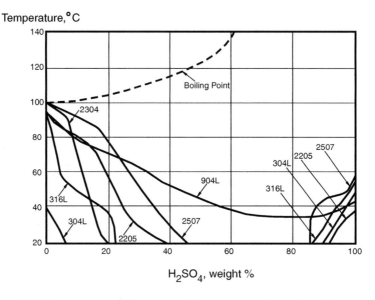

Figure 5.8 Iso-corrosion diagram, 0.1 mm/year, in naturally aerated sulphuric acid of chemical purity for three duplex stainless steels (2304, 2205 and 2507), and three austenitic stainless steels (304L, 316L and 904L).

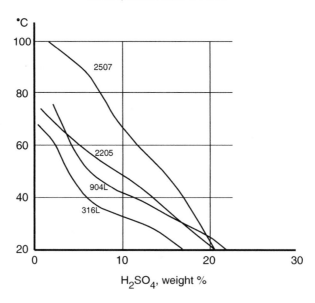

Figure 5.9 Iso-corrosion diagram, 0.1 mm/year, in sulphuric acid containing 2000 ppm chloride ions for two duplex stainless steels (2205 and 2507), and two austenitic stainless steels (316L and 904L).

In middle (20-90%) concentration sulfuric acid solutions, generally only the copper alloyed 25% Cr SDSS (e.g., S32550 or S32760) should be considered. Where they remain passive, they can be very useful materials. But the corrosion rate can increase by an order of magnitude or more with only a small change in conditions, e.g., a temperature increase of 20°F (6.7°C). See Figure 5.10.[6]

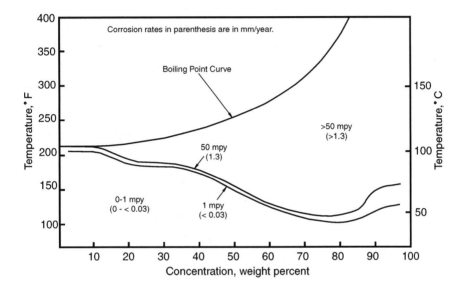

Figure 5.10　Corrosion resistance of Ferralium® 255 (S32550) in sulfuric acid iso-corrosion curve for 0.3 and 1.3mm/a.

In high (>90%) concentration sulfuric acid, there has been mixed experience with the S31803 (2205). Caution is therefore advised. However, the 25% Cr SDSS often give suitable performance and are used where high strength and erosion resistance is needed.

Phosphoric Acid

Wet process phosphoric acid typically contains significant amounts of impurities, with the chlorides and fluorides rendering the workhorse S31603 (type 316L) susceptible to corrosion. S31803 (2205) has clearly superior corrosion resistance in these types of environments and is used for the tank walls as well as other equipment in dedicated chemical tankers. It offers a lighter weight design due to its higher strength. The 25% Cr SDSS may find use where their higher strength and higher corrosion-erosion resistance and abrasion resistance offset their higher cost.

Nitric Acid

The molybdenum-free DSS (e.g., S32304) is useful in certain nitric acid solutions, offering superior corrosion resistance to S30403 (type 304L) and perhaps even S31002 (type 310L). Care must be taken to evaluate corrosion resistance of welds and end-grain attack. S31803 (2205) can be used in certain less aggressive acid applications, especially in heat exchangers where brackish water is used for cooling.

Other Inorganic Acids

As with the common austenitic stainless steels, DSS are generally not considered for hydrochloric acid service, except for handling very low concentrations. S31803 (2205) or the 25% Cr SDSS may then be considered. DSS is not generally considered for hydrofluoric acid service, although it has been used for handling stainless steel pickling acids, which are a mixture of nitric and hydrofluoric acids (up to 5%).

Organic Acids

Both S31803 (2205) and the 25% Cr DSS have very useful corrosion resistance in organic acids. They are commonly used in acetic and formic acid mixtures for distillation columns, heat exchangers, piping, etc.

Caustic Solutions

There does not appear to be much data available for pure caustic solutions, especially regarding caustic stress corrosion cracking (SCC). However, in typical kraft pulp mill digester liquors containing sodium hydroxide and sodium sulfate, the typically higher chromium contents of the DSS gives them far greater corrosion resistance than S30403 (type 304L) and S31603 (type 316L), with respect to both general corrosion and SCC.

Localized Corrosion

Chloride Stress Corrosion Cracking

The normally continuous ferrite phase in DSS gives them improved resistance, although not immunity, to chloride stress corrosion cracking. Table 5.7 gives a relative ranking of several stainless steels (duplex and austenitic) in the Wick test (modified ASTM C 192), which was run at 212°F (100°C) and in the Drop Evaporative test, where the estimated metal temperature is about 390°F (200°C).[7] The strong affect of temperature on relative SCC resistance can be seen. The 6% molybdenum austenitic stainless steel S31254 (254 SMO) is surprisingly good compared to even the superduplex stainless steel. DSS piping systems are commonly used to handle a variety of hot fluids in the CPI at locations near saltwater, where external SCC may result from chlorides accumulating under the insulation. However, caution is advised if the temperature exceeds certain ranges, e.g., with S31803 (2205) above about 250°F (120°C). Also, the type of the chloride needs to be examined before selecting an alloy, e.g., magnesium chloride which is found in seawater is more aggressive than sodium chloride.

Table 5.7 Relative Ranking of Some Stainless Steels
in the Wick Test and Drop Evaporation Test

Relative SCC Resistance Ranking	Wick Test, 212°F (100°C)	Drop Evaporation Test Estimated metal temperature about 392°F (200°C)
Highest	S31254, S32304	S31254
		N08904, S32750
	N08904, S31803	S32304
		S31803
Lowest	S30400, S31603	S30400, S31603

Chloride Pitting and Crevice Corrosion Resistance

The various DSS can be ranked by the standard test methods. They can also be ranked by their PREN values, which provides a comparative pitting resistance for these alloys based on their chemical composition (see Table 5.1a). It should be observed that most laboratory testing is based on fully annealed and water quenched specimens, not always indicative of real life conditions.

DSS are much more sensitive to improper heat treatment and thermal cycling than normal austenitic stainless steels, as well as to affects of welding (process, gases, incorrect heat input, heat tint, etc.). Figure 5.11 shows typical critical pitting temperatures of various alloys in a 1M NaCl solution, and critical crevice corrosion temperatures in ferric chloride. These rankings can be very useful in choosing replacement material where a known stainless steel has failed.

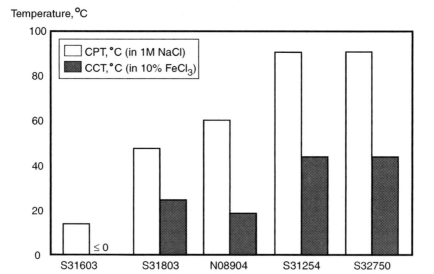

Figure 5.11 Typical critical pitting temperature (CPT) values in 1M NaCl, as measured using the Avesta Sheffield Pitting Cell and Critical crevice corrosion temperature (CCT) in 10% $FeCl_3$ [8] using PTFE multiple crevice washers, torque 0.28 N•m.

Corrosion in Brackish Water and Seawater

Duplex grades have found extensive use in heat exchangers where brackish water or seawater is used as a cooling medium. They are often chosen over S30403 (type 304L) or S31603 (type 316L) in brackish water because of their high resistance to chloride SCC. The same factors that affect choice of austenitic stainless steel grade (flow rate, pH, oxygen content, chlorination, contaminants, microbiological activity, etc.) need to be examined in order to choose the proper duplex alloy.[9]

Hydrogen Sulfide

Duplex alloys, because of their ferrite content, are somewhat sensitive to hydrogen sulfide stress corrosion cracking. Several of the duplex alloys are included in NACE MR0175-99 (Sulfide Stress Cracking Resistant Metallic Materials for Oilfield Equipment)[10]. For example, S31803 in the solution annealed condition is suitable in sour service provided that the hardness does not exceed 28 HRC . If further cold worked, S31803 is suitable for use up to 450°F (232°C) if the partial pressure of hydrogen sulfide does not exceed 0.3 psi (0.002 MPa), the yield strength does not exceed 160 ksi (1100 MPa), and the hardness does not exceed 36 HRC. Other limits are set for other DSS. New guidelines for suitability of all alloys, including DSS, are currently under review.

Guidelines for Selection

DSS are used primarily as problem solving alloys, where the more standard grades do not adequately perform. Often, the DSS is replacing existing equipment where problems have previously occurred, although as familiarity with this series of alloys increases, they are also being specified into new equipment design. Their higher strength allows lighter weight designs and thinner wall or cross-sectional thickness to be considered. Other characteristics may include higher fatigue strength, improved stress corrosion cracking resistance, high general corrosion resistance in certain media, and improved wear properties, both erosion and erosion-corrosion.

S31803 (2205) is often substituted for S31603 (type 316L) and S31703 (type 317L), while the super duplex alloys are most often substituted for the higher alloyed austenitics such as N08904 (904L) and the 6Mo austenitic stainless steels. S31803 (2205) has the highest availability of any duplex stainless steel around the world, with respect to product form, size, thickness, and quantity. Still, its availability is much lower than for S30403 (type 304L) and S31603 (type 316L) stainless steels, and availability may be an influential issue in alloy selection.

In any selection process, attention must be paid to potential issues relating to fabrication and welding. Although DSS processing and compositional controls have been optimized to overcome some of the earlier limitations associated with these materials, experienced fabricators need to be involved. Fabricators who are marginally competent to weld standard austenitic stainless steel grades should receive specific training on duplex stainless steels. Welding procedures and specifications should be carefully developed, including those required for the repair and later modification of equipment. Also, field welding procedures will often vary considerably from shop welding practices. In all cases, the proper training of the welders is essential. The cost of fabricating duplex stainless steel is not necessarily more than the costs associated with good quality fabrication of austenitic stainless steel.

When considering the substitution of an austenitic with a duplex stainless steel, caution should be used. Some general considerations are given below:

- Low temperatures: Chemistry and controlled processing for balanced ferrite/austenite ratios are necessary to minimize brittleness, especially the welds, in the temperature range -4 to -58°F (-20 to -50°C). Welding process selection is critical. DSS are rarely used below -58°F (-50°C).
- Elevated temperatures: Applications involving temperatures above 570°F (300°C) may result in embrittlement. Applications involving temperatures above 1200°F (650°C) may result in embrittlement and loss of corrosion resistance, even with short exposures.

- Hydrogen: Can result in hydrogen embrittlement, especially with high ferrite ratio imbalance.
- Hydrogen sulfide: DSS can be subject to sulfide stress corrosion and caution is warranted.

Summary

The use of duplex stainless steels is increasing at a rapid rate. They are not new, having been with us since the 1930s. However, there are more DSS grades available today, especially in the higher alloyed versions. Issues related to welding have been addressed but still require extra consideration. DSS are mostly used to replace austenitic stainless steels, especially where the latter have insufficient corrosion resistance or strength.

References

1.   J. Olsson, and M. Liljas, 60 Years of Duplex Stainless Steel Applications, Corrosion 94, Paper No. 395, NACE International, Houston, TX 1994.

2.   ASM Handbook Vol. 19 Fatigue and Fracture, "Fatigue and Fracture Properties of Duplex Stainless Steels," pg. 757-768, ASM International, Materials Park, OH 1996.

3.   Stainless Steels, Lacombe, et. al., Editors, Chapter 18 "The Duplex Stainless Steels," p. 624, Les Editions de Physiques, Les Ulis Cede A, France 1993.

4.   M. Moll, The world of duplex stainless steels, Duplex Stainless Steels 97, 5th World Conference, KCI Publishing, Zutphen, The Netherlands, 1997, p. 723ff.

5.   Avesta Sheffield Corrosion Handbook for Stainless Steels, Corrosion Tables pg. 59-60, Avesta Sheffield AB, Sweden 1994.

6.   Ferralium® Alloy 255, Data Sheet H2005B, Haynes International, Kokomo, IN.

7.   Anvig Andersen, et. al, SCC of Stainless Steel under Evaporative Conditions, Corrosion 98, Paper No. 251, NACE International, Houston, TX 1998.

8.   Materials Technology Institute, manual #3, "Corrosion Test Methods"

9.   B., Wallén, Corrosion of duplex stainless steels in seawater, Duplex Stainless Steels 97, 5th World Conference, KCI Publishing, Zutphen, The Netherlands, 1997, p. 59ff.

10.   Standard Material Requirements MR0175 (Latest Edition) Sulfide Stress Cracking Resistant Metallic Materials for Oilfield Equipment, NACE, Houston, TX.

General References

Proceedings, Duplex Stainless Steels 97, 5th World Conference, 2 vols., published by KCI Publishing, Zutphen, The Netherlands, 1997. These are the collected papers from the latest conference, containing a large body of information on metallurgy, corrosion, fabrication and manufacturing, and welding. Previous conference proceedings also hold much valuable and relevant information.

Chapter

6

SUPERAUSTENITIC STAINLESS STEELS

Dr. I.A. Franson and Dr. J.F. Grubb

Allegheny Ludlum Corporation

Brackenridge, Pennsylvania

Introduction

The definition of *superaustenitic stainless steel* is not precise. Over the past decade or two this term has been used primarily in reference to the new austenitic alloys containing 6% molybdenum. However, numerous other new austenitic alloys have been developed which also fit the "superaustenitic" description in that they are more highly alloyed and offer significantly better resistance to corrosion than the AISI 300 series austenitic stainless steels. These newer alloys contain less nickel than the nickel-based alloys. Cost of the superaustenitic alloys, therefore, is generally intermediate to the 300 series austenitic grades and the nickel-based alloys, while corrosion resistance, in some cases, challenges that of higher nickel alloys.

The superaustenitic stainless steels will include those commercial Fe-Cr-Ni-Mo-N austenitic alloys which do not have AISI 300 series designations, which have iron as their largest constituent, which contain less than 40% nickel, and which are used primarily for their aqueous corrosion resistance. Even with this rather narrow definition, there will be overlap. For instance, N08800 (alloy 800; Fe-21Cr-32Ni), which fits the superaustenitic definition (except it has no molybdenum addition), is included in the chapter on nickel-based alloys because it has been classified as part of this family of alloys. Other alloys which may fit the superaustenitic alloy composition ranges may also be excluded because they are principally used for high temperature applications. In any case, only those superaustenitic alloys which have UNS (Unified Numbering System) designations and which are included in ASTM specifications are discussed.

In the early 1990s, the ASTM adopted a new definition of stainless steels, consistent with other international standard specification organizations. Where previously a grade had to have more than 50% iron to be a stainless steel, the new definition requires that the iron be the largest element by weight percent. A number of grades, typically designated N08XXX, and presently included in the B specifications, would be S3XXXX stainless steels in the A specifications, if newly introduced. These grades are "grandfathered" in the B specifications, i.e. they will be retained in these specifications for an extended period of time because of user drawings and qualified procedures. However, no new N08XXX grades will be accepted in the B specifications, and the existing N08XXX grades will be individually sponsored in the A specifications when there is producer/user interest. Existing N08XXX grades will retain their UNS number to indicate their history.

Superaustenitic Alloy Development

Although many of the superaustenitic stainless steels are relatively new, the earliest developments date back more than 60 years. For instance, the alloy now known as N08904 (904L) was developed in France in the 1930s for use with sulfuric acid.[1] Hot workability problems with these early fully austenitic alloys initially inhibited their commercialization. The discovery that rare earth additions were effective in improving hot workability led to commercial production of wrought N08020 (alloy 20) in 1951.[2] This alloy also found primary application in the handling of sulfuric acid.

Existing superaustenitic alloys continued to be improved as experience with these materials was gained. N08020, for instance, was modified by introduction of columbium to provide stabilization against intergranular corrosion. Then nickel content was increased for improved resistance to stress corrosion cracking, resulting in the current N08020 (20Cb-3®) stainless steel used widely in the handling of sulfuric acid environments.[2]

Perhaps the most profound influence on the development of the superaustenitic alloys was the introduction of new refining technologies, such as the AOD (Argon Oxygen Decarburization) and VOD (Vacuum Oxygen Decarburization) processes in the 1970s. These refining processes allowed efficient and controlled reduction of carbon content to very low levels which eliminated the need for stabilizing elements. Close control of alloy composition was also made possible. Reduction of sulfur and other residual elements to low levels proved to be very beneficial to hot processing of superaustenitic alloys, eliminating the need for rare earth element additions. These benefits of the AOD and other similar refining processes made production and processing of new superaustenitic alloys (and other austenitic and ferritic stainless steels) possible and efficient. An example is N08028 (Sanicro28® alloy), which contains higher chromium and nickel (along with 3.5% molybdenum) than the then-existing superaustenitic steels. This alloy was specifically developed for phosphoric acid service.[1] Later in the 1980s a modification of this alloy, N08031 (VDM Alloy 31™) which contained 6.5% molybdenum and 0.2% nitrogen, was introduced.[1]

The first fully commercial superaustenitic alloy containing 6% molybdenum, N08366 (AL-6X™), was produced in 1969,[3] with the aid of calcium plus cerium additions to overcome hot workability difficulties.[4]

N08366 was specifically developed to provide sufficient pitting and crevice corrosion resistance for handling seawater. However, the high alloy content, particularly molybdenum, made N08366 prone to sigma phase formation which degraded corrosion resistance. Product form, thus, was limited to light gauges. The principal use of N08366 was for thin gage power plant condenser tube where seawater was used for cooling. It was used very successfully for a dozen years until it was replaced by the nitrogen-alloyed N08367 (AL-6XN®).[5]

Influence of Nitrogen

Introduction of nitrogen into the superaustenitic stainless steels in the 1970s and 80s had a profound effect on the utility of these alloys. Unlike carbon, nitrogen can be added in concentrations up to above 0.5% without precipitation of chromium compounds from austenite, which might lead to sensitization and intergranular corrosion. Nitrogen is an austenite stabilizer and provides the further benefit of acting to retard formation of sigma phase and other intermetallic phases. Nitrogen also provides a significant increase in strength properties and adds to the pitting and crevice corrosion resistance of the superaustenitic stainless steels.[1,6-8]

S31254 (254 SMO®), a 6Mo superaustenitic steel containing about 0.2% nitrogen, was introduced in 1976.[1] N08367 6Mo superaustenitic alloy, also containing 0.2% nitrogen, was brought to the marketplace in the early 1980s. The potent effect of the nitrogen addition allowed a full range of product forms, including heavy plate, to be produced and welded free of sigma phase precipitation, unlike the nitrogen-free N08366 material. The nitrogen stabilized 6Mo superaustenitic alloys have found broad application in many environments in addition to seawater.[9,10]

Superaustenitic alloys with up to 0.5% nitrogen emerged in the late 1980s and early 1990s. Through proper balance of chromium and manganese, nitrogen solubility is increased, permitting AOD processing and continuous casting of these alloys.[6] S32654 (654 SMO®) contains 7.3% molybdenum, 0.5% nitrogen, and 3% manganese, in addition to high chromium and nickel content. Because of higher chromium, molybdenum, and nitrogen, this alloy offers improved pitting resistance over the earlier 6Mo superaustenitic alloys. It finds applications in more severe environments and at higher temperatures than the 6Mo alloys can withstand.

Future Development

It is difficult to predict what directions future developments of superaustenitic alloys might take. The most highly alloyed superaustenitic stainless steels already contain high levels of chromium and molybdenum near known limits for iron-base alloys. Addition of more than 7% molybdenum, for instance, may lead to structural instability, even in the presence of high nitrogen. Although not as efficient as molybdenum, tungsten additions appear to add to pitting resistance while suppressing precipitation of intermetallic phases. Although nitrogen contents might be increased further through optimization of alloying elements, including manganese, the cost of producing and fabricating these higher nitrogen alloys might be sufficient to cause them to lose their competitive advantage compared to nickel-based alloys.

Superaustenitic Alloys

Compositions

Superaustenitic stainless steels which have UNS designations and which are included in ASTM specifications are listed in Table 6.1. These are Fe-Cr-Ni-Mo alloys. Many contain copper and about two-thirds also contain nitrogen as deliberate alloying elements. Those alloys with the highest specified nitrogen content out of necessity tend to have higher manganese contents.[8] For the most part, the superaustenitic stainless steels are refined to low levels of carbon, 0.02% or less. Those alloys with higher specified carbon content are stabilized against sensitization with titanium or columbium. In the case of the N08024 (20 Mo-4™) and S34565 (4565S) alloys, columbium is specified in addition to low carbon content. S31266 (URB-66™), alone, also contains a 2% tungsten addition.

Chromium in the superaustenitic stainless steels ranges from 19% to 28%. Nickel varies from 17 to 40%, and molybdenum varies from 2% up to about 8%. Copper is present from no deliberate addition up to 4%. Nitrogen ranges from residual values to deliberate additions of about 0.15% up to 0.5%

Table 6.1 Chemical Compositions of Superaustenitic Stainless Steels[a] (Weight %)

UNS No.	Alloy	C	Mn	P	S	Si	Cr	Ni	Mo	N max.	Cu max.	Fe	Other Elements
S31050	310MoLN	0.030	2.00	0.030	0.010	0.50	24.0-26.0	21.0-23.0	2.0-3.0	0.10-0.16	---	Bal.	---
S31254	254 SMO®	0.020	1.00	0.030	0.010	0.80	19.5-20.5	17.5-18.5	6.0-6.5	0.18-0.22	0.50-1.00	Bal.	---
S31266	B-66	0.030	2.00-4.00	0.035	0.020	1.00	23.0-25.0	21.0-24.0	5.0-7.0	0.35-0.60	0.50-3.00	Bal.	W 1.00-3.00
S32050	SR50A	0.030	1.50	0.035	0.020	1.00	22.0-24.0	20.0-22.0	6.0-6.8	0.24-0.34	0.40	Bal.	---
S32654	654 SMO®	0.020	2.00-4.00	0.030	0.005	0.50	24.0-25.0	21.0-23.0	7.0-8.0	0.45-0.55	0.30-0.60	Bal.	---
S34565	4565S™	0.030	5.00-7.00	0.030	0.010	1.00	23.0-25.0	16.0-18.0	4.0-5.0	0.40-0.60	---	Bal.	Cb 0.10 max.
N08020	20Cb-3®	0.07	2.00	0.045	0.035	1.00	19.0-21.0	32.0-38.0	2.0-3.0	---	3.0-4.0	Bal.	(Cb+Ta) 8 x C - 1.00
N08024	20 Mo-4®	0.03	1.00	0.035	0.035	0.50	22.5-25.0	35.0-40.0	3.5-5.0	---	0.50-1.50	Bal.	Cb 0.15-0.35
N08026	20 Mo-6®	0.03	1.00	0.03	0.03	0.50	22.0-26.0	33.0-37.2	5.0-6.7	0.10-0.16	2.00-4.00	Bal.	---
N08028	Sanicro 28®	0.030	2.50	0.030	0.030	1.00	26.0-28.0	30.0-34.0	3.0-4.0	---	0.6-1.4	Bal.	---
N08031	Nicrofer® 3127hMo	0.015	2.0	0.020	0.010	0.3	26.0-28.0	30.0-32.0	6.0-7.0	0.15-0.25	1.0-1.4	Bal.	---
N08320	20 modified	0.05	2.5	0.04	0.03	1.00	21.0-23.0	25.0-27.0	4.0-6.0	---	---	Bal.	Ti 4 x C min.
N08366	AL-6X™	0.035	2.00	0.040	0.030	1.00	20.0-22.0	23.5-25.5	6.0-7.0	---	---	Bal.	---
N08367	AL-6XN®	0.030	2.00	0.040	0.030	1.00	20.0-22.0	23.5-25.5	6.0-7.0	0.18-0.25	0.75	Bal.	---
N08700	JS700®	0.04	2.00	0.040	0.030	1.00	19.0-23.0	24.0-26.0	4.3-5.0	---	0.50	Bal.	(Cb+Ta) 8 x C - 0.40
N08825	825	0.05	1.0	0.03	0.03	0.5	19.5-23.5	38.0-46.0	2.5-3.5	---	1.5-3.0	bal.	Ti 0.6-1.2
N08904	904L	0.020	2.00	0.045	0.035	1.00	19.0-23.0	23.0-28.0	4.0-5.0	0.10	1.0-2.0	Bal.	---
N08925	INCO® 25-6Mo	0.020	1.00	0.045	0.030	0.50	19.0-21.0	24.0-26.0	6.0-7.0	0.10-0.20	0.8-1.5	Bal.	---
N08926	INCO® 25-6Mo, 1925hMo	0.020	2.00	0.030	0.010	0.50	19.0-21.0	24.0-26.0	6.0-7.0	0.15-0.25	0.5-1.5	Bal.	---
N08932	URSB-8™	0.020	2.00	0.025	0.010	0.40	24.0-26.0	24.0-26.0	4.5-6.5	0.15-0.25	1.0-2.0	Bal.	---

a. Maximum unless otherwise specified.

Specification

Basic ASTM product form specifications covering the superaustenitic stainless steels are listed in Table 6.2. As noted previously, many of the alloys are covered in ASTM "A" specifications pertaining to stainless steels, while others are included in ASTM "B" specifications for nickel alloys. Specifications other than those listed exist for some of the alloys. These can be located by contact with producers of the individual alloys or through search of the index of the various ASTM specification volumes.

Mechanical Properties

Minimum mechanical properties for the various superaustenitic alloys, from the various applicable ASTM plate, sheet, and strip specifications, are given in Table 6.3.

Those alloys with no nitrogen additions exhibit properties similar to those for 300 series austenitic stainless steels. However, in general, yield and tensile strength values are higher than for standard austenitic grades owing to the higher alloy content and, particularly, the presence of nitrogen in the superaustenitic grades.

Alloys with no nitrogen addition typically exhibit yield strengths of 28 to 35 ksi (193 to 241 MPa) and tensile strengths of 71 to 80 ksi (490 to 550 MPa). Addition of 0.2% nitrogen typically raises yield and tensile strengths to 44 and 95 ksi (300 and 655 MPa), respectively, representing a 33% increase in yield strength and 24% increase in tensile strength compared to the nitrogen-free alloys. The alloys with 0.5% nitrogen typically display 61 ksi (420 MPa) yield strength and 110 ksi (760 MPa) tensile strength, representing 85% increase in yield strength and 45% increase in tensile strength compared to nitrogen-free alloys. Nitrogen, thus, is a potent strengthener. Its influence is significantly stronger on yield strength than on tensile strength.

In any event the nitrogen-containing superaustenitic stainless steels are significantly stronger than the conventional 300 series austenitic grades and, therefore, provide higher allowable design stresses while exhibiting excellent elongation values typical of austenitic stainless steels.

Table 6.2 ASTM Specifications - Superaustenitic Stainless Steels

UNS No.	Alloy	Plate Sheet Strip	Bar Wire	Welded Tube	Welded Pipe	Billets Bar	Seamless Tube/Pipe	Miscellaneous
S31050	310MoLN	A 240/A 240	A 479/A 479M	A 249/A 249M	A 312/A 312M	A 479/A 479M	A 213/A 213M A 312/A 312M	---
S31254	254 SMO	A 240/A 240M	A 276 A 479/A 479M	A 249/A 249M A 269	A 312/A 312M A 358/A 358M A 409/A 409M A 813/A 813M A 814/A 814M	A 276 A 479/A 479M	A 312/A 312M A 269	A 182/A 182M A 193/A 193M A 194/A 194M A 403/A 403M
S31266	B66	A 240						A 182/A 182M
S32050	SR50A	A 240	A 479/A 479M		A 312/A 312M A 358/A 358M	A 479/A 479M	A 312/A 312M	
S32654	654 SMO	A 240/A 240M	A 479/A 479M	A 249/A 249M	A 312/A 312M		A 312/A 312M	
S34565	4565S	A 240/A 240M	A 479/A 479M	A 249/A 249M	A 358/A 358M	A 479/A 479M	A 312/A 312M	A 182/A 182M A 403/A 403M
N08020	20Cb-3	B 463 A 240/A 240M	B 473	B 468	B 464 B 474	B 472	B 729	B 366 B 471 B 462 B 475
N08024	20Mo-4	B 463	B 473	B 468	B 464 B 474	B 472	B 729	B 462 B 471 B 475
N08026	20Mo-6	B 463	B 473	B 468	B 464 B 474	B 472	B 729	B 462 B 471 B 475
N08028	Sanicro 28	B 709					B 668	

Table 6.2 ASTM Specifications - Superaustenitic Stainless Steels, (continued)

UNS No.	Alloy	Plate Sheet Strip	Bar Wire	Welded Tube	Welded Pipe	Billets Bar	Seamless Tube/Pipe	Miscellaneous
N08031	Nicrofer 3127HMo	B 582 B 625	B 649	B 626	B 619		B 622	B 564 B 581
N08320	20 modified	B 620	B 621	B 626	B 619		B 622	B 366 B 621
N08366	AL-6X	B 688	B 691	B 676	B 675	---	B 690	---
N08367	AL-6XN	B 688 A 240/A 240M	B 691	B 676	B 675 B 804	B 472	B 690	B 462 B 564
N08700	JS700	B 599	B 672	---	---	---	---	---
N08904	904L	B 625 A 240/A 240M	B 649	B 674 A 249/A 249M A 269	B 673 A 312/A 312M A 358/A 358M	---	B 677 A 312/A 312M A 269	---
N08925	INCO 25-6Mo	B 625	B 649	B 674	B 673	---	B 677	---
N08926	INCO 25-6Mo, 1925hMo	B 625 A 240/A 240M	B 649	B 674	B 804 B 673	B 472	B 677	B 366
N08932	URSB-8	B 625	---	---	---	---	---	---

Table 6.3 Mechanical Properties of Superaustenitic Stainless Steels

UNS No.	Alloy	Minimum Mechanical Properties (Plate, Sheet, and Strip)					Max. Hardness[a]		ASTM Spec.
		Tensile Strength		0.2% Yield Strength		% Elongation	HB	HRB	
		ksi	MPa	ksi	MPa				
S31050	310MoLN	84[b] / 78[c]	580 / 540	39 / 37	270 / 255	25.0 / 25.0	217	95	A240
S31254	254 SMO	100[d] / 95[e]	690 / 655	45 / 45	310 / 310	35.0 / 35.0	223	96	B 625
S31266	B-66	109	750	61	420	35.0	---	---	A240
S32050	SR50A	98	675	48	330	40.0	250	---	A240
S32654	654 SMO	109	750	62	430	40.0	250	---	A240
S34565	4565S	115	795	60	415	35.0	241	100	A240
N08020	20Cb-3	80	551	35	241	30.0	217	95	B 463
N08024	20Mo-4	80	551	35	241	30.0	217	95	B 463
N08026	20Mo-6	80	551	35	241	30.0	217	95	B 463
N08028	Sanicro 28	73	500	31	215	40	---	70/90	B 709
N08031	Nicrofer 3127hMo	94	650	40	276	40	---	---	B 625
N08320	20 modified	75	517	28	193	35.0	---	95	B 620
N08366	AL-6X	75	515	35	241	30	212	95	B 688
N08367	AL-6XN	100[d] / 95[e]	690 / 655	45 / 45	310 / 310	30.0 / 30.0	240	100	B 688
N08700	JS700	80	550	35	240	30.0	---	75/90	B 599
N08904	904L	71	490	31	215	35.0	---	70/90	B 625
N08925	INCO 25-6Mo	87	600	43	295	40	---	---	B 625
N08926	INCO 25-6Mo, 1925hMo	94	650	43	295	35	---	---	A240

Table 6.3 Mechanical Properties of Superaustenitic Stainless Steels (continued)

| | | Minimum Mechanical Properties (Plate, Sheet, and Strip) | | | | | | Max. Hardness[a] | | ASTM |
| | | Tensile Strength | | 0.2% Yield Strength | | | | | | |
UNS No.	Alloy	ksi	MPa	ksi	MPa	% Elongation		HB	HRB	Spec.
N08932	URSB-8	87	600	44	305	40		---	---	B 625

a. Hardness values are often shown for information only. Check individual specifications.

b. Less than ¼ inch.

c. Greater than ¼ inch.

d. Sheet/Strip.

e. Plate.

Physical Properties

The superaustenitic stainless steels are stable, single-phase austenitic (face-centered cubic) alloys. They are non-magnetic. Permeability of all of these alloys in the annealed condition is less than 1.01. To illustrate the stability of these alloys, consider that magnetic permeability (μ) of N08367 is 1.003 @ 200H in the 65% cold-worked condition as well as in the annealed condition.[11] Representative physical property ranges for the superaustenitic stainless steels are given in Table 6.4. Values for the same property for the same alloy varied significantly in some cases among sources. Therefore, the reader should contact manufacturers for data for specific alloys of interest.

Table 6.4 Superaustenitic Stainless Steels,
Representative Physical Properties[a]

Property	Range	Average
Density, (20°C), (gm/cm^3)	7.98 - 8.2	8.05
Thermal Expansion (20-100°C)(10^{-6}·K^{-1})	14.3 - 16.5	15.2
Electrical Resistivity (20°C), (μ·Ω·cm)	80 - 108	93
Thermal Conductivity (20°C), (W·m^{-1}·K^{-1})	11.4 - 16	12.9
Specific Heat (20°C), (J·kg^{-1}·K^{-1})	440 - 510	480
Modulus of Elasticity (20°C), (GPa)	186 - 200	193

a. Reader is urged to contact manufacturers of alloy of interest for specific data.

Average density, which was fairly consistent for all the superaustenitic alloys, is about 8.05 gm/cm^3 (0.291 lb/in.^{3}) which is similar to values for the 300 series austenitic stainless steels[12] and is lower than typical values for the nickel-based alloys[13] (see also, Chapters 4 and 7).

Thermal expansion values (20-100°C) average 15.2 x 10^{-6} K^{-1} which is lower than values for 300 series stainless steels and higher than values for nickel alloys. Thermal conductivity of the superaustenitic stainless steels averaged 12.9 W·m^{-1}·K^{-1} which is lower than values for 300 series steels and higher than values for nickel alloys. Room temperature specific heat (average 480 J·kg^{-1}·K^{-1}), is similar to values for 300 series stainless steels and higher than values for nickel-based alloys. Electrical resistivity averaged 93 micro-ohm·cm which is higher than values for 300 series alloys and lower than nickel-based alloys.

Elastic moduli for the superaustenitic steels and 300 series stainless steels are similar (average 193 GPa or 28 x 10^6 psi). Nickel alloys show somewhat higher values.

The toughness of the superaustenitic stainless steels, like that of other austenitic alloys, is high. These alloys typically exhibit Charpy V-notch impact values at room temperature of about 200 J·cm^{-2} (145 ft-lb). This excellent toughness extends to low temperatures also. Impact values at liquid nitrogen temperatures [-320°F (-196°C)] for instance, are generally reported to be in excess of 100 J·cm^{-2} (>73 ft-lb).

Physical Metallurgy

The alloying elements in the superaustenitic stainless steels must be in solid solution in order to maintain optimum corrosion resistance and fabricability of these alloys. The superaustenitic alloys are generally low in carbon to prevent precipitation of carbides which would lead to chromium depletion and sensitization of weld heat-affected zones (HAZ). The presence of nitrogen in most of these alloys further inhibits carbide precipitation so that sensitization from this source is typically not a problem. Precipitation of intermetallic compounds, particularly sigma (σ), but also chi (χ) or Laves phases, in these superaustenitic alloys can result in depletion of chromium and molybdenum in adjacent areas, which can serve as sites for pitting and in some cases intergranular corrosion.

High nickel contents minimize and high chromium plus molybdenum contents enhance precipitation of sigma and chi phases in the Fe-Cr-Ni-Mo alloy system at elevated temperatures. Manganese increases stability of sigma phase, but high nitrogen content effectively inhibits precipitation of this intermetallic phase,[14] as does rapid cooling following appropriate elevated temperature solution annealing. Rapid cooling may be effective in preventing the precipitation of sigma phase in sheet gages. However, nitrogen is necessary to prevent sigma phase precipitation in plate material, where these high cooling rates cannot be achieved. In plate, sigma (and/or chi) phase may still be present in small amounts at the center of the cross section where solidification-induced segregation makes the material more susceptible to sigma phase formation. If the cross-section of the material is not exposed to the corrosive environment, and these centerline phases do not come near the surface of the plate, corrosion resistance is not likely to be affected.[15]

Exposure of the superaustenitic stainless steels to temperatures where sigma phase and other intermetallic compounds may precipitate at grain boundaries and throughout the grains can lead to degradation of toughness as well as corrosion resistance.

Precipitation of sigma (or chi) phase is typically most rapid at about 1560°F (850°C) and becomes progressively slower at lower temperatures until effectively ceasing at 1000°F (540°C). The sigma phase precipitation may have a degrading effect on toughness as illustrated for the N08367 alloy in Figure 6.1 by way of an isotoughness [50 ft-lb (68J) Charpy V-notch impact energy] curve following elevated temperature exposures. In the annealed condition, unaffected by thermal exposure, the N08367 alloy typically exhibits Charpy V-notch impact value of about 150 ft-lb (200J) and a CPT of about 176°F (80°C).[11]

Resistance to pitting corrosion is most profoundly affected by precipitation caused by exposure to a lower temperature, as illustrated by the isopitting [CPT of 86°F (30°C) in ASTM G 48, Practice A test] curve in Figure 6.1. Note that loss of corrosion resistance occurs with shorter exposures than loss of toughness up to a temperature of about 1470°F (800°C).

Postweld or homogenizing heat treatment, if necessary, should be a full solution anneal at high enough temperatures to dissolve the sigma and/or chi phases, followed by rapid cooling to below 1000°F (540°C) to prevent reprecipitation. The appropriate solution annealing temperature for each alloy is slightly different and is often stipulated as part of ASTM specifications.

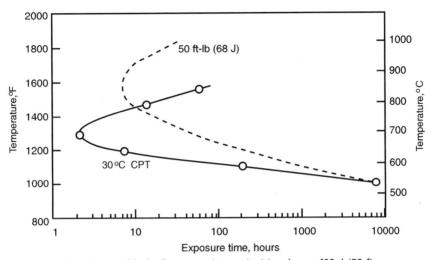

Upper curve (dashed) represents constant toughness [68 J (50 ft-lb)] as defined by room temperature Charpy V-notch impact tests. Lower curve (solid) represents constant Critical Pitting Temperature (CPT) of 86°F (30°C) as defined by ASTM G 48, Practice A test procedures.

Figure 6.1 Effect of thermal exposure on loss of toughness and pitting resistance of N08367 superaustenitic alloy.[31]

Corrosion Properties

The superaustenitic stainless steels are all distinctly more corrosion resistant than the 300 series austenitic stainless steels. However, because of the significant differences in alloying elements in these alloys, there is significant variation in corrosion resistance among the various superaustenitic stainless steels. The most resistant of these alloys are comparable to nickel-based alloys in many harsh environments.

Intergranular Corrosion

As was pointed out previously, most of the superaustenitic alloys are low carbon ("L") compositions, resistant to sensitization and intergranular corrosion. Presence of nitrogen in most of the alloys further inhibits carbide precipitation. Those alloys with higher maximum specified carbon, i.e., N08320, N08020, and N08700, also contain either titanium or columbium (niobium) which preferentially ties up carbon, leaving the chromium in solution and available for corrosion resistance. All of the superaustenitic alloys, therefore, are resistant to weld heat-affected zone sensitization. However, prolonged exposure of these alloys to temperatures in excess of 1000°F (540°C) could result in precipitation of sigma phase at grain boundaries, possibly making them susceptible to intergranular corrosion.

Pitting and Crevice Corrosion

Stainless steels, including the austenitic grades, are subject to pitting and/or crevice corrosion in chloride or other halide environments. High temperature and acid conditions exacerbate pitting. Crevice corrosion occurs under less severe conditions than those for pitting, especially at very tight crevices.

Resistance to these localized forms of corrosion improves with increases in chromium, molybdenum, and nitrogen. Tungsten, too, has been shown to add to pitting and crevice corrosion resistance of the superaustenitic stainless steels.[16] The PREN (Pitting Resistance Equivalent Number) has been widely used to gauge relative pitting and crevice corrosion resistance of the austenitic alloys. One representative PREN formula[1] is:

$$PREN = \%Cr + 3.3 \times \%Mo + 30 \times \%N$$

This calculation illustrates the strong influences of molybdenum and nitrogen in adding to pitting resistance. The contribution of tungsten to PREN has been shown to be half that of molybdenum.[16] The PREN is useful for a general ranking of similar grades for use in chloride environments, but small variations in PREN are not likely to be significant. However, the higher the calculated PREN number, the greater is the alloy's resistance to pitting and crevice corrosion.

---

[1] Many PREN formulas have been proposed. The "%C + 3.3 x %Mo" portion is generally accepted. The nitrogen coefficient, however, is more contentious, with values of 22 and 30 often widely cited.

PREN numbers for the superaustenitic stainless steels, calculated from mid-range values of chromium, molybdenum, and nitrogen (Table 6.1), are given in Table 6.5. These calculations suggest that even the least pitting-resistant superaustenitic alloy, N08020 with PREN = 28.3, offers somewhat better pitting and crevice corrosion resistance than S31603 (type 316L) at PREN = 25.2. Most of the alloys are significantly more resistant than S31603 or other 300 series stainless steels. The most pitting-resistant alloy, S32654 at PREN = 64.3, has better pitting resistance than the nickel-based N06625 (alloy 625) at PREN = 51.2 and reportedly approaches that of other high performance nickel-based alloys.

Table 6.5 Relative Pitting Resistance of Superaustenitic
Stainless Steels as Rated by PREN Number

| UNS No. | Typical Mid-Range Values, Wt.% | | | | PREN |
	Cr	Mo	N	W	%Cr + 3.3 x %Mo + 30 x %N
N08020	20	2.5	--	--	28.3
N08700	21	4.6	--	--	36.2
S31050	25	2.5	0.13	--	37.2
N08024	24	4.3	--	--	38.2
N08320	22	5.0	--	--	38.5
N08028	27	3.5	--	--	38.6
N08904	21	4.5	0.05	--	38.9
N08366	21	6.5	--	--	42.5
N08925	20	6.5	0.15	--	46.0
S31254	20	6.2	0.20	--	46.5
N08026	24	5.8	0.13	--	47.0
N08926	20	6.5	0.20	--	47.5
N08367	21	6.5	0.22	--	49.1
N08932	25	5.5	0.20	--	49.2
S32050	23	6.4	0.29	--	52.8
S34565	24	4.5	0.50	--	53.9
N08031	27	6.5	0.20	--	54.5
S31266	24	6.0	0.47	2.0	58.9*
S32654	24.5	7.5	0.50	--	64.3

*Note: PREN includes W contribution = ½ Mo

Although the PREN number does not reveal an absolute level of performance of these superaustenitic alloys in chloride–(or other halide–) containing solutions, it does provide a relative ranking of expected performance. Many studies have shown a direct correlation of PREN with performance in pitting tests, such as the ASTM G 48 ferric chloride tests.[1,16-18] Correlations have also been made between PREN and Critical Crevice Corrosion Temperature (CCT)[16] as well as Critical Pitting Temperature (CPT).[1,17,18] Experience has shown that alloys with PREN values greater than 40, e.g., the 6Mo alloys, are required for adequate resistance to ambient temperature seawater. For more severe conditions, such as higher temperature seawater service with crevices present, for instance at pipe flanges, or plate-frame heat exchangers, alloys with higher PREN values, such as S31266 or S32654, may have an advantage.

Stress Corrosion Cracking (SCC)

The 300 series austenitic stainless steels, particularly those with low nickel content, such as S30400 (type 304) and S31600 (type 316) are quite susceptible to chloride stress corrosion cracking (SCC) when high stresses are present and temperature exceeds about 120°F (50°C). Resistance to SCC in austenitic alloys increases with increasing nickel content.[19] The presence of 3% or more molybdenum also has a strong, synergistic beneficial effect.[20,21]

The superaustenitic stainless steels, therefore, with higher nickel and molybdenum contents than conventional 300 series austenitic stainless steels, offer substantially improved resistance to SCC, often superior to duplex alloys. The N08926 (1925hMo), with 25% nickel and 6.5% molybdenum, has been shown to exhibit superior resistance to SCC in laboratory tests compared to N08028 (alloy 28) and N08825 (alloy 825), both of which have higher nickel but lower molybdenum content. The explanation for this superior performance of the 6Mo alloys compared to the higher nickel but lower molybdenum alloys, may be that SCC often initiates in pits which are stress risers. The greater pitting resistance of the 6Mo alloys (PREN = 47.5) compared to N08028 (PREN = 38.6) or N08825 (PREN = 31.4), translates into fewer sites at which SCC initiation could occur.[17]

The superaustenitic stainless steels, thus, as a group, offer greatly improved resistance to SCC compared to the 300 series austenitic stainless grades with few cases being reported.[2,17] Performance in many severe environments has been excellent. These alloys may not be immune to SCC in very harsh chloride environments.

Data generally show the superaustenitic stainless steels to fail in several hundred to a thousand hours in the boiling 42% magnesium chloride test, compared to only a few hours for S30400 stainless steel. Therefore, very hot, aggressive acid chloride environments should be investigated, particularly if the superaustenitic stainless steel components will be used in a highly stressed condition.

General Corrosion

The standard 300 series austenitic stainless steels provide resistance to a wide variety of organic and inorganic environments. Because of higher chromium, nickel, and molybdenum contents, the superaustenitic alloys offer much resistance to a broader range of these environments, often equaling the resistance of more costly nickel-based alloys. Environments contaminated with chlorides or other halides, for instance, are good candidates for consideration of superaustenitic alloys because of their improved resistance to pitting, crevice corrosion, and/or stress corrosion cracking.

Organic Acid

Milder organic acids, such as acetic acid, are handled well by S31603 austenitic stainless steel. More aggressive organic acids,[11] however, such as hot formic acid or acetic-formic acid mixtures, may produce unacceptable corrosion rates on S31603. Presence of chlorides, too, can exacerbate corrosion rates. The high chromium, and in particular, molybdenum in superaustenitic alloys, provide much better resistance to these aggressive organic acid environments. This is illustrated by data given in Table 6.6.[11] The resistance of the 6Mo N08367 superaustenitic alloy is comparable to that of the N10276 (C-276) nickel alloy in boiling 45% formic acid.

Phosphoric Acid

Excellent resistance to the mineral acids is also available within the superaustenitic alloy family.[11] N08700 (JS700®), for instance, was developed specifically for service in wet phosphoric acid plants. Data given in Table 6.6 illustrate excellent resistance of the superaustenitic N08904 and N08367 alloys to corrosion by boiling 20% phosphoric acid, which is equivalent to that provided by nickel alloy N10276.

Table 6.6 Corrosion of Austenitic Alloys in Boiling
Formic and Phosphoric Acids[11]

| UNS No. (Alloy) | Corrosion Rate, mm/y (mpy) | |
	45% Formic	20% Phosphoric
S31603 (type 316L)	0.6 (23.4)	0.02 (0.60)
N08904 (904L)	0.2 (7.7)	0.01 (0.47)
N08367 (AL-6XN)	0.06 (2.4)	0.006 (0.24)
N10276 (C-276)	0.07 (2.8)	0.009 (0.36)

Comparison of iso-corrosion diagrams from various sources illustrates that the 4.5% Mo and 6% Mo superaustenitic alloys provide useful resistance to boiling [about 240°F (115°C)] 60% phosphoric acid and to 85% phosphoric acid to about 203°F (95°C). N08028 and the other higher chromium, nickel, and molybdenum superaustenitic alloys can handle 70% phosphoric acid to the boiling point [about 248°F (120°C)] and 100% acid to about 207°F (97°C). This performance is comparable to the resistance of nickel alloy N10276.

Phosphoric acid produced by the wet method invariably contains chlorides and fluorides and other impurities which make the acid considerably more aggressive, eliminating the possibility of using 300 series austenitic grades. The high chromium N08028 provides better resistance than superaustenitic alloys with 4.5 and 6% molybdenum,[22,23] although these materials also find use in phosphoric acid production. N08031, comparable in chromium to N08028, but higher in molybdenum (6%), is reported to offer greater resistance to halide- and sulfide-contaminated phosphoric acid than N08028.[24]

Sulfuric Acid

The superaustenitic alloys offer resistance to a much broader range of sulfuric acid environments than 300 series austenitic stainless grades. As was pointed out previously, N08904 was developed in the 1930s specifically for sulfuric acid applications. Similarly, alloy 20, later modified to N08020 (20Cb-3®), was developed in the early 1950s for sulfuric service. Compared to S30400 and S31600 austenitic stainless steels, the superaustenitic alloys offer superior resistance to corrosion by sulfuric acid environments, due to their higher chromium, nickel, molybdenum, and, in some cases, copper contents.

Comparison of manufacturer's corrosion data for superaustenitic and S31600 alloys for sulfuric acid illustrates the effects of various alloying elements[25] (Table 6.7). The temperature capability of N08031 in 20% and 50% sulfuric acid is better than that of nickel alloy N10276. N08031 also shows significantly lower corrosion rate in 80% sulfuric acid at 176°F (80°C) than nickel alloy N10276.

Presence of chlorides in sulfuric acid also increases corrosivity. For instance, N08367 is useful to about 140°F (60°C) in 30% sulfuric acid with a corrosion rate of about 0.1 mm/a (4 mpy). The presence of 200 ppm chloride lowers the temperature for this corrosion rate to about 104°F (40°C), and 2000 ppm chloride lowers it further to about 68°F (20°C).[11] The influence of chloride contaminant in 10% sulfuric acid on N08028 was shown to be small in comparison to the effect experienced by the higher nickel, lower chromium N08825.[23]

Overall, the superaustenitic stainless steels offer excellent resistance to sulfuric acid. Resistance to corrosion increases as total alloy content increases. N08031 with highest chromium (27%), highest nickel (31%), and 6.5% molybdenum plus 1.2% copper appears to provide the best resistance of this group of alloys.

Table 6.7 Sulfuric Acid - Approximate Maximum Use Temperature, °F (°C)a, b, c

UNS No. (Alloy)	Ni	Cr	Mo	Cu	20% H$_2$SO$_4$	50% H$_2$SO$_4$	80% H$_2$SO$_4$
S31600 (316)	12	17	2.5	--	86 (30)	--	--
N08904 (904L)	25	21	4.5	1.5	140 (60)	104-122 (40-50)	104 (40)
N08367 (AL-6XN)	25	21	6.5	--	158-176 (70-80)	104-122 (40-50)	104 (40)
N08020 (20Cb-3)	35	20	2.5	3.5	158-176 (70-80)	140 (60)	130 (55)
N08028 (Sanicro 28)	31	27	3.5	1.0	158-176 (70-80)	140 (60)	130 (55)
S32654 (654 SMO)	22	25	7.5	0.5	194 (90)	140 (60)	--
N08031 (Nicrofer 3127hMo)	31	27	6.5	1.2	219 (104)	225 (107)	167 (75)

a.   For a corrosion rate of 0.1 mm/y or less.
b.   For a given alloy and environment, the corrosion rate increases with increasing temperature.
c.   When comparing different alloys in a given environment, higher maximum use temperatures indicate greater corrosion resistance.

Nitric Acid

Nitric acid is strongly oxidizing and is, therefore, much less aggressive than sulfuric and, particularly, hydrochloric acids. For instance, S30403 austenitic stainless steel has usefulness in boiling nitric acid to about 50% concentration and in excess of 90% acid at room temperature, with a corrosion rate of 0.1 mm/y (4 mpy) or less.[27]

Increased chromium content improves resistance to nitric acid while presence of molybdenum has a negative effect. Some superaustenitic alloys, therefore, exhibit higher corrosion rates in the Huey Test (boiling 65% nitric acid) than S30403, as shown in Table 6.8.[11,23,24] The high chromium content (27%) in N08028 and N08031 alleviates the negative effect of molybdenum. These high chromium-molybdenum superaustenitic alloys are also useful in 20% HNO_3-4%HF pickling solutions at 140°F (60°C),[23] as well as in other highly oxidizing mixed acid solutions.[26]

Table 6.8 Huey Test–Nitric Acid [11, 23,24]

UNS No. (Alloy)	%Cr	%Mo	Corrosion Rate, mm/y (mpy) Boiling 65% Nitric
S30403 (304L)	18	---	0.17 (6.8)
N08904 (904L)	21	4.5	0.39 (15.2)
N08367 (AL-6XN)	21	6.5	0.67 (26.2)
N08028 (Sanicro 28)	27	3.5	0.06 (2.4)
N08031 (Nicrofer 3127hMo)	27	6.5	0.10 (3.9)

Hydrochloric Acid

The 300 series austenitic stainless steels offer very limited resistance to hydrochloric acid. S31600, for instance, is limited to about 1% HCl at room temperature and about 0.5% HCl at 140°F (60°C).[1,23]

The more highly alloyed superaustenitic alloys provide a distinct advantage. Comparison of manufacturer's corrosion data[1,11,23] shows room temperature capability increases from N08020 and N08904

(2% HCl) to 6Mo alloys–S31254 and N08367–and N08028 (3 to 4% HCl) to S32654 and N08031 (8% HCl). N08031 also is reported to provide a corrosion rate of 0.13 mm/y (5 mpy) or less in 3% HCl to a temperature of 176°F (80°C).[26] These data show clearly that for chemical processes where dilute hydrochloric acid may be found, the superaustenitic alloys may well be cost effective alternatives to high nickel alloys. Caution should be used to avoid situations where concentration could occur, such as in wet-dry interfaces.

Other Environments

The promotional literature for the various superaustenitic stainless steels also points out the usefulness of these materials in other environments. N08367 alloy and N08028, for instance, have both been shown to exhibit low corrosion rates in hot 43-45% caustic and caustic/salt solutions.[11,23] Several of the alloys are approved by inclusion in NACE MR0175[28] for use in sour gas environments.[28] S31050 (310MoLN) superaustenitic steel performs well in urea synthesis.[27]

Fabrication Characteristics

The superaustenitic stainless steels, like the 300 series austenitic grades, are readily fabricated into complex shapes. Rules applicable to the metallurgically stable 300 series alloys, such as S30400,[29] also apply to the superaustenitic stainless steels. All of the superaustenitic alloys are stable, i.e., they do not form martensite on cold deformation, as do some of the lean 300 series alloys. The superaustenitic steels all experience the high work hardening rates and excellent ductility common to all stable austenitic grades. The significant difference for most of the superaustenitic alloys is their higher yield and tensile strengths in comparison to 300 series stainless grades. Greater strength translates into greater power requirements during forming and machining of the superaustenitic alloys, which could be significantly greater than the 300 series alloys and several times that of carbon steel. In this respect, the superaustenitic alloys are similar to the nickel-based alloys of similar strength.

Cold Forming

There is a dearth of information concerning cold forming of the superaustenitic alloys. However, their stable structure and good ductility indicate these alloys can be blanked, pierced, bent, and drawn like their stable 300 series relatives. Strain hardening exponent (n), average strain ratio (R_m) and Limiting Draw Ratio (LDR) for N08367 alloy strip, for instance, are very similar to those values for the stable S30500 (type 305) austenitic stainless steel.[30] Power requirements, however, will generally be greater unless advantage can be taken of their greater strength and corrosion resistance to reduce section thickness. Die wear also can be expected to be greater. Springback on press-brake forming or tube bending will be greater, particularly with the more highly alloyed and nitrogen-containing superaustenitics. Compensating die changes, thus, may be needed to accommodate these differences. Heat treatment after cold forming is generally not required.

Hot Forming

Austenitic stainless steels, including the superaustenitic grades, are easily hot formed into many useful shapes, such as seamless tube and pipe, flanges and fittings, bolts and other fasteners, dished heads, etc. However, the range of temperature over which hot forming of the superaustenitic alloys is practicable is, generally, narrower than that for the common austenitic stainless steels such as S30400 or S31600. A suggested hot forming temperature range for these alloys is 2100 to 2250°F (1150 to 1230°C). Hot shortness will become a problem at higher temperatures.

Severe (catastrophic) oxidation accompanied by deep pitting may also occur at these high temperatures with superaustenitic alloys which contain substantial amounts of molybdenum. Below 2100°F (1150°C), second phases may precipitate and cause degradation of corrosion resistance and mechanical properties at room temperature as illustrated in Figure 6.1.[31] Therefore, a final solution anneal heat treatment above 2100°F (1150°C), followed by rapid cooling, is generally necessary. A tenacious scale forms during high temperature forming and annealing of the superaustenitic steels. Scale formation and associated chromium depletion of the base metal can have a very detrimental effect on corrosion resistance.[15] It is important, therefore, that the scale and depleted surfaces are completely removed by pickling or grinding.

Machinability

Machinability of austenitic stainless steels in general is considered to be relatively difficult because these alloys exhibit a relatively high work hardening rate and are very tough and ductile.[32] In spite of these physical characteristics, these steels are readily machined into a multitude of parts daily. The superaustenitic stainless steels with mechanical properties similar to 300 series austenitic grades have similar machinability. However, the strength of the superaustenitic grades is generally considerably greater, while work hardening rate and ductility are similar to the 300 series. Considerable energy, therefore, is required in removing a chip from these higher strength superaustenitic alloys resulting in tools that run hotter with greater tendency to forming a built-up edge, leading to reduced tool life. Chips are stringy and tend to tangle.

Tools and work pieces need to be very rigid to avoid chatter. It is generally a good idea to avoid using tools at more than about 75% of rated capacity. Feeds should be heavy enough to ensure that cutting edges get beneath previous work-hardened zones. Tools will require more frequent sharpening than when used on the 300 series alloys. Proper sulfurized and/or chlorinated lubricants are necessary for optimum results. All traces of cutting fluids must be removed prior to welding, annealing, or use in corrosive service. These comments pertain to well annealed material. Presence of carbides, nitrides, and especially sigma phase particles will increase tool wear.

Welding

All of the superaustenitic stainless steels are weldable using a variety of welding techniques. In general, few matching composition filler metals are available for joining these alloys. Notable exceptions are

N08020 and N08904 for which matching filler metals are available. Normally, more highly alloyed nickel-based filler metals are recommended. As-cast superaustenitic weld deposits of matching composition are highly segregated (cored) because of their high alloy content. Consequently the use of the more highly alloyed (higher molybdenum) nickel alloy filler metals brings the corrosion resistance of the weld deposit up to the same level as the base metal.

Autogenous welds, if made, need to be given a full anneal to eliminate the cast structure and dissolve second phases. Welding of these superaustenitic alloys is discussed in more detail in Chapter 8: "Weld Fabrication of Nickel-Containing Materials."

Applications

The superaustenitic stainless steels are used to provide resistance to corrosive environments that are too severe for the 300 series austenitic stainless steels. The superaustenitic alloys all have higher nickel content which, together with their molybdenum content, provides much greater resistance to stress corrosion cracking (SCC) in the presence of chlorides than the 300 series alloys. Pitting and crevice corrosion resistance, too, are generally better in the superaustenitic alloys because of higher chromium, molybdenum, and nitrogen content, as reflected in PREN numbers. For these reasons, the superaustenitic alloys have been used widely in place of 300 series alloys in a broad variety of equipment where pitting and/or crevice corrosion or stress corrosion cracking was present or likely to be present.

The availability of standard product forms such as sheet, strip, plate, tube, pipe, fittings, bar, billet and castings, and the similarity in fabrication requirements to standard austenitic alloys, have allowed this substitution to be made, in most cases, without too much difficulty.

The higher alloy (chromium, nickel, molybdenum, and copper) content of the superaustenitic alloys also provides substantially improved resistance to general corrosion by a wide variety of environments, particularly acids, in comparison to the 300 series austenitic stainless steels. The superaustenitic alloys thus provide the user with a number of alternatives for chloride or acid environments that are too severe for 300 series stainless steels and where more costly nickel-based alloys might otherwise be used.

The use of superaustenitic alloy in fabricated structures is illustrated by a distillation column used in processing of tall oil shown in Figure 6.2. The 146 ft (44.5 m) column, its internals, pipe and fittings, and associated heat exchangers were fabricated using N08367. Numerous other examples of use of this 6Mo stainless steel and all of the other superaustenitic stainless steels can be obtained from producers of these materials.

Figure 6.2 Distillation column used in processing of tall oil.
(Courtesy Union Camp Corporation).

Water

Presence of chloride ion in water can cause great problems for 300 series austenitic stainless steels. Chloride levels in excess of 1000-2000 ppm can result in crevice corrosion on S31600, for instance. Lesser amounts may cause stress corrosion cracking if temperatures exceed about 140°F (60°C) and residual stresses are present. The presence of sulfate-reducing and other strains of bacteria in water, particularly under stagnant conditions, has resulted in microbiologically influenced corrosion (MIC) on S30400, S31600, and even more highly alloyed 300 series austenitic alloys, in many cases resulting in failure. The superaustenitic alloys have been widely used to solve these pitting/crevice corrosion, stress corrosion cracking, and MIC problems in the power industry, chemical process industries, food and pharmaceutical industries, and pulp and paper industries.

All of the superaustenitic stainless steels, because of their high nickel and molybdenum contents, are highly resistant to stress corrosion cracking in the presence of chlorides. This has resulted in their wide use in shell-and-tube heat exchangers,[9] plate-and-frame heat exchangers, jacketing for steam heated vessels, and other applications where chlorides are present along with elevated temperature and residual stresses from forming operations or welding. An example of such an application is the selection of N08028 as the primary pressure boundary material in water walls and heat exchanger bundles in syngas coolers.[33]

Resistance to pitting and crevice corrosion varies with the chromium, molybdenum, and nitrogen content of the alloys, i.e., the PREN number, as shown in Table 6.5. Alloys with PREN number less than 30, such as N08020, offer only marginally better resistance to pitting and crevice corrosion than S31600 (PREN 24) stainless steel. Alloys with PREN values of 36 to 40, such as N08904 or N08028, are substantially more resistant to pitting and crevice corrosion than S31600 and can be used in brackish waters with chloride content in the 5,000-10,000 ppm level. Depending on temperature and crevice conditions these alloys have been used in seawater but with mixed success. N08904, for instance, suffered severe crevice corrosion at tube-to-tubesheet joints, on gasket surfaces and on weld seams of nozzles of seawater cooled heat exchangers.[34]

Seawater

Better choices for seawater service are the 6Mo and more highly alloyed superaustenitic alloys with PREN values of 46 or higher (Table 6.5). These alloys have been used for over 20 years in a wide variety of seawater applications, including nuclear and fossil power plant steam surface condenser tubing and tubesheets, piping and heat exchangers for nuclear power plant service water systems,[35] offshore oil/gas production platform piping systems, heat exchangers, ballast water systems, and fire protection systems.[36] The 6Mo-superaustenitic stainless steels are also used in reverse-osmosis and flash distillation desalination plants,[26,34] primarily in the form of piping.

It has been suggested that the 6Mo-superaustenitic alloys be limited to 95°F (35°C) in continuously chlorinated seawater to avoid crevice corrosion.[36] Higher temperature applications require the superaustenitic alloys with PREN values of 60 or greater (Table 6.5). S32654, for instance, has been shown to resist corrosion in chlorinated seawater to 113°F (45°C), with excursions to 140°F (60°C).[37]

Pulp and Paper

The 6Mo and other superaustenitic alloys with PREN above 46 have found broad application in the pulp and paper industry. These alloys have been found suitable for use in C- and D-stage bleach washers, effluent coolers, piping, paper machine headboxes, black liquor recovery boilers, reheaters, filter drums, and scrubbers and other applications.[1,11,25] The superaustenitic grades with PREN greater than 60 (S32654 and S31266, Table 6.5) also extend the use of these alloys to pulp mill bleach plant environments that may be too severe for the 6Mo alloys. These alloys demonstrate similar resistance to localized corrosion as high molybdenum nickel-based alloys in aggressive D-stage bleach plant environments,[17,37] and are resistant to the uniform corrosion experienced by nickel-based alloys when pH is 4 or higher.

Microbiologically Influenced Corrosion

Microbiologically influenced corrosion (MIC) has been responsible for failure of S30400 and S31600 austenitic stainless steels and other alloys in biologically active fresh and saline waters. The 6Mo-superaustenitic alloys have been shown to perform well in these waters.[37,38] For instance, no failures attributable to MIC have been observed during more than 25 years of service of 6Mo (N08366) superaustenitic alloy power plant surface condenser tubes in biologically active fresh and salt waters.

Phosphoric Acid

As pointed out in the Corrosion Properties section, the superaustenitic stainless steels offer excellent resistance to phosphoric acid environments with corrosion rates which often rival those of nickel-based alloys. Consequently, these alloys have been employed widely in contact with phosphoric acid, often in heat exchangers where resistance to phosphoric acid is combined with resistance to chlorides in cooling water.

Corrosion resistance to phosphoric acid improves dramatically with increased chromium content.[27] Alloys with 20-22% chromium, such as N08904 and N08700, are used widely in wet process phosphoric acid applications. The superaustenitic alloys with 27% chromium, i.e., N08028 and N08031, offer improved corrosion resistance. These alloys provide resistance to halide- and sulfide-contaminated phosphoric acid comparable to that offered by more costly nickel alloys. These high chromium superaustenitic alloys also provide exceptional resistance to erosion-corrosion in phosphoric acid digestion slurries. They are, therefore, used in mixers.

Other wet process and superphosphoric acid production applications include heat exchangers (often replacing graphite), plate components, piping systems, impellers and pumps, and agitators/stirrers.

Sulfuric Acid

Data given in the Corrosion Properties section and Table 6.7 illustrate that the superaustenitic stainless steels offer excellent resistance to sulfuric acid. Resistance was shown to increase with total alloy content. N08031, with highest alloy content (27 chromium, 31 nickel, 6.5 molybdenum, 1.2 copper), provides exceptional resistance among this group of alloys.

Applications for superaustenitic stainless steels include acid distribution systems,[17] acid coolers, and piping[27] in production plants and a broad range of equipment handling sulfuric acid in other process plants. For instance, N08031 is mentioned as having been used for heat exchangers associated with sulfuric acid pickling baths.[25] Where seawater cooling is used for acid coolers, the superaustenitic alloys with sufficiently high PREN number should be considered.

Flue gas desulfurization (FGD) systems have consumed large quantities of superaustenitic alloys. Table 6.8 presents guidelines for selection of alloys for FGD environments.[39] The severity of the environment increases with decrease in pH and increase in the amount of halide ion present, as well as increasing temperature.

Table 6.9 illustrates that 6Mo superaustenitic alloys fill the gap between the S31726 (317LMN) austenitic stainless steel and nickel-based alloys such as N10276 or N06022. Although not included in the tests which formed the background for Table 6.8, N08031 or S32654 could be used under some of the conditions indicated to require nickel-based alloys.

The superaustenitic alloys have been used for many FGD system components, including absorbers and internals, spray systems and piping, outlet ducts and breeching, reheaters, stack liners, etc. In most cases solid plate or sheet product has been used in fabricating the components. However, in some cases, superaustenitic stainless steel liners have been used to clad ductwork, employing the "wallpaper" technique.[40]

Table 6.9 Alloy Selection Guide Flue Gas Desulfurization Environments(39)

	5,000 ppm Cl⁻			10,000 ppm Cl⁻			50,000 ppm Cl⁻		
	140°F	150°F	160°F	140°F	150°F	160°F	140°F	150°F	160°F
pH 4	S31726	S31726	6Mo	S31726	S31726	6Mo	6Mo	6Mo	N10276 N06022
pH 2	6Mo	6Mo	6Mo	6Mo	6Mo	6Mo	6Mo	6Mo	N10276 N06022
pH 1	6Mo	6Mo	N10276 C-22	6Mo	6Mo	N10276 N06022	N10276 N06022	N10276 N06022	N10276 N06022
pH 0.5	6Mo	N10276 N06022	N10276 N06022	6Mo	N10276 N06022	N10276 N06022	N10276 N06022	N10276 N06022	N10276 N06022

Hydrochloric Acid

Process equipment is often exposed to hydrochloric acid as a by-product of chemical reactions or as a consequence of condensation of chlorine laden gases. Because hydrochloric acid is highly corrosive, 300 series austenitic stainless steels have only very limited applicability in these conditions. Stress corrosion cracking often accompanies the general corrosion experienced by these alloys. The superaustenitic stainless steels offer a cost effective alternative to high nickel alloys for many conditions where dilute hydrochloric acid is present, providing much improved general corrosion resistance and stress corrosion cracking resistance.

The superaustenitic stainless steels have found use in heat exchangers, piping systems, process vessels, vapor recovery systems, and other process equipment where dilute hydrochloric acid is present. The 6Mo superaustenitic grades have been used in condensing heat exchangers and wet scrubbing systems associated with waste incineration systems.[37,41] These grades have also been applied to heat exchangers incorporated into high-efficiency home heating systems, an example shown in Figure 6.3, where acid condensates, including hydrochloric acid form.

Figure 6.3 Heat exchanger formed from N08367 6Mo superaustenitic alloy sheet for use in high efficiency home heating furnace. (Courtesy Coleman Corporation).

Where 6Mo superaustenitic grades and N08028 prove to offer marginal resistance, superaustenitic grades with a PREN number of 52 or higher (Table 6.5), such as S32654 or N08031 and similar alloys, will provide the user with sufficient resistance to perform well in many hydrochloric acid applications.[37, 42]

Hydrofluoric Acid

Even low concentrations of hydrofluoric acid (HF) cause active general corrosion of S31600 austenitic stainless steel at room temperature. N08904 remains passive up to about 20% HF and the N08028 remains passive in even 40% HF at room temperature.[23]

The most highly alloyed superaustenitic alloys offer significantly better resistance to HF than the conventional 6Mo superaustenitic grades.[37] The superaustenitic stainless steels, therefore, particularly the highest alloy superaustenitics, have utility in dilute HF acid environments.

Applications include heat exchangers, vessels, piping systems, and scrubber components[26] where very dilute HF, perhaps in combination with dilute HCl, is present. The high chromium N08028 exhibits significantly lower corrosion rate than N08904 or N08020 in 20% nitric-4% hydrofluoric acid pickling liquor.[23]

Similar results were observed for these alloys in 10%HNO$_3$-3% HF pickling liquor (ASTM A 262, Practice D) at 160°F (70°C)[43] The high chromium superaustenitic stainless steels thus have utility in some HNO$_3$-HF pickling liquor applications, particularly at ambient temperature.

Nitric Acid

The superaustenitic alloys offer little advantage over 300 series austenitic stainless steels, particularly S31008 (alloy 310S) with 25% chromium and low carbon content (0.015%).[27] Exceptions are N08028 and N08031, both of which contain 27% chromium and 3.5 and 6.5% molybdenum, respectively. The high resistance to stress corrosion cracking and pitting which these alloys offer in addition to resistance to nitric acid, makes them well suited for nitric acid coolers operating with high chloride cooling waters.[23]

Urea Synthesis

S31050 is used in carbamate condensers, scrubbers, and strippers in the manufacture of urea.[27] Where chloride-induced stress corrosion cracking is a concern, N08028 should be considered. This superaustenitic alloy is reported to offer similar resistance to ammonium carbamate in urea synthesis as S31050.

Organic Acids

S31603 austenitic stainless steel is used widely in organic acid environments, particularly acetic acid. However, stronger acids, such as formic acid, often require alloys with higher chromium and molybdenum contents. The superaustenitic alloys, therefore, offer excellent resistance to organic acids.[11, 23] These alloys find applications which involve formic or acetic-formic acid mixtures in equipment such as heat exchangers, vessels, columns, and internals.

Sour Gas Environments

Many of the superaustenitic alloys have demonstrated resistance to sour (hydrogen sulfide), chloride-containing environments. Several of the alloys, e.g. S31254, N08020, N08367, N08926, N08024 and N08028, are included in NACE Standard MR0175-99[28]. These alloys, therefore, find uses in piping and heat exchangers in oil refineries and deep, hot, sour gas wells where hydrogen sulfide and chlorides are present.[23]

The reader is advised to consult the latest edition of NACE Standard MR0175 (which is updated each year) for information regarding limitations on the superaustenitic alloys in these sour environments.

Other Applications

The preceding discussion of specific applications for the superaustenitic stainless steels was intended to provide the reader with a flavor of the broad usefulness of these alloys. In addition to the applications mentioned, many others exist.

The superaustenitic grades will provide resistance where the 300 series austenitic stainless steels have inadequate pitting, crevice corrosion, and stress corrosion cracking resistance. These alloys will also perform well in mineral and organic acid environments too severe for the 300 series alloys, and, in some cases, nickel-based alloys or titanium. Because of this good resistance to corrosion, the superaustenitic alloys find many applications in such industries as chemical processing, food and/or pharmaceutical processing, oil and gas production and refining, pulp and paper, electric power, and others facing severe environments.

The reader is encouraged to contact producers of the superaustenitic alloys for more information concerning specific applications or for coupons for testing.

Summary

There are more than twenty Fe-Cr-Ni-Mo austenitic alloys in ASTM specifications that are not included among the 300 series austenitic alloys and do not have sufficient nickel to be a nickel-based alloy. Many of these alloys also contain nitrogen as an alloying element. Some also contain copper or higher manganese. These are the "superaustenitic" stainless steels. These alloys provide a desirable combination of mechanical and physical properties and fabricability in combination with superb corrosion resistance. The superaustenitic stainless steels offer substantially improved resistance to stress corrosion cracking and pitting/crevice corrosion in comparison to the 300 series austenitic stainless steels. Within the broad composition range of the superaustenitic alloys, resistance to organic and mineral acids is also exceptional, often rivaling that of nickel-based alloys. The superaustenitic alloys, therefore, provide the user with cost effective alternatives to nickel-based alloys. As a consequence, the superaustenitic alloys have found many applications worldwide, within many industries, where they perform well. The use of these versatile stainless steels is expected to continue to grow as new applications are defined and, perhaps, as new alloys continue to be developed.

References

1.   M. Liljas, "Development of Superaustenitic Stainless Steels," *Acom. 2-1995*, Avesta Sheffield, Avesta, Sweden, (1995), p.1.

2.   *Carpenter 20Cb-3 Stainless Steel*, Carpenter Technology Corporation, Reading, PA (1980), p. 3.

3.   H. E. Deverell and J. R. Maurer, "Stainless Steels in Seawater," presented at NACE CORROSION/77, March 14-18, 1977, San Francisco, CA, Paper No. 97, p.3.

4.   U.S. Patent 4,043,838, August 23, 1977, "Method of Producing Pitting-Resistant, Hot-Workable Austenitic Stainless Steel," H. E. Deverell, Assigned to Allegheny Ludlum Industries, Inc.

5.   J. R. Maurer, "Improving a Stainless Condenser Tube Alloy with a Small Addition of Nitrogen," Joint ASME/IEEE Power Generation Conference, Milwaukee, WI, AM. Soc. Mech. Engrs., Paper No. 85-JPGC-Power-38, 1985.

6.   B. Wallen, J. Liljas, and P. Stenvall, "A New High Nitrogen Superaustenitic Stainless Steel for Use in Bleach Plant Washers and Other Aggressive Chloride Environments," CORROSION/92, Paper No. 322, NACE, April 26-May 1, 1992, Nashville, Tennessee, (1992).

7.   J. R. Kearns, "The Effect of Nitrogen on the Corrosion of Austenitic Stainless Steels Containing Molybdenum," *J. Of Materials for Energy Systems*, Vol. 7, No. 1, (1985), p. 17.

8.   *High Manganese, High Nitrogen Austenitic Steels*, Proc. of two conferences: Material's Week '87, ASM International, Cincinnati, OH, October, 10-15, 1987; Material's Week '92, ASM International, Chicago, IL, November 2-4, 1992, R. A. Lula, Ed.

9.   I. A. Franson and J. R. Maurer, "An Advanced Fe-Ni-Cr-Mo-N Austenitic Alloy for Utility Heat Exchanger Service," *Proc. Of Int'l. Conf. On Advances in Material Technology for Fossil Power Plants*, Sept. 1-3, 1987, Chicago, IL, ASM International, Paper No. 8706-007, (1987), p. 519.

10.   R. M. Davison and J. D. Redmond, "Practical Guide to Using 6Mo Austenitic Stainless Steel," *Material Performance*, Vol. 27, (12), (1988), p. 39.

11.   *AL-6XN® Alloy*, Allegheny Ludlum Corp., Pittsburgh, Pennsylvania, (1995), p. 18.

12.   *Steel Products Manual, Stainless Steels and Heat-Resisting Steels*, Iron and Steel Society, Warrendale, Pennsylvania, Nov. 1990.

13.   W. L. Mankins and S. Lamb, "Nickel and Nickel Alloys," *Metals Handbook, Tenth Ed., Vol. 2, Properties and Selection: Nonferrous Alloys and Special Purpose Materials*, ASM International, Metals Park, Ohio, (1990), p. 441.

14.   M. O. Speidel and P. J. Uggowitzer, "High Manganese, High Nitrogen Austenitic Stainless Steels: Their Strength and Toughness," Proc. of two conferences: Material's Week '87, ASM International, Cincinnati, OH, October, 10-15, 1987; Material's Week '92, ASM International, Chicago, IL, November 2-4, 1992, R. A. Lula, Ed., p. 135.

15. J. F. Grubb and J. R. Maurer, "Correlation of the Microstructure of a 6% Molybdenum Stainless Steel with Performance in a Highly Aggressive Test Medium," Paper No. 300, CORROSION/95, NACE International, March 27-31, 1995, Orlando, Florida, 1995.

16. J. Charles, et. al., "A New High Nitrogen Austenitic Stainless Steel with Improved Structure Stability and Corrosion Resistance Properties," Creusot-Loire Industries (1995).

17. F. E. White, "Superaustenitic Stainless Steels, Cost Effective Alternative to Nickel-Base Alloys," *Stainless Steel Europe*, Vol. 4 (21), October (1992), p. 58.

18. R. D. Kane, "Super Stainless Steels Resist Hostile Environments," *Advanced Materials and Processes*, Vol., (7) (1993), p. 16.

19. H. R. Copson, "Effect of Composition on Stress Corrosion Cracking of Some Alloys Containing Nickel," *Physical Metallurgy of Stress Corrosion Fracture*, T. N. Rhodin, Ed., Interscience (1959), p. 242.

20. M. O. Speidel, "Stress Corrosion Cracking of Stainless Steels in NaCl Solutions," *Met Trans. A*, Vol. 12A(5), (1981), p. 779.

21. M. Ueda, et. al., "A New Austenitic Stainless Steel Having Resistance to Stress Corrosion Cracking," *Nippon Steel Technical Report*, Overseas No. 2, January (1973), p. 66.

22. U. Heubner, et. al., *Nickel Alloys and High-Alloy Special Stainless Steels*, Expert Verlag, Sindelfingen, (1987), p. 67.

23. S. Bernhardsson, "Corrosion Performance of a High-Nickel Alloy," *Materials and Design*, Vol. 10(4), July-August, (1989), p. 186.

24. M. Rockel and F. E. White, "Superaustenitic Stainless Steels: Cost Effective Solutions to the Corrosion Problems of the Chemical Process Industries," in *Proc. Conf. on Use of Special Steels, Alloys and New Materials in Chemical Process Industries*, Lyon, France, 22-23 May, 1991, *Bull Cercle Etud. Metaux*, Vol. 16(1), May (1991), p. 13.6 - 13.13.

25. D. C. Agarwal, F. E. White, and W. R. Herda, "The 6% Mo Superaustenitics - The Cost Effective Alternative to Nickel Alloys," *CORROSION REVIEWS*, Special Issue on Application of Stainless Steels for Corrosion Control in Industrial Systems, Vol. 11(3,4), (1993), pp. 65-82.

26. M. B. Rockel, W. Herda and U. Brill, "Two New Austenitic Special Stainless Steels with High Molybdenum Content and Nitrogen Additions," *Proc. International Conf. On Stainless Steels*, Chiba, Japan, ISIJ, (1991), p. 78.

27. H. Tornbloom and M. Tynell, "Experience with Some High Alloyed Stainless Steels in the Chemical Process Industries," Paper No. 606, presented at CORROSION/89, NACE, April 17-21, 1989, New Orleans, Louisiana, U.S.A.

28. NACE Standard MR0175, "Sulfide Stress Cracking Resistant Metallic Materials for Oilfield Equipment," NACE International, Houston, Texas, U.S.A.

29. "Forming," in *Stainless Steels, ASM Specialty Handbook*, J. R. Davis and Associates, editors, ASM International, Materials Park, Ohio, U.S.A., (1994), p. 257.

30. I. A. Franson, "Factors Affecting Formability of Stainless Steels," *Proceedings of FABTECH INTERNATIONAL 1995 Conference*, Soc. Manufacturing Engineers, October 9-12, 1995, Rosemont, Illinois, (1995) p. 1049.

31. J. F. Grubb, "High Temperature Aging of a 6% Mo Superaustenitic Stainless Steel." *Proceedings International Conference: Stainless Steels '96*, June 3-5, 1996, Düsseldorf, Germany, Verein Deutscher Eisenhuttenleute (1996), p. 367.

32. "Machining," in *Stainless Steels, ASM Specialty Handbook*, J. R. Davis and Associates, editors, ASM International, Materials Park, Ohio, USA, (1994), p. 319.

33. B. Grooters and W. T. Bakker, "Materials for Water Walls of Syngas Collers," presented at the 1996 Gasification Technologies Conference, Mark Hopkins Inter-Continental, San Francisco, California, October 2-4, 1996, as reported in *Materials and Components in Fossil Energy Applications*, DOE/EPRI, No. 126, February 1, 1997, p. 5.

34. S. Narain and S. H. Asad, "Corrosion Problems in Low-Temperature Desalination Units," *Materials Performance*, Vol. 31, (4), (1992), p. 64.

35. R. S. Tombaugh and J. R. Maurer, "Retubing and Bundle Replacement for Life Extension of Nuclear Service Water System Heat Exchangers," International Joint Power Generation Conference, San Diego, California, October, 1991.

36. P. Lovland, "Super Stainless Steels: Review of Norwegian Offshore Practice," *Stainless Steel Europe*, November, 1993, p. 28.

37. J. Olsson, "Avesta UNS S32654, A New Superaustenitic Stainless Steel for Harsh Environments", ACOM 1-1995, Avesta Shelfield, Avesta, Sweden (1995), p1.

38. C. W. Kovach and J. D. Redmond, "High Performance Stainless Steels and Microbiologically Influenced Corrosion," *Proceedings of International Conference: Stainless Steels '96*, June 3-5, 1996, Düsseldorf, Germany, Verein Deutscher Eisenhuttenleute (1996), p. 198.

39.   I. A. Franson and P. Whitcraft, "Resistance of 6% Mo Superaustenitic Alloys and Other Stainless Steels and High Nickel Alloys to Severe FGD Environments," presented at 34th Annual Liberty Bell Corrosion Course, NACE International, Northeast Region Meeting and Exhibition, September 9-11, 1996, Philadelphia, Pennsylvania, USA.

40.   D. Felker, C. E. Bailey, and W. L. Mathay, "Sheet Lining of FDG Components - A Low Cost Approach to the Use of High Nickel Alloys," Paper No. 23, in *Solving Corrosion Problems in Air Pollution Control Equipment*, Proceedings of the 1992 NACE Air Pollution Seminar, November 17-19, 1992, Orlando, Florida, USA.

41.   G. Sorell, "Corrosion-Resistant Alloys for Incinerator Air Pollution Control Equipment," Paper No. 31, in *Solving Corrosion Problems in Air Pollution Control Equipment*, Proceedings of the 1992 NACE Air Pollution Seminar, November 17-19, 1992, Orlando, Florida, USA.

42.   D. C. Agarwal, G. K. Grossman, and R. Kircheimer, "Two New Alloys in Waste Incineration Systems - UNS#N06059 and UNS #N08031," Paper No. 210, CORROSION/93, NACE International, 1993.

43.   J. F. Grubb, "Pickling and Surface Chromium-Depletion of Corrosion-Resistant Alloys," *Proceedings of International Conference on Stainless Steels*, 1991, Chiba, Japan, ISIJ, (1991), p. 944.

Chapter

7

NICKEL-BASED ALLOYS

J. R. Crum and E. Hibner

INCO Alloys International Inc., Huntington, WV

N. C. Farr, and D. R. Munasinghe

INCO Alloys Ltd., Hereford, England

The term *nickel alloys* and *nickel-based alloys* often have different meanings depending upon the context and user. In this book the term nickel alloys will refer to a general family of nickel containing alloys. Nickel-based alloys are considered to be those in which the principle constituent is nickel, though it may be less than 50% of the total composition, which is consistent with the ASTM B02.07 committee scope.

Regarding the designation of alloys referenced in this chapter, the first time an alloy is mentioned in the text, the trade name will appear in parentheses followed by its UNS number, as shown in the following example:

Example: MONEL® alloy 400 (N04400)

Thereafter the material will be referred to by its UNS Number without the trade name.

Nickel-containing copper-based alloys have been used for tools, weapons, and coins since the beginning of civilization.[1] The first useful nickel based-alloy was MONEL® alloy 400 (N04400) which was developed in 1905. It was found to have high strength and good corrosion resistance in a variety of aqueous environments. The NIMONIC® (Ni-Cr-Ti) alloy system, with improved high temperature properties, was developed shortly thereafter, while Paul D. Merica discovered the use of aluminum and titanium to produce precipitation hardening in nickel-based alloys. High strength age hardened MONEL alloy K-500 (N05500) was introduced in 1924. Further work with the addition of molybdenum to a nickel base in the 1920s led to the development of the HASTELLOY® series of Ni-Mo and Ni-Cr-Mo alloys with exceptional corrosion resistance in reducing environments. The Ni-Cr-Fe alloy INCONEL® alloy 600 (N06600) was developed in 1931. This material, which was initially used for dairy equipment, eventually became the primary component of nuclear power plant steam generators. The higher chromium INCONEL alloy 690 (N06690) was later introduced as an improved material for this application. Due to a shortage of nickel in the 1950s, the lower nickel INCOLOY® alloy 800 (N08800) was developed to replace N06600 in domestic appliance applications.

Also during the 1940s alloy NIMONIC[2] alloys 80 and 80A (N07080) were developed for advanced aircraft applications. In the early 1960s and 70s, INCONEL alloys 718, 625, and 617 (N07718, N06625, and N06617), with exceptional high temperature strength and corrosion resistance, were invented. Though developed for aircraft engines, these materials have found wide use in other applications requiring high temperature strength and/or corrosion resistance. Mechanically alloyed (MA) nickel-based materials were also developed for critical engine applications.

Specialized nickel-based alloys were developed for use in the chemical process industry as well. In the 1950s INCOLOY alloy 825 (N08825) and Carpenter® 20Cb-3 (N08020) (Ni-Cr-Fe-Mo-Cu alloys), were developed with improved aqueous corrosion resistance, especially in sulfuric acid. In the 1970s improved melting techniques helped make possible HASTELLOY alloy C-276 (N10276), a higher purity version of HASTELLOY alloy C (N10002) and HASTELLOY alloy B-2 (N10665), an improved version of the Ni-Mo HASTELLOY alloy B (N10001).

In the 1980s demand for higher performance materials in pollution control and the chemical process industry led to the development of HASTELLOY alloy C-22™ (N26022). In the following decade Nicrofer® 5923hMo-alloy 59 (N06059), INCONEL alloy 686 (N06686), and the HASTELLOY C-2000 (N06200) were introduced to provide even better localized and general corrosion resistance in oxidizing and reducing acid media by additions of increased chromium, molybdenum, tungsten, or copper.

The increasing use of nickel alloys in natural gas production led to the development of high strength age-hardenable INCOLOY alloy 925 (N09925) and the age-hardened alloy 625 (N06625) variations, INCONEL alloy 725 (N07725) and Carpenter 625PLUS® (N07716). INCOLOY alloys 800H (N08810) and 800HT (N08811) were developed for higher temperature strength in petrochemical refinery applications. Alloys such as INCONEL alloy 601 (N06601), HASTELLOY alloys 214 (N07214) and HR-160® (N12160), and Nicrofer alloys 603GT™ and 45TM™ (N06045) have improved oxidation resistance in many applications.

Older and newer nickel alloys can be divided into different families depending on elemental additions made to the nickel base. The more common nickel alloys in these families are shown schematically in Figure 7.1. For each alloy, the elemental additions made to a nickel base to arrive at this alloy are indicated. Tables 7.1 summarizes the chemical compositions.

DEVELOPMENT OF WROUGHT NICKEL ALLOYS

□ represents solid solution material and

○ represents precipitation-hardenable material

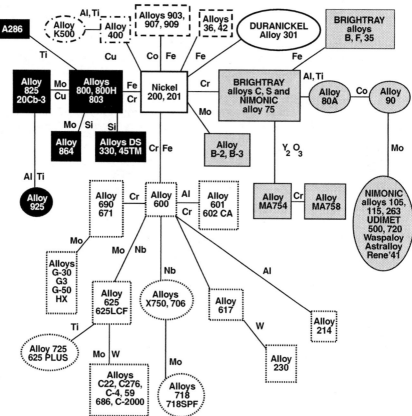

Figure 7.1 Development of wrought nickel alloys.

Table 7.1 Chemical Composition of Nickel and Nickel-Based Alloys

UNS No.	Common Name	Chemical Composition
S35045	Alloy 803	Ni 32.0-37.0 Cr 25.0-29.0 C 0.06-0.010 Al 0.15-0.60 Ti 0.15-0.6 Fe rem
S35135	Alloy 864	Ni 34 Cr 21 Mo 4.2 C 0.03 Si 0.8 Ti 0.6 Fe rem
N02200	Nickel 200	Ni 99.0 min C 0.15 max Cu 0.25 max Fe 0.40 max Mn 0.35 max S 0.010 max Si 0.35 max
N02201	Nickel 201	Ni 99.0 min C 0.02 max Cu 0.25 max Fe 0.40 max Mn 0.35 max S 0.010 max Si 0.35 max
N02211	Nickel 211	Ni 93.7 min C 0.02 max Cu 0.25 max Fe 0.75 max Mn 4.25-5.25 S 0.015 max Si 0.15 max
N04400	Alloy 400	Ni 63.00-70.00 C 0.3 max Cu rem Fe 2.50 max Mn 2.00 max S 0.024 max Si 0.50 max
N05500	Alloy K-500	Ni 63.0-70.0 Al 2.30-3.15 C 0.25 max Cu rem Fe 2.00 max Mn 1.50 max S 0.01 max Si 0.50 max Ti 0.35-0.85
N06002	Alloy X	Ni rem C 0.05-0.15 Co 0.5-2.5 Cr 20.5-23.0 Fe 17.0-20.0 Mn 1.00 max Mo 8.0-10.0 P 0.40 max S 0.030 max Si 1.00 max W 0.20-1.0
N06003	80Ni-20Cr	Ni rem C 0.15 max Cr 19-21 Fe 1.0 max Mn 2.5 max S 0.01 max Si 0.75-1.6
N06004	65Ni-15Cr	Ni 57 min C 0.15 max Cr 14-18 Fe rem Mn 1.0 max S 0.01 max Si 0.75-1.6
N06022	Alloy C-22 Alloy 622	Ni rem C 0.015 max Cr 20.0-22.5 Fe 2.0-6.0 Co 2.5 max Mn 0.50 max Mo 12.5-14.5 P 0.02 max S 0.02 max Si 0.08 max V 0.35 max W 2.5-3.5
N06025	Alloy 602CA	Cr 25, Fe 9, Al 2.2, Y Trace, Zr, Ti, rem Ni
N06030	Alloy G-30	Ni rem C 0.03 max Cb 0.30-1.50 Co 5.0 max Cr 28.0-31.5 Cu 1.0-2.4 Fe 13.0-17.0 Mn 1.5 max Mo 4.0-6.0 P 0.04 max S 0.02 max Si 0.8 max W 1.5-4.0
N06045	Alloy 45TM	Ni 45 min, Cr 26-29, Co 29, Fe 21-25, Si 2.5-3.0
N06050	HX-50	Ni rem, C 0.015 max, Mn 1.0 max, Cr 19.0-21.0, 8.0-10.0 Mo, Fe 15.0-20.0, Si 1.0 max, Co 2.5 max, W 1.0 max
N06059	Alloy 59	Ni rem Al 0.1-0.4 C 0.010 max Co 0.3 max Cr 22.0-24.0 Fe 1.5 max Mn 0.5 max Mo 15.0-16.5 P 0.015 max S 0.005 max Si 0.10 max
N06075	Alloy 75	Ni rem C 0.08-0.15 Cr 18.0-21.0 Cu 0.50 max Fe 5.00 max Mn 1.00 max Si 1.00 max Ti 0.20-0.60
N06230	Alloy 230	Ni rem Al 0.20-0.50 B 0.015 max C 0.05-0.15 Cr 20.0-24.0 Co 5.0 max Fe 3.0 max La 0.005-0.05 Mn 0.30-1.00 Mo 1.0-3.0 P 0.03 max S 0.015 max Si 0.25-0.75 W 13.0-15.0
N06455	Alloy C-4	Ni rem C 0.015 max Co 2.0 max Cr 14.0-18.0 Fe 3.0 max Mn 1.0 max Mo 14.0-17.0 P 0.04 max S 0.03 max Si 0.08 max Ti 0.70 max
N06600	Alloy 600	Ni 72.0 min C 0.15 max Cr 14.00-17.00 Cu 0.50 max Fe 6.00-10.00 Mn 1.00 max S 0.015 max Si 0.50 max
N06601	Alloy 601	Ni 58.0-63.0 Al 1.0-1.7 C 0.1 max Cr 21.0-25.0 Cu 1.0 max Fe rem Mn 1.0 max S 0.015 max Si 0.50 max

Table 7.1 Chemical Composition of Nickel and Nickel-Based Alloys (Continued)

UNS No.	Common Name	Chemical Composition
N06617	Alloy 617	Ni 44.5 min Al 0.80-1.50 B 0.006 max C 0.05-0.15 Co 10.0-15.0 Cr 20.0-24.0 Cu 0.50 max Fe 3.00 max Mn 1.00 max Mo 8.00-10.0 S 0.015 max Si 1.00 max Ti 0.60 max
N06625	Alloy 625	Ni rem Al 0.40 max C 0.10 max Cb 3.15-4.15 Cr 20.0-23.0 Fe 5.0 max Mn 0.50 max Mo 8.0-10.0 P 0.015 max Si 0.50 max Ti 0.40 max
N06626	Alloy 625LCF	Ni rem, 0.03 max C, 0.15 max Si, 5.0 max Fe, 3.15-4.15 Nb, 8.0-10.0 Mo, 20.0-23.0 Cr, 0.02 max N, 0.4 max Al, 0.40 max Ti, 1.0 max Co
N06686	Alloy 686	Ni rem, max 0.01 C, 5 max Fe, 0.02-0.25 Ti, 3.0-4.4 W, 19.0-23.0 Cr, 15.0-17.0 Mo
N06690	Alloy 690	Ni 58.0 min C 0.05 max Cr 27.0-31.0 Cu 0.50 max Fe 7.0-11.0 Mn 0.50 max S 0.015 max Si 0.50 max
N06985	Alloy G-3	Ni rem C 0.015 max Cr 5.0 max Cr 21.0-23.5 Cu 1.5-2.5 Fe 18.0-21.0 Mn 1.0 max Mo 6.0-8.0 P 0.04 max S 0.03 max Si 1.0 max W 1.5 max other Cb+Ta 0.50 max
N07001	Waspaloy®	Ni rem Co 12-15 Cr 18-21 Mo 3.5-5.0 Al 1.2-1.6 Ti 2.75-3.25
N07031	Pyromet® 31	Ni rem C 0.04 Cr 22.5 Fe 15.0 Mo 2.0 Al 1.4 Ti 2.3 Cu 0.5 B 0.005
N07080	Alloy 80A	Ni rem Al 1.0-1.8 B 0.008 max C 0.10 max Co 2.0 max Cr 18.0-21.0 Cu 0.2 max Fe 3.0 max Mn 1.0 max S 0.015 max Si 1.00 max Ti 1.8-2.7 P 0.045 max
N07090	Alloy 90	Ni rem Al 0.8-2.0 C 0.13 max Co 15.0-21.0 Cr 18.0-21.0 Fe 3.0 max Mn 1.0 max Si 1.5 max Ti 1.8-3.0
N07214	Alloy 214	Ni rem Al 4.0-5.0 B 0.006 max C 0.05 max Co 2.0 max Cr 15.0-17.0 Fe 2.0-4.0 Mn 0.5 max Mo 0.5 max P 0.015 max S 0.015 max Si 0.2 max Ti 0.5 max W 0.5 max Y 0.002-0.040 Zr 0.05 max
N07500	UDIMET® 500	Ni 53 Co 18.5 Cr 18 Mo 4 Ti 3 Al 3
N07716	Alloy 625Plus	Ni 57.0-63.00 Al 0.35 max C 0.03 max Cr 19.0-22.0 Fe rem Mn 0.20 max Mo 7.00-9.50 P 0.015 max S 0.010 max Si 0.20 max Ti 1.00-1.60
N07718	Alloy 718	Ni 50.0-55.0, Cr 17.0-21.0, Cu 0.30 max, Fe rem, Mo 2.8-3.3, Al 0.2-0.8, max C .08, max Mn 0.35, max Si 0.35, max Co 0.015, Ti 0.05-1.15, Cb 4.75-5.50
N07719	Alloy 718SPF	Ni 50.0-55.0, Cr 17.0-21.0, Cb 4.75-5.25, Ti 0.65-1.15, Mo 2.80-3.30, Al 0.20-0.80, Fe Bal
N07725	Alloy 725	Ni 55.0-59.0 Al 0.35 max C 0.03 max Cr 19.0-22.5 Cb 2.75-4.00 Fe rem Mn 0.35 max Mo 7.00-9.50 P 0.015 max S 0.010 max Si 0.20 max Ti 1.00-1.70
N07750	Alloy X-750	Ni 70.0 min Al 0.40-1.0 C 0.08 max Cb 0.70-1.20 Cr 14.0-17.0 Cu 0.5 max Fe 5.0-9.0 Mn 1.0 max S 0.01 max Si 0.50 max Ti 2.25-2.75

Table 7.1 Chemical Composition of Nickel and Nickel-Based Alloys (Continued)

UNS No.	Alloy Name	Chemical Composition
N07751	Alloy 751	Ni 70.0 min Al 0.90-1.5 C 0.10 max Cb 0.70-1.20 Cr 14.0-17.0 Cu 0.50 max Fe 5.00-9.00 Mn 1.00 max S 0.01 max Si 0.50 max Ti 2.00-2.60
N07754	Alloy MA754	Ni rem Al 0.2-0.5 C 0.05 max Cr 19.0-23.0 Fe 2.5 max Ti 0.3-0.6 Other Y₂O₃ 0.5-0.7
N08020	Alloy 20Cb-3	Ni 32.00-38.00 C 0.07 max Cb(8xC-1.00) Cr 19.00-21.00 Cu 3.00-4.00 Fe rem Mn 2.00 max Mo 2.00-3.00 P 0.045 max S 0.035 max Si 1.00 max
N08330	Alloy 330	Ni 34.0-37.0 C 0.08 max Cr 17.0-20.0 Cu 1.00 max Fe rem Mn 2.00 max P 0.03 max Pb 0.005 max S 0.03 max Si 0.75-1.50 Sn 0.025 max
N08800	Alloy 800	Ni 30.0-35.0 Al 0.15-0.60 C 0.10 max Cr 19.0-23.0 Cu 0.75 max Fe rem Mn 1.5 max S 0.015 max Si 1.0 max Ti 0.15-0.60
N08810	Alloy 800H	Ni 30.0-35.0 Al 0.15-0.60 C 0.05-0.10 Cr 19.0-23.0 Cu 0.75 max Fe rem Mn 1.5 max S 0.015 max Si 1.0 max Ti 0.15-0.60
N08811	Alloy 800HT	Ni 30.0-35.0 Al 0.15-0.60 C 0.06-0.10 Cr 19.0-23.0 Cu 0.75 max Fe 39.5 min Mn 1.5 max S 0.015 max Si 1.0 max Ti 0.15-0.60 Other Al+Ti 0.85-1.20
N08825	Alloy 825	Ni 38.0-46.0 Al 0.2 max C 0.05 max Cr 19.5-23.5 Cu 1.5-3.00 Fe rem Mn 1.0 max Mo 2.5-3.5 P 0.03 max S 0.03 max Si 0.5 max Ti 0.6-1.2
N09925	Alloy 925	Ni 38.0-46.0 Al 0.10-0.50 C 0.03 max Cb 0.50 max Cr 19.5-23.5 Cu 1.50-3.00 Fe 22.0 min Mn 1.00 max Mo 2.50-3.50 S 0.03 max Si 0.50 max Ti 1.90-2.40
N10276	Alloy C-276	Ni rem C 0.02 max Co 2.5 max Cr 14.5-16.5 Fe 4.0-7.0 Mn 1.0 max Mo 15.0-17.0 P 0.030 max S 0.030 max Si 0.08 max V 0.35 max W 3.0-4.5
N10665	Alloy B-2	Ni 69 Mo 28 Fe 2 Co 1max Cr 1 max Mn 1 max
N10675	Alloy B-3	Ni 65 Mo 28.5 Co 3 max W 3 max Cr 1.5 Fe 1.5
N12160	Alloy HR160	Cr 28 Fe 2 Co 29 Si 2.75 rem Ni
---	Alloy 671	Ni Rem Cr 48 C 0.05 Ti .35
---	Alloy DS	C 0.10, Mn 0.8-1.5, Si 1.9-2.6, Cr 17.0-19.0, Cu 0.50, Ti 0.20, S 0.03, Fe rem
---	Alloy PE16	Ni 43.5 C 0.06 Cr 16.5, Mo 3.3 Ti 1.2 Al 1.2 Bal Fe
---	Alloy 70	Ni rem C 0.1 max Si 1.0 max Mn 1.0 max Al 0.3-1.0 Cr 18-21 Fe 24-26 Nb 1.0-2.0 Ti 1.0-2.0 Co 2.0 max
---	Alloy 101	Ni rem, C 0.05 Si 0.50 Mn 0.5 Al 1.2-1.6 Ti 2.8-3.2 Cr 23.5-25.0 Co 18.0-20.5 Fe 0.5 max Mo 1.2-1.7 Nb 0.7-1.20
---	NIMONIC 86	Ni rem Cr 25.0 Mo 10.0 C 0.05 Ce 0.03
---	NIMONIC 105	Ni rem C 0.12 max Si 1.0 max Mn 1.0 max Cr 14.0-15.7 Co 18-22 Mo 4.5-5.5 Ti 0.5-1.5 Al 4.5-4.9 Cu 0.2 max Fe 1.0 max B 0.003-0.010 Zr 0.15 max
---	UDIMET 720	Ni 56 Cr 16 Co 14.7 Ti 5 Mo 3 Al 2.5 W 1.25

Table 7.2 American Cross Referenced Specifications - Nickel and Nickel-Based Alloys

UNS No.	Common Name	SAE/AMS	Military	ASTM	ASME	AWS
N02200	Alloy 200	---	---	B 160, B 161, B 162, B 163, B 366, B 725, B 730	SB-160, SB-161, SB-162, SB-163	A5.11 (ENi-1) A5.14 (ENi-1)
N02201	Alloy 201	5553	---	B 160, B 161, B 162, B 163, B 366, B 725, B 730	SB-160, SB-161, SB-162, SB-163	A5.11 (ENi-1) A5.14 (ENi-1)
N04400	Alloy 400	4544, 4574, 4575, 4675, 4730, 4731, 7233	T-1368, T-23520, N-24106	B 127, B 163, B 164, B 165, B 366, B 564, F 96, F 467 (400), F468 (400)	SB-127, SB-163, SB-164, SB-165, SB-564	A5.11 (ENiCu-7) A5.11 (ENiCrMo-3) A5.14 (ERN.Cu-7) A5.14 (ERNiCrMo-3)
N05500	Alloy K-500	4676	N-24549	F 467 (500), F 468 (500)	---	A5.11 (ENiCu-7) A5.11 (ENiCrMo-3) A5.14 (ERNiCu-7) A5.14 (ERNiCrMo-3)
N06022	Alloy C-22 Alloy 622	---	---	B 366, B 564, B 574, B 575, B 619, B 622, B 626	SB-366, SB-564, SB-574, SB-575, SB-619, SB-622, SB-626, SFA-5.14 (ERNiCrMo-10)	A5.14 (ERNiCrMo-10) A5.11 (ENiCrMo-10)
N06030	Alloy G-30	---	---	B 366, B 581, B 582, B 619, B 622, B 626	SB-366, SB-581, SB-582, SB-619, SB-622, SB-626, SFA-5.14 (ERNiCrMo11)	A5.14 (ERNiCrMo-11)
N06950[a]	Alloy G-50	---	---	NACE MR0175	---	---
N12160	HR160	---	---	B 366, B 435, B 564, B 572, B 619, B 622, B 626	---	---
N06050	HX-50	---	---	A 608(HX-50)	---	---
N06059[a]	Alloy 59	---	---	B 564, B 574, B 575, B 619, B 622, B 626	---	---

Table 7.2 American Cross Referenced Specifications - Nickel and Nickel-Based Alloys (continued)

UNS No.	Common Name	SAE/AMS	Military	ASTM	ASME	AWS
N06455	Alloy C-4	---	---	B 366, B574, B 575, B 619, B622, B 626	SB-366,SB-574, SB-575, SB-619, SB-622, SB-626, SFA-5.14 (ERNiCrMo-7)	A5.14 (ERNiCrMo-7)
N06600	Alloy 600	5540, 5580, 5665, 5687, 7232	T-23227, N-23228, N-23229	B 163, B 166, B 167, B 168, B 516, B 517, B 564	SB-163, SB-166, SB-167, SB-168, SB-564	A5.14 (ERNiCr-3) A5.14 (ERNiCrCoMo-1) A5.11 (ENiCrFe-2) A5.11 (ENiCrCoMo-1)
N06601	Alloy 601	5715, 5870	---	B 166, B 167, B 168	---	A5.14 (ERNiCrFe-10)
N06617	Alloy 617	---	---	---	SFA-5.14 (ERNiCrCoMo-1)	A5.14 (ERNiCrCoMo-1) A5.11 (ENiCrCoMo-1)
N06625	Alloy 625	5401, 5402, 5581, 5599, 5666, 5837, 7490	E-21562 (EN625, RN625)	B 366, B 443, B 444, B 446, B 704, B 705	SB-443, SB-444, SB-446, SFA-5.14 (ERNiCrMo-3)	A5.14 (ERNiCrMo-3) A5.11 (ENiCrMo-3)
N06626	Alloy 625LCF	5599, 5879	---	B 443	SB-443	A5.14 (ERNiCrMo-3) A5.11 (ENiCrMo-3)
N06686a	Alloy 686	---	---	B 564, B 574, B 575, B 619, B 622, B626, B 751, B 775, B 829	SB-564, SB-574, SB-575, SB-619 SB-622, SB626	A5.11 (ENiCrMo-14) A5.14 (ERNiCrMo-14)
N06690	Alloy 690	---	---	B 163, B 166, B 167, B 168	SB-163, SB-166,SB-167, SB-168	A5.11 (ENiCrFe-7) A5.14 (ERNiCrFe-7)
N06985a	Alloy G-3	---	---	B 581, B 582, B 619, B 622, B 626	SB-581, SB-582, SB-619, SB-622, SB-626, SFA-5.14 (ERNiCrMo-9)	A5.14 (ERNiCrMo-9)
N07716a	Alloy 625Plus	---	---	B 805	---	A5.14 (ERNiCrMo-15)
N07725a	Alloy 725	---	---	B 805	---	A5.14 (ERNiCrMo-15)

Table 7.2 American Cross Referenced Specifications - Nickel and Nickel-Based Alloys (continued)

UNS No.	Common Name	SAE/AMS	Military	ASTM	ASME	AWS
N08020[a]	Alloy 20Cb-3	---	---	A 265, B 366 B 462, B 463 B 464 B 468, B 471, B 729	---	A5.4 (E320 & E320LR) A5.9 (ER320 & ER320LR)
N08800	Alloy 800	5766, 5871	---	B 163, B 366, B 407, B 408, B 409, B 514, B 515, B 564	SB-163, SB-407, SB-408, SB-409, SB-564	A5.11 (ENiCrFe-2) A5.11 (ENiCrCoMo-1) A5.14 (ERNiCr-3) A5.14 (ERNiCrCoMo-1)
N08825 [a]	Alloy 825	---	---	B 163, B 423, B 424, B 425, B 564, B 704, B 705	SB-163, SB-423, SB-424, SB-425	A5.11 (ENiCrMo-3) A5.14 (ERNiCrMo-3)
N09925	Alloy 925	---	---	---	---	A5.14 (ERNiCrMo-15)
N10276	Alloy C-276	---	---	B 366, B 564, B 574, B 575, B 619, B 622, B 626, F 467 (276), F 468 (276)	SB-366, SB-564, SB-574, SB-575, SB-619, SB-622, SB-626, SFA-5.14 (ERNiCrMo-4)	A5.14 (ERNiCrMo-4) A5.11 (ENiCrMo-4)

a. Other cross reference specification: NACE International Std. MR0175

Table 7.2 lists the basic specification or standard number. Since these standards are constantly being revised, they are presented herein as a guide and may not reflect the latest revision.

Future developments such as ever increasing pollution control requirements in all industries will affect nickel alloy usage. Also, growing polymer and oil and gas industries will continue to push the limits of existing alloys. As a result, the nickel-based alloy system is continuously evolving, with new materials appearing almost every year to provide improved properties for specific applications in old, established industries and in completely new applications.

Applications

Nickel-based alloys are used for a wide range of applications requiring aqueous corrosion resistance, high temperature strength, or both. The chemical composition of these alloys are given in Table 7.1.

Corrosion resistance may be required in aqueous and/or high temperature environments where a variety of attack mechanisms are possible. Also, resistance to mechanical damage by many mechanisms such as fatigue or creep rupture is needed. As a consequence, nickel alloys have often been used to replace steels or stainless steels which have proved inadequate for a particular service.

Water, Seawater, and Atmospheric Applications

The corrosive nature of water can best be described by the level of impurities contained therein. Using this guideline, water can be divided into four main categories: distilled or high purity water, fresh water, brackish water and seawater. Relative corrosivity increases in this respective order.

The main corrosion mechanisms encountered with common engineering materials in contact with waters are pitting and stress corrosion cracking (SCC). Dissolved oxygen and chlorides are the usual major impurities which contribute to these failure mechanisms.

Nickel-based alloys are resistant to chloride SCC, but some alloys may pit or crevice corrode in seawater.

In potable water systems, N04400 or N04405 (MONEL alloy R-405) are used for specialized applications such as valve seats, while sheaths for electrical resistance water heaters of various types are made of N08800 and N08825. N04400 is also used in brackish water applications and various heat exchangers using steam or water to heat or cool a process fluid. N04400 is used in seawater for sheathing and N05500 for bolts and shafts. Though usually resistant, N04400 and N05500 may pit or crevice corrode in stagnant conditions.

As conditions become more corrosive, with increasing chloride content or higher temperature, a Ni-Cr-Mo alloy may be required to resist attack. Alloys such as N08825, N06985 (HASTELLOY alloy G-3), N06625, N07725, N07716, and N10276, in order of increasing molybdenum content, are commonly used in brackish or seawater applications where less expensive materials are susceptible to pitting or crevice corrosion. Ni-Cr-Mo alloys are usually resistant to seawater unless a tight crevice condition exists, such as underneath a nonmetallic gasket or marine deposits etc. In these situations, only the most highly alloyed materials, such as alloys N10276, N26022, N06686, or Nicrofer 5923hMo – alloy 59 (N06059), resist attack.

Typical applications for nickel alloys in marine environments are pumps, valves, propeller shafts, exhaust systems, piping systems, tanks, heat exchangers, fasteners, bellows, and desalination equipment.

Atmospheric corrosion is also influenced by the level of impurities contained in the atmosphere, including chlorides and sulfur compounds. Even in the most severe coastal exposure sites, however, nickel alloys are resistant to attack with corrosion rates <0.0025 mm/y (<0.1 mpy). N02200 (Nickel 200) and N04400 develop a thin gray-green patina when exposed outdoors. The more resistant Ni-Cr-Mo alloys will remain bright and shiny even after extended atmospheric exposure. The most common atmospheric application for nickel alloys is the use of N04400 for architectural service as roofs, gutters, and flashings[3].

Chemical Process Industry

A wide range of liquid, gas, and solid service environments are found in the chemical process industry (CPI) ranging from water to strong acids or bases or a multitude of organic compounds at temperatures from ambient to 2300°F (1260°C). Corrosion failures may be caused by one or more of many corrosion mechanisms.

Many severe environments which cause failure of nonmetallic materials or stainless steels can be contained by a nickel alloy. The choice of nickel alloy is often determined by the oxidizing or reducing nature of the particular environment.

Alloys high in chromium resist general corrosion attack in oxidizing solutions while high nickel or molybdenum content is better for reducing solutions. Nickel alloys containing both chromium and molybdenum are useful, though not necessarily optimum, for both environments. Alloys typically used for general corrosion resistance in each environmental condition are listed in Table 7.3. Examples of reducing and oxidizing environments are also given. Temperatures above about 212°F (100°C) may limit the use of nickel alloys in some strong acids.

In sulfuric acid, which can be either oxidizing or reducing depending on concentration, alloys containing both chromium and molybdenum are usually more resistant. Small additions of copper are also helpful. A combination of chromium and molybdenum additions is also useful in nickel alloys used for phosphoric acid service. High levels of molybdenum are especially beneficial in hydrochloric acid solutions, which are reducing in nature.

Table 7.3 Alloy Selection Based on
Oxidizing or Reducing Conditions

Solution Condition	Alloys	Typical Solution
Oxidizing	Ti 50Ni-50Cr R20033 (alloy 33) N06030 (G-30) N06690 (690) S31000 (310)	Chromic Acid Nitric Acid Conc. Sulfuric Peracetic Acid Oxidizing Salts
Oxidizing or Reducing	N06059 (59) N06686 (686) N 06200 (C-2000) N06022 (C-22) N06022 (622) N10276 (C-276) N06455 (C-4) N06625 (625) N06985 (G-3) N08825 (825)	Halogenated Organic Acids Flue Gas Condensates
Reducing	Zr N10665 (B-2) N10675 (B-3) N10629 (B-4) N02200 (200) N04400 (400)	Caustic Soda Halogen Acids Dilute Sulfuric Phosphoric Acid Alkaline Salts

In many inorganic and organic chemical processes, stainless steels are usually the most cost effective material while nickel alloys are only used for the more extreme conditions where general corrosion, pitting, or SCC cause failure of the traditional stainless steels. Pitting and crevice corrosion can be significant problems in many CPI processes involving halide salts or chlorinated solvents[4]. The use of Ni-Cr alloys containing molybdenum or molybdenum and tungsten are useful in resisting this localized attack. Resistance to localized corrosion generally increases with increasing molybdenum content in alloys such as N08825, N06985, N06625, N06059, and HASTELLOY alloys C-4 (N06455) and C-2000 (N06200). A high level of localized corrosion resistance is also obtained in Ni-Cr alloys containing both molybdenum and tungsten, such as alloys N10276, N06022, and INCONEL alloy 686 (N06686).

Resistance to SCC in chloride and caustic solutions increases with increasing nickel content. This is a deciding factor in the recommendation of nickel alloys for many aqueous applications. The effect of nickel content on chloride SCC resistance is shown in Figure 7.2.[4] Certain nickel alloys have proven performance in specific applications where SCC and/or general corrosion are of concern. N02200, for example, is often used in highly basic solutions such as encountered in evaporator tubing for caustic soda production, while N04400 is often used for handling salts.

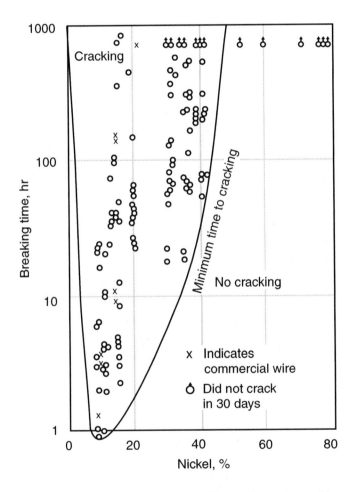

Figure 7.2 Stress corrosion cracking of iron-chromium nickel wires in boiling 42% magnesium chloride.

In most aqueous applications, material strength is a much less important consideration than corrosion resistance. In high temperature applications above about 1000°F (540°C) high temperature strength, creep resistance, and oxidation resistance become more important. Alloys such as N06600, N08800, N08810, N08811, N08330, N06617 and RA330® (N06230) are often used when these properties are needed. Though alloys N08800, N08810, and N08811 have the same base composition in common, the different high temperature strength and creep properties of these materials are due to differences in heat treatment and carbon, titanium, and aluminum ranges, as shown in Table 7.4.[5]

Table 7.4 Comparison of UNS N08800, N08810, and N08811

UNS No. (Alloy)	C	Al+Ti	Annealing Temperature	ASTM Grain Size
N08800 (800)	0.1 max.	0.30 - 1.20	1800°F (982°C)	--
N08810 (800H)	0.05 - 0.10	0.30 - 1.20	2100°F (1149°C)	≤5
N08811 (800HT)	0.06 - 0.10	0.85 - 1.20	2100°F (1149°C)	≤5

Pulp and Paper

Though paper mills have historically relied upon other materials of construction for many processes including traditional stainless steels, nickel alloys play a crucial role in some pulp and paper applications. N06600 has performed well in batch or continuous pulp digester operations for over 30 years. Likewise, N05500 has performed well in doctor blade applications where corrosion and wear resistance are needed. In low pH, high chloride environments such as bleached pulp washers, N06985 and N10276 have been shown to resist attack. In chlorine dioxide washer service nickel-based alloys have had mixed performance.

INCONEL Welding Electrode 112 (AWS A5.11 ENiCrMo-3) and INCONEL Filler Metal 625 (A5.14 ERNiCrMo-3) have been used widely to join austenitic stainless steels seeing service in the pulp and paper industry. This welding product provides an over matching weld metal which is more resistant to corrosive attack than the base metal, thus avoiding preferential weld attack. These filler metals are also used for weld overlays, along with HASTELLOY Welding Electrode C-22 (AWS A5.11 ENiCrMo-10) and HASTELLOY Filler Metal C-22 (AWS A5.14 ERNiCrMo-10), to produce corrosion resistant surfaces on some types of equipment.

Natural Gas Production

Natural gas is one of the world's most abundant sources of energy. In the past, drilling and production tubulars were steel only, while stainless steels and nickel alloys were used only for valves or instrumentation. Today corrosion-resistant alloy (CRA) tubulars made from nickel alloys are needed because of:

1.   deeper wells involving higher temperatures, up to 500°F (260°C), and pressures,
2.   enhanced recovery methods such as steam or CO_2 injection,
3.   increased weight considerations, especially offshore, and
4.   the need for greater corrosion resistance in wells containing hydrogen sulfide (H_2S), carbon dioxide (CO_2), and chlorides.

Material selection is especially critical for sour gas wells–those containing H_2S partial pressure ≥ 0.05 psi (0.0003MPa). Generally, the need for higher corrosion resistance and increased strength increases with well depth as temperature, pressure, and CO_2, chloride, and H_2S levels also increase. As these conditions become more severe, tubular material selection goes from C-Mn or Fe-Cr steels used for shallow, *sweet* (no H_2S) wells, duplex (austenitic-ferritic) stainless steel, N08825 or N09925, to N06985 or HASTELLOY alloy G-50 (N06950), for sour well service. Under the most severe conditions, N10276 has been used.

In general, resistance to sulfide SCC, hydrogen embrittlement, and localized corrosion in these environments increases with increasing nickel, chromium and molybdenum content. These materials are cold worked or age hardened to specified levels in order to obtain the strength needed to support the weight of several thousand meters of tubing and withstand the intense pressure.

Material selection for downhole and wellhead accessories such as hangers, valves, pumps, packers, and wire lines is also important. For many of these components age-hardenable alloys are used to obtain the needed strength in heavier cross-sections which cannot be strengthened by cold work. Nickel alloys commonly used for these applications include: N05500, N09925, N07725, N07716, N07718, and INCONEL alloy X-750 (N07750).

As with tubulars, these components must resist SCC. The potential for SCC becomes greater with higher temperature and concentrations of H_2S and the presence of chloride ions and elemental sulfur. Lower temperature hydrogen embrittlement and sulfide stress cracking (SSC) are also potential failure mechanisms, which are promoted by galvanic corrosion, acidizing operations, or dissolved H_2S.

Alloy strength is another factor. As strength increases, environmental cracking susceptibility increases. In order to obtain the optimum level of strength, toughness, and cracking resistance maximum hardness levels are specified for each alloy in NACE MR0175 (see Table 7.5). Typical nickel alloy mechanical properties for oil-country applications are listed in Table 7.6

Selection of corrosion resistant alloys (CRA) for oil field applications can be a complex procedure. If done improperly, it can lead to mistakes and misunderstandings of the performance of a CRA in a specific sour gas environment.

Different methods are used by individuals and companies when choosing a CRA for specific service environments.[6] A recognized selection procedure is to review the literature for corrosion data that, in general, apply to anticipated field conditions. Then a group of alloys is selected that represents a range of alternatives. A test program, simulating the subject field environment, is sometimes initiated. A final CRA selection is made for a specific application based on test results and an economic analysis of cost effective alternatives. While more detailed testing and analysis is often required, guideline tables and diagrams are often used before extensive efforts are made to make a final alloy selection for a specific oil field application.

Table 7.5 NACE MR0175-99 Maximum Hardness Levels

UNS No.	Condition	HRC Maximum
N04400, N04405	All	35
N05500	Hot Worked and Age-Hardened Solution-Annealed Solution-Annealed and Aged-Hardened	35
N06002, N06625	All	35
N06007 N06250 wrought N06255 wrought N06600 N06975 wrought	All	35
N06950	All	38
N06985	All	39
N07718	Solution-Annealed Hot Worked Hot Worked and Aged	35
	Solution-Annealed and Aged Cast, Solution-Annealed and Aged	40
N07716, N07725	Annealed and Aged	40
N07750	Solution-Annealed Solution-Annealed and Aged Hot Worked Hot Worked and Aged	35
	Cold Worked and Aged for Springs	50
N08800	All	35
N08825	All	35
N09925	Solution-Annealed Cold Worked	35
	Annealed and Aged	38
	Cold Worked and Aged	40
	Hot-Finished and Aged	40
N10276	Solution-Annealed Solution-Annealed and Cold-Worked	35
	Cold-Worked and Unaged for Service Over 250°F (121°C)	45

Table 7.6 Representative Mechanical Properties of Nickel Alloys
for Oil-Country Applications[a]

UNS No.	Material Condition	Yield Strength ksi	Yield Strength MPa	Tensile Strength ksi	Tensile Strength MPa	% El.	Hardness
N04400	Annealed	31.3	216	78.6	542	52	60 HRB
	Cold Worked	93.7	646	108.8	716	19	20 HRC
N05500	Aged	97.5	672	152.5	1051	25	28 HRC
N06950	Cold Worked	147.3	1029	136.0	952	24.1	30 HRC
N06625	Annealed	69.5	479	140.0	965	54	95 HRB
	Cold Worked	125.7	867	150.4	1037	30	33 HRC
N07716	Annealed & Aged	133	917	186	1282	32	37 HRC
N07718	Aged	159.0	1096	191.5	1320	20	40 HRC
N07725	Aged	132.9	916	183.3	1264	28	36 HRC
	Cold Worked & Aged	163.3	1126	189.6	1307	15	38 HRC
N07750	Aged	132.8	916	188.0	1296	27	34 HRC
N08825	Annealed	47.0	324	100.0	690	45	85 HRB
	Cold Worked	114.0	786	130.5	900	15	28 HRC
N09925	Annealed & Aged	113.0	779	176.0	1214	26	36 HRC
	Cold Worked	129.0	889	140.0	965	17	32 HRC
	Cold Worked & Aged	153.0	1055	176.0	1214	19	---
N06985	Annealed	41.4	285	99.3	685	54	83 HRB
	Cold Worked	119.7	825	141.1	973	18	28 HRC
N10276	Annealed	52.0	359	110.4	761	64	83 HRB
	Cold Worked	156.9	1082	172.5	1189	17	35 HRC

a. Properties represent various product forms. Tubular goods are supplied to specified minimum yield strengths that may differ from values in this table.

Corrosion data for cold-worked austenitic alloys including alloy 825 (N08825), alloy 28 (N08028), alloy 25-6MO (N08926), alloy 625 (N06625), alloy C-276 (N10276) and age-hardened alloys such as alloy 925 (N09925), alloy 718 (N07718) and alloy 725 (N07725) are presented. Tables 7.7 through 7.15 list sour oil patch environments from a literature review in which cold-worked oil country tubular goods (OCTG's) and age-hardened CRA's have either been recommended or where corrosion testing has validated their use. Results are generally based on stress corrosion cracking and sulfide stress cracking hydrogen embrittlement data.

For references [7] and [8], H_2S limits are based on the presence of a significant concentration of Cl^- salts in the aqueous phase. It is recognized that alloys exposed to environments with little or no Cl^- may be able to tolerate higher H_2S partial pressures. Appropriate testing and/or available test data are required to identify these environments.

Most alloys have displayed corrosion resistance at 230°C, depending on the chloride concentration, H_2S content, and the presence of elemental sulfur.[9] The cold-worked alloys ranked by corrosion resistance are:

N10276 > N06625 and N06985 > N08825 > N08028 > N08926

The precipitation hardened alloys ranked by corrosion resistance are:

N07725 > N09925 > N07718.

Tables 7.7 through 7.15 show service environments where the literature has indicated acceptable corrosion resistance for the austenitic alloys listed.

The precipitation hardened alloys are used at different strength levels depending on the application, but generally N09925 is used at 758 MPa (110 ksi) minimum yield strength and N07725 and N07718 are used at 827 MPa (120 ksi) minimum yield strength.

The cold-worked solid solution alloys are also used at various strengths. Of the standard OCTG's, N08825 and N08028 are used at 758 MPa (110 ksi) minimum yield strength and N06985 and N10276 are used at 862 MPa (125 ksi) minimum yield strength. See the manufacturers for mechanical properties for specific grades available.

Ultimately, it is the user's responsibility to establish the acceptability of an alloy for a specific environment. This chapter presents data from a literature review intended for use in selecting materials for corrosive sour oil patch environments. A group of alloys that represents a range of alternatives can be selected for testing in an environment simulating the field environment under study. A final CRA selection is made for a specific application based on test results and an economic analysis of cost effective alternatives.

The manufacturers of equipment and components will also have a data bank of previous service recommendations which can be an excellent aid in determining the candidate alloys for particular service conditions.

Table 7.7 Environments in Which N08825
Has Been Reported as Acceptable

Reference	#2	#3	#3	#3	#5	#5
Cl⁻ (ppm)	151,750	Any	Any	Any	100,000	150,000
pH	---	---	---	---	3.5	3.5
Temperature (°C)	200	175	220	230	205	205
H_2S (MPa)	6.0	1.4	0.7	0.2	0.69	0.69
CO_2 (MPa)	---	Any	Any	Any	2.76	2.76
S°	---	0	0	0	0	0

Table 7.8 Environments in Which N06625
Has Been Reported as Acceptable

Reference	#2	#3	#3	#3
Cl⁻ (ppm)	151,750	Any	Any	Any
pH	---	---	---	---
Temperature (°C)	200	230	190	150
H_2S (MPa)	6.0	1.0	3.5	Any
CO_2 (MPa)	---	Any	Any	Any
S°	---	0	0	0

Table 7.9 Environments in Which Alloy C-276
Has Been Reported as Acceptable

Reference	#2	#3	#3	#6	#7	#7
Cl⁻ (ppm)	Any	Any	Any	151,750	121,400	121,400
pH	---	---	---	3.1	3.0	3.1
Temperature (°C)	260	205	230	230	230	230
H_2S (MPa)	66.0	Any	1.0	0.83	6.9	0.5
CO_2 (MPa)	---	Any	Any	Any	4.8	4.8
S°	---	0	0	Yes	Yes	0

Table 7.10 Environments in Which N08926
Has Been Reported as Acceptable

Reference	#2	#2	#2	#3	#3	#8
Cl⁻ (ppm)	121,400	12,140	12,140	Any	Any	60,700
pH	---	---	---	---	---	3.3
Temperature (°C)	250	200	150	150	170	120
H_2S (MPa)	0.0	0.14	0.27	0.3	0.1	0.7
CO_2 (MPa)	---	---	---	Any	Any	1.4
S°	---	---	---	0	0	0

Table 7.11 Environments in Which N08028
Has Been Reported as Acceptable

Reference	#2	#2	#3	#3	#3
Cl⁻ (ppm)	37,634	15,175	Any	Any	Any
pH	---	---	---	---	---
Temperature (°C)	100	204	175	220	230
H_2S (MPa)	0.5	1.31	1.4	0.7	0.2
CO_2 (MPa)	---	---	Any	Any	Any
S°	---	---	0	0	0

Table 7.12 Environments in Which N06985
Has Been Reported as Acceptable

Reference	#3	#3	#3	#9
Cl⁻ (ppm)	Any	Any	Any	151,750
pH	---	---	---	3.3
Temperature (°C)	230	190	150	220
H_2S (MPa)	1.0	3.5	Any	2.1
CO_2 (MPa)	Any	Any	Any	2.1
S°	0	0	0	0

Table 7.13a Environments in Which N07718
Has Been Reported as Acceptable

Reference	#4	#4	#4	#4	#4	#4
Cl⁻ (ppm)	151,750	151,750	60,700	151,750	151,750	151,750
pH	3.13	3.13	3.19	3.13	3.13	3.14
Temperature (°C)	148.9	148.9	148.9	121.1	148.9	148.9
H_2S (MPa)	0.34	2.76	2.76	0.69	0.69	0.34
CO_2 (MPa)	4.83	2.76	1.38	4.83	4.83	4.83

Table 7.13b Environments in Which N07718
Has Been Reported as Acceptable

Reference	#4	#2	#3	#3	#3	#3
Cl⁻ (ppm)	151,750	91,050	Any	Any	Any	Any
pH	3.15	---	---	---	---	---
Temperature (°C)	148.9	150	175	205	220	230
H_2S (MPa)	0.17	1.4	1.4	1.0	0.7	0.2
CO_2 (MPa)	4.83	---	Any	Any	Any	Any
S^o	---	---	0	0	0	0

Table 7.14a Environments in Which N09925
Has Been Reported as Acceptable

Reference	#4	#4	#4	#4	#4	#4	#4
Cl⁻ (ppm)	151,750	151,750	60,700	151,750	91,050	151,750	151,750
pH	3.13	3.13	3.19	3.13	3.13	3.13	3.14
Temperature (°C)	148.9	148.9	148.9	121.1	148.9	148.9	148.9
H_2S (MPa)	0.34	2.76	2.76	0.69	0.69	0.69	0.34
CO_2 (MPa)	4.83	2.76	1.38	4.83	4.83	4.83	4.83
S^o	0	0	0	0	0	0	0

Table 7.14b Environments in Which N09925
Has Been Reported as Acceptable

Reference	#4	#2	#3	#3	#3	#3	#9
Cl⁻ (ppm)	151,750	91,050	Any	Any	Any	Any	300,000
pH	3.15	---	---	---	---	---	3.1
Temperature (°C)	148.9	150	175	205	220	230	182
H₂S (MPa)	0.17	1.4	1.4	1.0	0.7	0.2	6.8
CO₂ (MPa)	4.83	---	Any	Any	Any	Any	2.9
S°	0	---	0	0	0	0	0

Table 7.14c Environments in Which N09925
Has Been Reported as Acceptable

Reference	#9	#9	#9	#9
Cl⁻(ppm)	99,000	Saturated	Condensed	63,000
pH	5.0	3.3	3.4	3.2
Temperature (°C)	177	199	105	190
H₂S (MPa)	6.2	2.3	0.3	2.5
CO₂ (MPa)	3.1	1.5	0.9	3.3
S°	0	0	0	0

Table 7.15 Environments in Which N07725
Has Been Reported as Acceptable

Reference	#3	#3	#3	#10	#10	#10	#10
Cl⁻(ppm)	Any	Any	Any	100,000	250,000	250,000	151,750
pH	---	---	---	3.3	3.0	3.0	3.1
Temperature (°C)	230	190	150	220	205	175	175
H₂S (MPa)	1.0	3.5	Any	1.4	4.1	8.3	2.1
CO₂ (MPa)	Any	Any	Any	1.4	4.8	4.8	4.8
S°	0	0	Yes	Yes	0	0	Yes

Petrochemical and Refinery Processing

High nickel alloys have found widespread application for process equipment in these industries from refining of the crude stock through to manufacture of products such as ethylene, ammonia, and vinyl chloride. Corrosion problems in the early stages of refining where the crude may be heated to (700°F) (370°C) are usually associated with contaminants, mainly salt, napthenic acid, and various sulfur compounds. N04400, either as a lining or clad plate, is particularly useful for the upper parts of the fractionating tower where hydrolysis of chlorides and chlorinated solvents leads to formation of dilute hydrochloric acid causing rapid wastage of carbon steel and SCC of conventional stainless steels. The alloy is also well suited for fractionating trays, bubble caps and condensers, as well as overhead condensers and transfer piping.

For removal of sulfur, nitrogen, and metallic contaminants from petroleum feedstock in hydrocracking and hydrodesulfurizing (HDS) units, N08800 and N08801 are used for fired heater tubing, transfer piping, and various other components. Due to a higher titanium content, the latter alloy can be well stabilized and thereby is highly resistant to polythionic acid cracking encountered in HDS units. In other applications, N08825 and N06625 have also been found to be resistant to polythionic acid cracking.

The most serious corrosion problems in the catalytic cracker units, used for converting heavy fractions to gasoline, occur in the catalyst regenerator. Since temperatures reach 1200°F (650°C) or higher, N08810 and N08811 are used for critical parts such as the plenum chamber, cyclone support rods, and catalyst grid supports.INCONEL alloy 625LCF® (N06626) is also used for expansion bellows in these units because of its high resistance to thermal and mechanical fatigue at temperatures up to about 1200°F (650°C.)

Alkylation is another process presenting serious corrosion problems, especially hydrofluoric acid alkylation. While the extreme reactivity of hydrofluoric acid excludes most commonly available materials, N04400 is suitable for acid regeneration and dehydration columns, tar neutralization towers, trays, pumps, and valves, as well as piping and fittings. When exposed to hot, moist, and highly-aerated HF vapors, however, stress relief of equipment is usually recommended to avoid any possibility of SCC.

The alternative process of sulfuric acid alkylation is generally less aggressive through maintenance of a high concentration of the sulfuric acid catalyst. Nevertheless, corrosion resistant alloys, like N08825 are needed for valves and acid feed nozzle systems. For flare stack tips, N08800 and INCOLOY alloy DS have been used successfully, offering sufficient oxidation and carburization resistance, strength and resistance to thermal fatigue.

Petrochemical processing also imposes severe demands on material reliability and often necessitates the use of nickel alloys. In ethylene production, for example, pyrolysis tube skin temperatures can reach as high as 1925 - 2100°F (1050 - 1150°C) and high creep strength is of utmost importance along with carburization resistance. Although cast heat resisting alloys, such as HP and HX (N06002) alloys, have tended to dominate ethylene applications, newer designs incorporating much shorter residence times and smaller diameter tubes have seen a resurgence of wrought alloys, like N08810 and N08811. Moreover, further improvements in process efficiency have been attained with the advent of internally finned tube technology[17] and newer alloys like INCOLOY alloy 803 (S35045). This alloy has a higher chromium and nickel content than N08810, thereby providing enhanced strength, oxidation, and carburization resistance. N08810 and N08811 are also often the preferred materials for convection tubes, transfer piping, and quench systems.

Over the years N08810 type has proved the workhorse for components of steam-hydrocarbon reformers in hydrogen and ammonia production, typical examples being catalyst tubing, manifolds, heaters, and internals in secondary reformers. The alloy is also the preferred choice for pigtails which accommodate thermal expansion differentials between the reformer tubes and headers. Apart from providing the desired corrosion resistance, N08810 generally maintains sufficient ductility after long time exposure to enable 'nipping' and thereby isolate any overheated or failed reformer tubes.

In the production of vinyl chloride monomer, ethylene dichloride is thermally cracked at temperatures up to about 1200°F (650°C). N08800 has better corrosion resistance than stainless steels such as S32100 (type 321) and is also less susceptible to chloride SCC during shutdowns. N06600 is generally considered to have even better resistance than N08800 and stainless steels, especially when traces of free chlorine are present. Extreme care in decoking is necessary, however, since operating temperatures are relatively close to the breakaway corrosion temperatures of all the alloys.

In the oxychlorination stage, where ethylene is converted to dichlorethylene, both N08825 and N08800 have been used with success.

Two processes are involved in the production of styrene-alkylation and dehydrogenation. For the former, N06625 is recommended for the reactor exchangers, coolers, and piping systems, where the aluminum chloride catalyst may hydrolyze to acid chlorides causing pitting problems. In the dehydrogenation stage, alloys N08810 and N08811 are the preferred choice for the steam superheaters and transfer piping where high temperature strength is required.

N06230 and N06617, albeit much more costly materials, have also been considered for various applications in the petrochemical industry where high creep rupture strength is required.

Nickel-Based Wrought Superalloys for Gas Turbine Engineering

Since the development of the Ni-Cr high temperature alloys for use in the earliest jet engines, nickel-based wrought super alloys have found many applications such as compressor blades, combustor fabrications, turbine casings, and discs. One of the first alloys to be developed for gas turbine applications was NIMONIC alloy 75 (N06075), a simple 80-20 Ni-Cr alloy with an addition of about 0.3% titanium and low carbon content. By 1944, NIMONIC alloy 80A (N07080) had been introduced, with improved alloy performance due to increases in the titanium and aluminum contents plus general improvements in production technique.

Nickel-based alloys offer virtually no strength advantage over steels or titanium at lower temperatures, which have the advantages of lower price and lower density, respectively. Consequently, at the cool end of corrosion-free compressors, the use of high nickel alloys for blading cannot normally be justified. However, as the air passes through the compressor, its temperature, and that of the adjacent parts of the engine, are raised up to 1100°F (593°C) in some cases. While the strength of steel (and also titanium) falls rapidly as the temperature increases above 570-750°F (300-400°C), there is little loss of properties with the nickel-based alloys up to 1100°F (593°C). As a result, the materials are little affected by creep at temperatures below about 1050°F (550°C), whereas steels are subject to creep at comparable or even lower temperatures. These nickel-based alloys include N07718 and NIMONIC alloy 90 (N07090).

Sheet materials are used in a variety of applications in gas turbines. The major ones produced in nickel-based alloys include combustion chambers, casings, liners, transition ducts, exhaust ducts, afterburner parts, and bearing housings. Sheet components generally have the basic similarity of purpose in containing and directing gas at high temperatures and pressure. The essential material requirements include oxidation resistance, creep strength, and thermal and mechanical creep strength while retaining a degree of formability and weldability. The high temperature oxidation resistance of nickel alloys is superior to that of stainless steels. Typical sheet alloys include NIMONIC alloy 263 (N07263), N06617, and HAYNES® 230™ (N06230). For exhaust applications, N06626 is used increasingly due to its resistance to low cycle fatigue. Existing super-plastic forming (SPF) technology, for the manufacture of complex, light-weight fabrications, is now being adopted at elevated operating temperatures with the use of INCONEL alloy 718SPF™ (N07719).

The rotor blade probably operates under more rigorous conditions of stress and temperature than any other engine component. High stresses are produced by rotation at blade-tip speeds of up to 1280 ft/s (390 m/s). Under operating conditions in the turbine, the predominant mode of failure for blades is by creep rupture. The basic members of the NIMONIC series of alloys are used for blading applications and were developed from N06075. The first of the precipitation hardened alloys was N07080, followed by alloys containing extra solid solution-hardening and precipitation-hardening elements. The most highly alloyed member in the group is NIMONIC alloy 115, a complex Ni/Cr/Co/Mo/Ti/Al alloy. Figure 7.3a illustrates the temperature capability of wrought nickel-based superalloys, whereas Figure 7.3b shows their relative creep rupture properties. The trend within the industry has been toward higher purity materials utilizing single crystal technology and investment casting techniques, which allows blades to be internally cooled with compressor air, and to operate in higher temperature environments. This drive toward hotter turbine inlet temperatures is to improve engine efficiency.

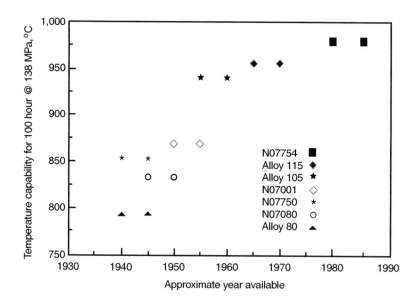

Figure 7.3a Temperature capability of
wrought nickel-based superalloys.

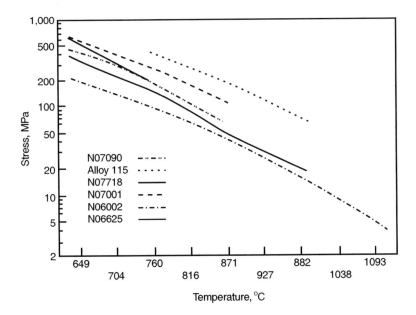

Figure 7.3b Nickel-based wrought superalloys –
creep rupture properties (1,000 hour).

The main functions of a casing are to contain and direct the gas flow, to locate the non-rotating elements, and to provide a means of external location and support for the engine as a whole. As turbines are built in units that produce from almost zero to about 100 MW, their casings vary in size from a few centimeters to several meters in diameter. Manufacturing methods include casting, forging, and fabricating from sheet and forged rings, or formed and welded rings. A common method of producing casings for aero engines is to forge or roll a ring or casing section from which the casing shape is machined. Typical casing alloys include INCOLOY alloy 909 (N19909), INCONEL alloy HX (N06002) and Hastelloy X (N06263).

Compressor and turbine discs operate at high rotational speeds. Their failure in service would be catastrophic, both in terms of aircraft safety and the economics of land-based turbine operation. Material evaluation is critical. The choice is mainly dictated by the operating temperature and is usually confined to steels, titanium, or nickel-based alloys. When operating conditions become more severe and temperatures rise much above 932°F (500°C), nickel-based alloys are normally employed. These include N07718, INCOLOY alloy 901 (N09901), and N07001. To meet the desired levels of internal cleanliness and freedom from segregation, double and even triple melting refining methods are employed in the manufacture of some of these alloys.

Nickel alloys such as N06600 and N07718, are also utilized for engine and airframe bolting and fastener applications.

Nickel-Based Wrought Superalloys for Rockets, Missiles, and Spacecraft

Rocket engines often require unique properties in materials. They must resist very high temperatures and/or very low cryogenic temperatures while demonstrating very high strength and resistance to the environment. Additionally, controlled low expansion characteristics may be required at both low and high temperatures. In certain applications, high temperature strength is needed in combination with resistance to hydrogen embrittlement or surface oxidation. At cryogenic application temperatures, alloys must remain ductile while retaining a high fracture toughness and strength. Early rocketry originally used large amounts of N02200 tubing for the water cooling of rocket thrust chambers. Nickel-base alloys which regularly find applications in this industry include N07718, N19909 and N06002. Additionally, N04400 is used for valves and pumps in the handling of liquid oxygen due to its non-sparking characteristics.

For the very highest temperatures requiring extreme resistance to creep and corrosion, mechanically alloyed yttrium oxide dispersion strength (ODS) alloys are used for critical components in military aircraft gas turbines, stress bearing parts in industrial furnaces, and molten glass containment. INCONEL alloy MA754 (N07754) is now specified for the heat shield in the X-33 prototype space shuttle. It will replace the ceramic shield on the original space shuttle.

Power

Fossil fueled power plants often use N08800, N08810, or N08811 reheater or superheater tubes for high temperature strength and oxidation resistance when less expensive steel or stainless steel tubing experiences unacceptable life. Tubes using N08800 clad with INCONEL alloy 751 (N07751) 50Ni-50Cr have been used for some very severe conditions. Corrosion and oxidation of ashpit seals and drip lips can be prevented by use of alloys such as N08825, N06625, and N06601. N04400 feedwater heater tubes have been used successfully for over 40 years. The cold drawn stress relieved condition provides improved strength, though SCC of this product has occurred in some instances.

Flue gas desulfurization (FGD) pollution control systems for fossil plants rely heavily on nickel alloys for resistance to general corrosion, pitting, and crevice corrosion[18]. Absorption of sulfur dioxide gas and chloride contamination results in aggressive low pH, high chloride environments. Generally, nickel alloys are used in environments with lower pH, higher chloride content, and higher temperature where stainless steels and nonmetallic linings have failed. N10276 is the most commonly used nickel-base alloy in FGD applications. Other pitting and crevice corrosion resistant nickel alloys often used are N06022, N06059, and N06686. An FGD system schematic drawing is shown in Figure 7.4. In this figure, the most common areas of nickel alloy usage are indicated as: the wet/dry interface (A), the scrubbed gas reheater (E), the bypass duct (F), and the stack liner (G). These are the most aggressive areas of the system. In addition, these nickel alloys are sometimes used in the less aggressive jump (B) and spray zone (C-D) areas when extended life or increased reliability are required. Not shown are induced draft fans which are often constructed of N06625 or N10276. Corrosion resistant welding products such as HASTELLOY alloy C-22 welding electrode (AWS 5.11 ENiCrMo-10), INCO-WELD® welding electrode 686CPT™ (AWS A5.14 ENiCrMo-14) and alloy C-276 (AWS A5.11 EniCrMo-4) are used.

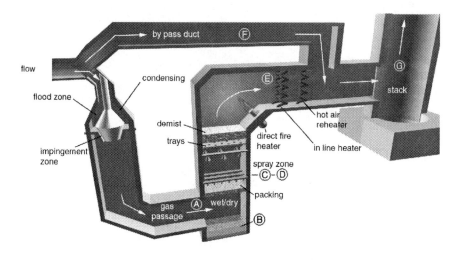

Figure 7.4 FGD system areas using nickel alloys.

The use of nickel alloys in FGD systems[19] is made more economical by use of alloy clad steel plate and alloy sheet linings over steel structures. The clad steel is used for new structures, while the sheet lining can be used for new units or repair of existing units[20,21].

Nuclear power plants have used large quantities of nickel alloys since the beginning of the nuclear power program. In pressurized water reactors (PWR), N06600 and N08800 are used for steam generator tubes, nozzles, and baffle plates. N06600 is now being replaced with the high chromium INCONEL alloy 690 (N06690) due to caustic and high purity water SCC failures of N06600. Age hardenable alloys N07750 and N07718 are also used for high strength bolting and springs in contact with primary water. These same nickel alloys are used to a lesser extent in boiling water reactors (BWR).

Radwaste evaporator systems use pitting resistant alloys such as N08825 and N10276. N06690 is also important in the glass vitrification of solid radwaste.

Advanced gas cooled reactors utilize N08800, N07080, PE16, and N06002.

INCOLOY alloy 908 (N09908) is being used for sheathing and piping in the International Thermonuclear Experimental Reactor program where special controlled expansion properties are required.

Electrical Resistance Alloys

Nickel-chromium-iron alloys are used for industrial and domestic electrical heating applications, typically in the form of wire or strip elements at temperatures up to 2100°F (1150°C). The main factors which determine alloy selection are the operating temperature, resistivity in the range of 110 to 120 µW•cm, the change of resistance with temperature defined by the temperature coefficient of resistance, mechanical strength at the operating temperature, resistance to embrittlement in service and resistance to high temperature oxidation and other forms of corrosion.

Applications include heating elements for industrial furnaces, domestic heating appliances, cookers, irons, kettles, and the like. In transportation, applications include dynamic braking resistors in locomotives which dissipate electrical energy when the train or tram needs to brake.

In applications where frequent switching of the unit is expected, BRIGHTRAY alloys C and S (N06003) (80/20 Ni-Cr alloy) with rare earth additions, produced exclusively in wire form are most suitable for up to 2100°F (1150°C).

For continuous operating conditions at temperatures up to 2100°F (1150°C), a conventional N06003 is more suitable. Elements made of this alloy are equally suited to oxidizing, reducing, or neutral atmospheres.

BRIGHTRAY alloy B (N06004) (60/16 Ni-Cr alloy, the balance being iron with some rare earth additions), can be used up to 2010°F (1100°C). The higher temperature coefficient of resistance of this alloy makes it suitable for less arduous operating conditions than the 80/20 type alloys.

Used in applications up to 1925°F (1050°C), a 37/18 Ni-Cr alloy, the balance being mainly iron with a small silicon addition, is suitable for applications where resistance to sulfidation and green rot, (alternate carburization and oxidation), is important.

These alloys are competing in many applications with ferritic Fe-Cr-Al alloys. This alloy system exhibits higher resistivity and lower density but usually at the expense of lower strength and increased likelihood of embrittlement leading to shorter component life.

All of the electrical resistance alloys oxidize when heated, forming a protective layer under continuous operating conditions. Under cyclic conditions there may be a tendency for the oxide to spall when the elements are cooled. This process repeated several times can lead to ultimate failure of the element. The addition of a small quantity of rare earth elements aids in the adherence of the oxide to the metal under cyclic conditions and vastly improves service life. The effect of these additions is seen in Figure 7.5.

Waste Incineration - Environmental conditions within waste combustion systems are extremely complex and vary enormously depending on the type of waste, operating conditions, and location within the system. Consequently many different materials, both non-metallic and metallic, may be needed to resist the variety of corrosion mechanisms operative over the entire temperature spectrum (Figure 7.6).

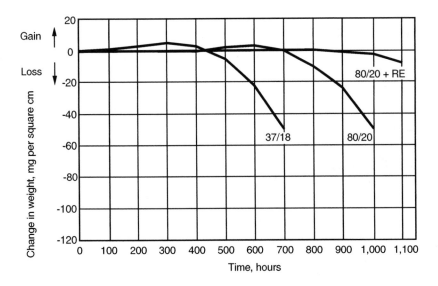

Figure 7.5 Comparative cyclic oxidation
properties at 1832°F (1000°C).

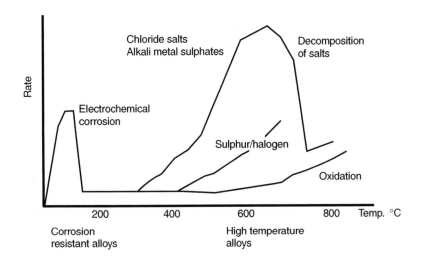

Figure 7.6 Effect of temperature on corrosion rate
in waste combustion systems.[22]

At low temperatures or during downtime periods, condensate or dewpoint corrosion can occur and conventional stainless steels are rarely resistant to this type of attack. Even higher nickel alloys, such as N08020 and N08825, may suffer pitting or crevice corrosion beneath deposits. For ultimate resistance, under more arduous halide conditions, Ni-Cr-Mo alloys as used in FGD systems are necessary. Although more costly, such alloys can prove cost effective on a life cycle basis.

For intermediate temperatures up to about 1100°F (600°C), N08825 has provided long-standing service as superheater tubing while N06625 welding consumables have been used for weld overlaying waterwall tubes. Under similar conditions, iron-rich alloys exhibited liquid phase corrosion caused by chlorine-rich phases.

At higher temperatures, sulfidation and halide attack tend to dominate, often initiated by the presence of alkaline and heavy metal salt deposits. Many of these salt mixtures have low melting points and cause accelerated wastage through fluxing type reactions. Countless materials have been used in waste incinerations for such applications as tube supports, shields, dampers, ducting, and thermowells. Alloys include traditional Fe-Ni-Cr alloys often with high Cr content. In sulfur-rich environments high nickel materials are not favored. N06600, N06601, N06617, N06625, N07214, and N06230 are generally better suited to chloride containing environments. Two common cobalt-containing alloys are HAYNES 188 (R30188) and HAYNES 556 (R30556). Two more recently developed alloys that have performed well in incinerator environments are HASTELLOY alloy HR-160 (N12160) and Nicrofer alloy 45™ (N06045). These alloys not only have high chromium contents but also have relatively high levels of silicon which further contributes to their good sulfidation resistance.

If fluorides (HF) are present, alloys offering the best corrosion resistance normally contain high levels of nickel and molybdenum, with low chromium content, such as Hastelloy alloy N (N10003) with 7% Cr. However, N06600 with 15% Cr also offers useful resistance to many high temperature fluoride environments. High chromium levels are normally detrimental because of the low melting points and poor protective qualities of chromium-rich fluoride compounds.

For long term disposal of radioactive waste, N06601 and N06690 are often the preferred materials of construction for glass vitrification-encapsulation processes. While near matching welding consumables are available for these alloys, N06072 (INCONEL filler metal 72) with an even higher chromium content of 44% has also been used with some success, particularly for overlaying areas exposed to more severe operating conditions.

Heating and Heat Treating

Nickel alloys find a multitude of applications in the thermal processing and heat treatment industry. The factors which must be considered in choosing an alloy for a specific application include: resistance to high temperature corrosion in a specific atmosphere, effects of thermal fatigue and thermal shock, high temperature strength and stability requirements, including creep rupture properties, along with ease of fabrication and repair of components.

The main high temperature corrosion mechanisms encountered are oxidation, halogenation, carburization, sulfidation, and nitridation. Several common applications are listed below and the most commonly used alloys are noted.

Radiant tubes made from alloys such as N08811, N06600, and N06601, offer prolonged service life, extended times between overhauls, and hence reduced maintenance costs; These provide efficient heat transfer without deformation during high temperature heavy furnace loading.

Furnace muffles in alloys, such as N06600 and N06025 (Nicrofer alloy 6025HT™) can be readily fabricated from sheet or plate and can provide several years of uninterrupted life.

Many burners ranging from small units in furnaces or domestic heating systems to large units on oil and gas platforms, use alloys such as N08811, alloy DS, N08330, N06601, and N06617 in their design.

Pit retorts which can operate at temperatures up to 2200°F (1200°C) provide cost savings due to the thinner sections which can be used in various atmospheres. Alloys commonly used for this application include N06601 and N06230.

Baskets and trays used for tempering, hardening, annealing, and quenching components made from wrought alloys are lighter than cast alternatives. They also offer better resistance to thermal fatigue and thermal cracking due to their lower thermal expansion characteristics and fine grained structures. Alloys commonly used for this application include alloys N08800, alloy DS, and N08330.

Mesh belts woven from wire in alloys such as N06600, N06601, alloy DS, N08330, and N07214 are commonly used in many industries for both thermal processing and sintering. Larger belts are used in the mining and refining industries for handling hot sinters and ashes. N08811 has been used in chain link belts for handling iron ore sinters.

N06601 and INCONEL alloy 601GC® (grain controlled) welded tubing are used as rollers in hearth furnaces in the handling of ceramic components. The tight oxide films, which are resistant to spalling, maintain a clean furnace environment with little or no contamination of the ceramic components. Likewise, N06600, N06601, and N06230 are used for their high temperature corrosion resistance as wire strand annealing tubes.

Molten Salt Corrosion

Corrosion by molten salts is important in the heat treating, metal processing, and metal refining industries, as well as nuclear and solar energy systems, where salts may provide media for heat transfer and energy storage. Usually corrosion proceeds by oxidation followed by fluxing and dissolution of the oxidized species in the melt, not unlike fuel ash corrosion. Consequently, salts containing little or no oxidant are generally not corrosive and metals are stable in the salt. Certain types of attack also show similarities to molten metal corrosion involving thermal gradient mass transfer and impurity reactions.

Chlorides of sodium, potassium, and barium are commonly used in neutral salt baths providing temperatures in the range of 1400-1800°F (760-980°C) suitable for annealing and normalizing steels. Alloys having a high nickel content with less than 20% chromium generally offer the best corrosion resistance, although somewhat erratic alloy behavior has sometimes been observed, possibly due to "aging" of the salt during usage, or the presence of impurities. N06600 has been used for protector tubes in the electrical heating of sodium chloride-potassium chloride salt baths up to 1470°F (800°C) and also for vessels holding molten zirconium tetrachloride (Zr Cl$_4$) in the production of zirconium.

For the lower heat treatment temperatures of 375-1100°F (190-590°C), molten salt mixtures of sodium or potassium nitrate and nitrites are used. Carbon or low alloy steels have proved adequate for containing these salts up to 800°F (430°C), but at higher temperatures austenitic stainless steels or high nickel alloys, such as N06600, are needed. In a mixture of sodium nitrate and potassium nitrate at 1250°F (675°C), corrosion resistance also improved with increasing nickel content in the alloy, although pure nickel suffered severe attack.

Molten fluorides and also hydrogen fluoride (see later) are among the most aggressive salts due to their strong fluxing action on chromium oxides, for which most heat resisting alloys rely on for protection. N10003 with 7% chromium has provided the best corrosion resistance in a number of nuclear applications involving temperatures up to 1380°F (750°C), but at lower temperatures stainless steels and N06600 become candidate materials for handling fluoride salts. Nickel 201 (N02201) has also generally shown good resistance to fluoride salts up to 1300°F (700°C).

The excellent resistance of nickel to aqueous caustic soda solutions persists to the molten anhydrous salt up to at least 1075°F (580°C),

Figure 7.7. Above 600°F (315°C), however, the low carbon grade version N02201 is the preferred choice to prevent graphitization and possible embrittlement. In molten caustic soda above 1075°F (580°C), nickel is subject to thermal gradient mass transfer, and in circulating systems corrosion rates may be many times higher than static conditions (Table 7.16).

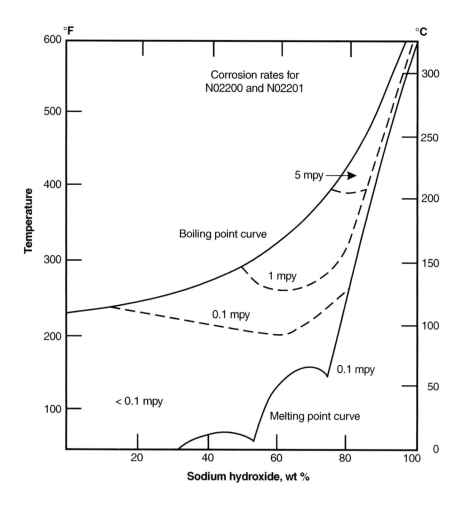

Figure 7.7 Corrosion rates for nickel N02200 (alloy 200) and low carbon nickel N02201 (alloy 201) in sodium hydroxide.[23]

N06600 (alloy 600) is not as resistant as nickel to anhydrous sodium hydroxide, but is usually favored when higher strength is required or sulfur-bearing contaminants are present, being less prone to formation of the low melting eutectic of nickel-nickel sulfide.

Table 7.16 Static Corrosion Rates of Nickel and
Nickel Alloys in Molten Caustic Soda

| Alloy | UNS No. | Corrosion Rate (mils/year) | | | |
		400°C (750°F)	500°C (932°F)	580°C (1076°F)	680°C (1256°F)
Nickel 201	N02201	0.9	1.3	2.5	37.8
Alloy C	N30002	N.T.	100.5	a	N.T.
Alloy D	---	0.7	2.2	9.9	b
Alloy 400	N04400	1.8	5.1	17.6	N.T.
Alloy 600	N06600	1.1	2.4	5.1	66.4
Duranickel alloy 301	N03301	1.7	3.2	10.4	40.7
Alloy 75	N06075	1.1	14.3	20.8 (pitted)	47.6

N.T. = not tested
a = Gained weight. Swollen outside surface largely oxide heavily corroded.[24]

Molten Metal Corrosion

In common with other metal systems, nickel and nickel alloys may undergo corrosion attack in molten metals by simple solution or inward diffusion of the molten phase, either uniformly on a broad front or selectively along grain boundaries. Variants of these basic interactions may also occur through mass transfer caused by the existence of temperature gradients or concentration gradients in the system, or by impurities such as carbon, oxygen, and nitrogen in the molten metal.

Although solution rates are strongly dependant on temperature, nickel and high nickel alloys in general have poor resistance to molten metals because of their relatively high solubility. Hence, they are not normally considered suitable for use as containment materials for handling molten aluminum, tin, magnesium, zinc, bismuth, or antimony. In molten lead, Ni-Cr, and Ni-Cr-Fe alloys have shown useful resistance up to about 1100°F (600°C) but are severely attacked at higher temperatures and in the presence of oxygen or lead oxide. Low melting point alloys of lead, tin, and bismuth are also corrosive and may cause cracking of nickel alloys in the stressed state due to intergranular penetration of the molten phase. While not immune to attack by molten copper and copper alloys, N06075 and N07080 have been used with success as dies or die inserts for casting of aluminum bronze at pouring temperatures of 2100°F (1150°C).

Corrosion rates of nickel and Ni-Cr alloys such as N06600 in molten sodium are suitably low at temperatures up to about 1200°F (650°C) but increase significantly above 1300°F (700°C) particularly under flowing conditions where temperature gradients may give rise to mass transfer effects (Table 7.17). High levels of oxygen in the sodium also tend to increase corrosion rates, whereas carbon may cause carburization, albeit less severe than for stainless steels. The corrosivity of potassium and sodium-potassium mixtures is generally comparable to that of sodium. However, molten lithium is far more aggressive and nickel-chromium alloys are subject to mass transfer at temperatures above 800°F (425°C).

Table 7.17 Corrosion Rates of Materials Exposed to Dynamic Sodium
(30-40 ft/s) Containing 25-30 ppm Oxygen at 650°C for 8 Weeks.[25]

Material	Loss in Metal section	
	mpy	mm/a*
Nickel	3.9	0.1
N06600	6.1	0.15
N07080	4.9	0.12
S31600	1.8	0.05

*Converted data.

Ni-Cr alloys, such as N06600 and N06075, are able to handle molten mercury up to 930°F (500°C) without any indication of intergranular attack or loss in strength. Ni-Cu alloys, such as N04400, on the other hand, are sensitive to stress cracking in mercury, even at moderate temperatures, and if adopted should be stress relieved without subsequent removal of the resulting surface oxide film.

Fuel Ash and Molten Salt Corrosion

Many of the corrosion reactions in gas turbines, boilers, superheaters, and waste incinerators involve ash and salt deposits derived from combustion products and impurities in the fuel such as S, Cl, Na, and V. Often these deposits are extremely aggressive and not only interfere with the development of protective oxide films but also cause fluxing or rapid dissolution of the underlying metal, particularly when present in the molten state.

So called "hot corrosion" of gas turbine components is linked with deposition of sodium sulfate and is usually more severe in land base turbines burning lower grade fuel oils or in marine situations where ingested salt may play a part. Alloys with less than about 10-15% chromium, designed to allow greater additions of the major age-hardening elements, titanium and aluminum, are particularly prone to attack and normally require coatings, such as those based on aluminides or MCrARY systems, for protection. A series of wrought alloys of the Nimonic type but with higher chromium contents were specifically developed to resist hot corrosion, but so far these alloys have found only limited application in gas turbines compared with cast alloy counterparts such as IN 738LC and IN 792 or cobalt-based alloys.

Chromium is also a key element in providing protective scales against fireside corrosion in the superheater sections of coal fired boilers where metal temperatures may reach up to 1200°F (650°C). Corrosion in this instance is caused by a liquid alkali-iron trisulfate phase forming on the tube surfaces. While low alloy Cr-Mo steels and austenitic types such as S31009 (type 310H), and S34709 (type 347H) are often suitable for superheaters and reheaters, the nickel-base INCONEL alloy 671 containing 50% Cr has provided significantly longer lives, particularly at higher temperatures, for example as tube supports.

Corrosion in oil-fired boilers is frequently more acute than in coal ash environments, especially with the use of low grade, high vanadium fuel oils. Vanadium pentoxide and alkali metal sulfates are the principal constituents responsible for the accelerated attack, forming a series of complex vanadates with low melting points in the range 1000-1240°F (535-672°C). Thus serious corrosion problems can be expected above 1050°F (565°C). Again alloys of high chromium content provide the best corrosion resistance, although a high nickel content also appears to be of benefit in environments having a high V/S ratio. Hence, 50 Ni-50 Cr alloys or INCOCLAD® 671 co-extruded tubing are generally well suited for superheaters and other similar applications where residual fuel ash corrosion is the overriding factor. For tube supports, spacers, and hangers, cast Ni-Cr alloys with chromium contents ranging from 35-60% may be used.

Alternatively metal spraying or weld overlaying with high chromium consumables may be used to provide protection to these molten slags. These techniques are especially suited for protection of critical areas which are particularly susceptible to attack. Moreover, in all these environments, significant improvements in component life can be achieved by reducing the design operating temperature or by using fuel additives, such as MgO, which raises the melting point of any deposits formed above the metal surface temperature.

Nickel Alloys for Automotive Applications

Nickel alloys are used in a variety of applications in the automotive industry due to their attractive high temperature mechanical properties and excellent corrosion resistance. The main applications for nickel alloys are in spark plug and glow plug manufacture, exhaust valve production, and components of the exhaust system such as manifolds, flexible coupling (bellows), catalytic converters, and mufflers.

Spark plugs provide an electrode gap across which a high voltage discharge can occur. This is typically composed of a round center electrode, consisting of a nickel-chromium alloy outer layer with a copper core, and a rectangular or rhomboidal section ground electrode in a Ni-Cr alloy. These components operate at temperatures of 750-1650°F (400-900°C). The nickel-chromium electrodes resist erosion due to the spark discharge and maintain a precise spark gap for longer times. Addition of silicon and manganese to these alloys improves corrosion resistance. In selecting a spark plug alloy, the thermal conductivity, electrical conductivity, corrosion resistance, and resistance to spark erosion should be considered.

Glow plugs are used as igniters in diesel engines and comprise a central hollow electrode typically in a Ni-Cr-Fe alloy such as N06601. Similar characteristics to spark plugs are required from the glow plug alloy, but it operates under more arduous conditions.

Engine exhaust valves operating at high temperature require high temperature strength, resistance to corrosion from the products of internal combustion, and good wear characteristics. The majority of automotive valves are produced in steel such as 21-4N, but the age-hardenable Ni-Cr alloys such as N07080, INCONEL alloy 751 (N07751), and in some cases N07090, are suitable for applications which require superior performance due to higher operating temperatures and harsher environments.

Automotive exhaust systems provide several applications for high nickel alloys. As the average engine operating temperature rises, there is a trend toward the use of fabricated manifolds in Ni-Cr-Fe and Ni-Fe-Cr alloys such as N06601 and N08810 instead of those made with the conventional ferritic heat resistant S40900 and alloy 444 stainless steel or cast manifolds. The light weight of the wrought manifold also allows the catalytic converter which is positioned close to the manifold to heat to its operating temperature quicker.

Flexible coupling or bellows provide a shock absorbing function for the exhaust system in its connection to the engine. The material used here needs to have excellent fatigue characteristics accompanied by good resistance to high temperature and salt corrosion. Steels such as S32100 are commonly used at present, but with more stringent requirements for emissions control worldwide, including the requirement for the exhaust systems to remain leak-free for 100,000 miles or 10 years, nickel alloys are being used in greater quantities. Ni-Cr-Fe alloys such as N06600 and N06601 are used for some applications. A major leap in bellows performance has been achieved from using the Ni-Cr-Mo alloy, alloy N06626. This alloy, which is especially well suited for bellows applications, provides improved low cycle thermal fatigue characteristics along with the excellent aqueous corrosion properties of the conventional N06625. INCOLOY alloy 864™ (S35135) was developed specifically for flexible coupling and other exhaust applications. This material provides much better performance than the stainless steels and at a lower cost than alloy N06626.

The bellows unit links the manifold to the catalytic converter, where again nickel alloys are used for several applications. An oxygen sensor sheathed in an alloy such as N06601, is positioned at the entrance to the catalyst. The catalyst unit usually comprises a ceramic monolith onto which the catalytic agent is coated. However, as the operating temperature of these units increases, there is an opportunity for using metal monolith catalysts. These are made by coiling fine, corrugated strip in alloys such as Fe-Cr-Al and, to a lesser extent Ni-Cr-Fe-Al alloys like N06601 and alloy N06025.

The ceramic monolith needs packing and positioning within a steel casing, and several nickel alloys provide the good high temperature corrosion resistance required for this application. Alloys such as N06601 and N07750 in the form of knitted mesh supports produced from fine wire prove to be very efficient. N06601 again finds several such uses for internal metal parts of the catalyst which also need to withstand the high temperature exhaust products.

Corrosion Properties in Aqueous and High Temperature Environments

Various corrosion mechanisms can cause failure of metals during atmospheric, aqueous, or high temperature service. Certain alloying additions are used in nickel-base alloys to provide resistance to these forms of corrosion. A summary of alloying additions and their effects on high temperature and aqueous corrosion resistance is provided in Table 7.18.

Table 7.18 Effects of Alloying Addition on Corrosion Resistance

Nickel	Carbon
<ul><li>Improves high temperature strength</li><li>Improves resistance to oxidation, nitridation, carburization, and halogenation</li><li>Detrimental to sulfidation</li><li>Improves metallurgical stability</li><li>Improves resistance to reducing acids and caustic</li><li>Improves resistance to stress corrosion cracking</li></ul>	<ul><li>High temperature strength</li></ul>
Aluminum	**Silicon**
<ul><li>Improves oxidation and sulfidation resistance</li><li>Age-hardening component</li></ul>	<ul><li>Improves oxidation and carburization resistance, especially under alternating oxidizing and reducing conditions</li></ul>
Iron	**Cobalt**
<ul><li>Improves economy of alloy</li><li>Controlled thermal expansion addition</li></ul>	<ul><li>Improves high temperature strength and oxidation resistance</li><li>Controlled thermal expansion addition</li></ul>
Rare Earth Elements	**Molybdenum**
<ul><li>Improves adherence of oxide layer</li></ul>	<ul><li>Improves high temperature strength</li><li>May reduce oxidation resistance</li><li>Improves resistance to pitting and crevice corrosion</li><li>Improves resistance to reducing acids</li></ul>
Copper	**Nitrogen**
<ul><li>Improves resistance to reducing acids and salts (Ni-Cu alloys)</li><li>Improves resistance to sulfuric acid.</li></ul>	<ul><li>Can improve high temperature strength by precipitation of stable nitrides</li><li>Improves pitting and crevice corrosion resistance</li><li>Improves metallurgical stability by acting as an austenitizer</li></ul>
Niobium	**Titanium**
<ul><li>Improves high temperature strength</li><li>Improves resistance to pitting</li><li>Carbide stabilization component</li></ul>	<ul><li>Detrimental to oxidation due to titanium oxides disrupting primary oxide scale</li><li>Age-hardening component</li><li>Carbide stabilization component</li></ul>
Sulfur	**Tungsten**
<ul><li>Improves machinability</li></ul>	<ul><li>Improves resistance to pitting and crevice corrosion</li><li>Improves high temperature strength</li></ul>
Chromium	**Yttrium Oxide**
<ul><li>Improves oxidation and sulfidation resistance</li><li>Below 18% can improve resistance to halogens or high temperature halides</li><li>Improves aqueous corrosion resistance</li></ul>	<ul><li>Improves high temperature strength</li><li>Improves grain size control</li><li>Improves resistance to oxidation</li></ul>

Aqueous Corrosion

General corrosion is the uniform thinning of a metal component due to corrosive attack. All nickel alloys form a passive oxide surface film when exposed to air which will usually prevent significant corrosion under milder conditions, such as atmospheric exposure and immersion in low temperature neutral solutions. The effectiveness of this oxide film in resisting corrosion in more aggressive environments is enhanced by the addition of alloying elements such as chromium and molybdenum.

As shown in Table 7.18, alloying with chromium is helpful for oxidizing solutions, while molybdenum additions increase resistance to reducing conditions. Additions of both chromium and molybdenum in the same alloy provide resistance to both oxidizing and to reducing conditions.

Localized corrosion (pitting and crevice corrosion) is a common failure mechanism for aqueous environments. The presence of a high level of chloride or other halide contamination along with acidic and oxidizing conditions, significantly increases pitting and crevicecorrosion attack. Alloying with chromium, molybdenum, and tungsten improve resistance to this form of corrosion. Thus, nickel alloys containing higher levels of these alloying elements, such as alloys N10276, N06022, N06686, and N06059 can resist these forms of attack under all but the most extreme environmental conditions. ASTM Test Method G 48 is often used to evaluate nickel alloys for resistance to localized corrosion.

Stress corrosion cracking (SCC) is the combined action of corrosion by a specific environment and tensile stress to produce a brittle fracture in an otherwise ductile material[27,28]. Cracking may be either intergranular or transgranular in nickel alloys. The most common contaminants causing SCC are chlorides and caustic. In general, both chloride and caustic SCC resistance of nickel alloys increases with increasing nickel content.

Ni-Fe-Cr alloys containing greater than about 40% nickel are often used to replace stainless steels which have failed by chloride SCC. At elevated temperatures above about 300°F (149°C), caustic SCC is possible in many nickel alloys, though N02200 (99% nickel) is resistant to even molten caustic. Hydrofluoric acid has been found to cause SCC of Ni-Cu and Ni-Cr-Fe alloys, and Ni-Cr-Fe alloys have failed by intergranular SCC in the high temperature, high purity water environments of nuclear steam generators. High temperature, hydrogen sulfide containing chloride environments may also cause cracking of nickel alloys.

Intergranular corrosion (or intercrystalline corrosion) is preferential corrosion at or along the grain boundaries of a metal. Intergranular attack (IGA) usually occurs in highly oxidizing environments. Ni-Fe-Cr or Ni-Cr-Fe alloys with high carbon levels, can be *sensitized* and made vulnerable to IGA as a result of chromium depletion, caused by grain boundary precipitation of chromium carbides during exposure at temperatures between about 1100°F (593°C) and 1400°F (760°C).[29] Minor additions of titanium or columbium act as carbide stabilizers by tying up the carbon to prevent unwanted grain boundary chromium carbide precipitation.

Highly alloyed Ni-Cr-Mo alloys may become susceptible to intergranular corrosion after exposure to temperatures between about 1200°F (649°C) and 1900°F (1038°C) due to grain boundary precipitation of intermetallic phases such as mu (μ). Higher purity, lower carbon and silicon is helpful but not entirely effective in preventing this precipitation. ASTM Test Methods A 262[30] and G28[31] are commonly used to evaluate nickel alloys for resistance to intergranular corrosion.

High Temperature Corrosion

Oxidation is the dominant mode of high temperature corrosion. Oxidation results in the formation of an oxide layer, the composition of which depends very much on the alloy chemistry. In many cases the oxide will be a combination of nickel oxide with oxides of elements such as chromium and aluminum. This layer may provide a barrier to further oxidation as it grows thicker. The rate of scale growth is generally proportional to the temperature.

The majority of nickel-based alloys used for high temperature oxidation resistance contain chromium, aluminum, or silicon which form protective oxide layers, Cr_2O_3, Al_2O_3, or SiO_2 usually above a thinner nickel oxide (NiO) layer. The Cr_2O_3 is stable up to about 1650°F (900°C), above which it begins to vaporize as CrO_3. The Al_2O_3 or SiO_2 layers are more stable at higher temperatures up to 2200°F (1200°C). Hence Ni-Cr-Fe-Al alloys such as N06601, N06025, and N07214 provide better oxidation resistance at higher temperatures than the Ni-Fe-Cr alloys such as N08811, N08330, and alloy DS.

It is also possible to have internal oxidation of certain alloy when oxygen diffuses into the alloy and combines to form oxides below the original metal surface. The amount of oxidation should be gauged by the depth of metal affected by internal oxidation in combination with the overall metal loss.

Small quantities of rare earth elements e.g. yttrium, cerium and lanthanum, improve the oxide adhesion to the metal under conditions of cyclic oxidation. This is the case with alloys such as the Ni-Cr electrical resistance alloys. Under cyclic conditions an oxide can form and then spall off when cooled to expose fresh metal which will again oxidize. The cumulative effect of this process is gradual metal loss from the component. The improved adherence of the oxide resulting from rare earth additions can hence extend service life of the alloy by preventing or reducing the spalling of the oxide layer. The cyclic oxidation resistance of a range of alloys is shown in Figure 7.8.

Sulfidation is a high temperature corrosion mechanism resulting in the formation of a sulfide scale which grows faster than oxide scales and which is typically more volatile and less protective than an oxide scale.

Sulfur is found in many fuels such as coal, gas, and oil. It leads to problems when these fuels are combusted, producing gases such as H_2S, SO_2, and SO_3. In pure nickel alloys this can be a serious concern under reducing conditions since sulfur will combine with nickel to form a low melting point (1200°F (650°C)) Ni/Ni$_3$S$_2$ eutectic which will result in rapid corrosion.

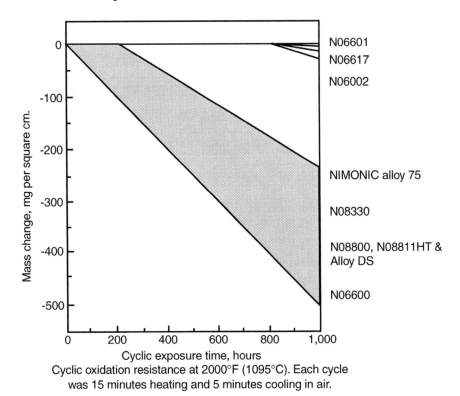

Cyclic oxidation resistance at 2000°F (1095°C). Each cycle
was 15 minutes heating and 5 minutes cooling in air.

Figure 7.8 Cyclic oxidation resistance at 2000°F (1095°C).
15 minutes heating and 5 minutes air cooling per cycle.[33]

The relative sulfidation resistance of nickel-based alloys is particularly dependant on the oxidizing potential, i.e., whether the environment is oxidizing or reducing, in nature.[24] Ni-Fe-Cr alloys with a high iron content tend to perform well under reducing conditions due in part to the Fe-FeS eutectic which melts at 1805°F (985°C). Alloys such as N08800, alloy DS, and N06045 tend to provide good resistance. Stainless steels such as S31000 and the Ni-Cr-Fe alloy N06601, due to its aluminum content will also provide near comparable resistance to corrosion. In general, high iron contents, aluminum or silicon additions, and chromium contents higher than 15% are required for good corrosion resistance under these conditions.

Under oxidizing-sulfidation conditions the corrosion rates are slightly lower than under reducing conditions, and alloys with a high iron content and over 20% chromium tend to perform best. When different amounts of chromium are added to nickel, varying types of scales form, depending on the chromium content. Like many sulfides, the sulfide layer on Ni-Cr alloys will be duplex in nature, comprising a mixture of sulfides of the elements present in the form of compact outer layer and a porous inner layer. Here again the Ni-Fe-Cr alloys such as N08800 and alloy DS and some stainless steels tend to perform well.

Halogenation - Nickel alloys are relatively more resistant to chlorination than iron-based alloys.[35] Unlike many of the metal halides which form on other metals, such as iron or aluminum, $NiCl_2$ has a relatively high melting point. Nickel alloys, such as Nickel 201, have the best resistance to chlorination in mainly chlorine atmospheres. The resistance to chlorination reduces as other alloying elements are added. However, alloys such as N06600 which has a relatively high nickel content and the 80/20 Ni-Cr alloys also provide reasonable resistance.

There are many cases where oxygen will also be present with chlorine. Under these conditions Ni-Cr-Al based alloys such as N07214 and N06601 provide the best high temperature chlorination resistance. There is no clear theory which explains this behavior. One can only speculate that an alumina scale of some description provides a significantly better barrier to chlorination.

Corrosion resistance of nickel alloys to fluorine-bearing environments follows a similar pattern to that for chlorine. High nickel alloys such as Nickel 201 provide the best resistance at temperature up to 930°F (500°C), along with alloys such as the Ni-Cu N04400. At higher temperature, alloys such as N06600 become more suitable, in many cases due to their strength retention at these temperatures. Higher chromium content alloys than N06600 are not as effective in hydrogen fluoride gases at elevated temperatures.

In other halogen vapors (Br, HBr, I, etc.), nickel alloys offer some advantage over stainless steel which may be prone to pitting or SCC under aqueous conditions.

Carburization - Material problems associated with carbonaceous atmospheres, such as CO, CH_4 and other hydrocarbons, mostly arise from inward diffusion of carbon and eventual internal carbide formation causing embrittlement and degradation of other mechanical properties.[35]

Additionally, local variations in thermal expansion coefficient and volume increase that accompany carburization may induce internal stresses sufficient to cause cracking or warping (distortion) on furnace tubes etc., particularly in cast heat resisting alloys.

Metal wastage is not usually a problem except for *metal dusting*, another form of carburization attack which occurs at relatively low temperatures of 800-1650°F (430-900°C) and manifests itself as pitting or general thinning.

Since nickel reduces the diffusivity of carbon in Fe-Ni-Cr alloys, high nickel alloys are generally more resistant to carburization as shown by the data in Figure 7.9. Chromium, aluminum, and silicon are also favorable alloying additions against carburization through formation and maintenance of stable compact oxide films which act as a barrier and help restrict ingress of carbon. Depending on oxidizing potential and carbon activity, such oxide scales may be developed under normal operating conditions, or under start-up, or sometimes by pretreatment. Alloys of relatively high aluminum content offering excellent resistance to carburization include N06617 (1.2% Al), N06025 (2.2% Al), and N07214 (4.5% Al). This latter alloy relies on an alumina scale for protection.

The benefits of silicon are generally observed where SiO_2 is able to form, either in the surface scale or as a subsurface reaction product.

Alloys having a relatively high silicon content include N12160 (2.75% Si) and alloy DS (2.3% Si). The latter alloy is particularly effective against "green-rot," a form of deep internal attack involving carburization/oxidation reactions and observed in baskets, trays, mesh belts, and other heat treatment fixtures exposed to atmospheres that are part reducing and part oxidizing.

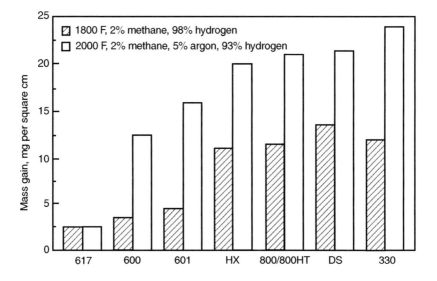

Figure 7.9 Resistance to gas carburization at 1800°F (980°C) and 2000°F (1090°C). Test duration of 100 hours.

<u>Nitridation</u> - Ammonia-bearing and nitrogen-base atmospheres are widely employed in the heat treatment, sintering, and metal processing industries and may cause nitridation of the materials of construction for retorts, muffles, belts, and jigs used in these processes. As with carburization, metal wastage is not normally involved, but nitrogen pick-up leads to internal nitride precipitation causing brittle behavior. Nickel is a very useful alloying element since it does not form stable nitrides and the diffusivity of nitrogen in nickel is low. On the other hand, elements such as chromium, iron, titanium, and aluminum readily form nitrides and are not generally useful alloying additions, at least in oxygen-free environments.

In ammonia-bearing environments various austenitic stainless steels have proved suitable for processing equipment, but under more hostile conditions of higher temperature and higher ammonia concentrations, nickel-base alloys are used, with N06600 often showing the best overall performance. Despite good nitriding resistance, however, the alloy is not normally considered for highly stressed components in ammonia synthesis because of possible hydrogen embrittlement.

Alloys such as N06601 (1.2% Al), generally tend to suffer from formation of internal chromium and aluminum nitrides, but even very low levels of oxygen or moisture in the atmosphere are known to retard nitridation by preferential oxide film formation. N07214 is also reported to be resistant to nitridation if a stable oxide film of alumina is present. N12160 with 29% Co and 2.75% Si also provides substantially better nitridation resistance to ammonia than stainless steels and a number of higher alloyed iron- and nickel-base materials. In the production of nitric acid, N06230 and N06617 have found application for catalyst support grids. Exposure to pure nitrogen tends to give rise to nitridation at temperatures above 1800°F (980°C), but N06600 still performs well.

Summary - Guidelines for Selection

Though the various nickel alloys were usually designed for specific applications, most have found multiple uses. In many cases, the same alloy may be used in both aqueous and high temperature environments in completely different industries. Thus it is difficult to define general rules for nickel alloy selection. Individual alloys used for various applications are summarized in Table 7.19. The effect of elemental additions on corrosion resistance are outlined in Table 7.18. It should be noted that many corrosion-resistant nickel alloys are also available in age-hardened versions, although the age-hardening process can reduce the corrosion resistance to some extent.

Most of the nickel-based alloys have in common their relative ease of fabrication and weldability. They can be fabricated by standard hot and cold deformation processes, though hot forming temperature range and work-hardening rate may vary. Annealing and age-hardening temperatures are also specific to each alloy. This information is available from suppliers. Information on welding and fabrication of nickel alloys is provided in Chapter 8 of this book.

This has been a brief overview of the more common corrosion-resistant nickel alloys and their more important applications. Details pertaining to specific alloy properties and performance data in various applications can be found in the referenced literature.

Tabular Summary of Alloys, General Properties, Applications, Compositions, and Specifications

Table 7.10 Corrosion Resistant Alloys, Properties and Typical Applications

UNS No.	Common Name	General Properties	Applications
N02200	Nickel 200	resistant to reducing acids, strong bases, high electrical conductivity	chemical processing, electronics
N02201	Nickel 201	same as Nickel 200 but also resists embrittlement at elevated temperature	chemical processing, electronics
N02211	Nickel 211	electrical	electronics
N04400	Alloy 400	resistant to a wide range of moderately aggressive environments, higher strength than nickel	chemical processing, power, marine
N05500	Alloy K-500	high strength, age hardenable, corrosion similar to alloy 400	pump shafts, impellers, springs, bolting, valve trim
N06002	Alloy HX	oxidation resistance, high temperature strength	thermal processing, aerospace and land based gas turbines
N06003	Alloy 80-20	cyclic oxidation resistance	resistance heating
N06004	Alloy 65-15	oxidation resistance	resistance heating braking resistors
N06022	Alloy C-22	resists highly aggressive oxidizing and reducing aqueous environments, resists localized corrosion	chemical process, pollution control, pulp and paper
N06022	Alloy 622		
N06025	Alloy 602CA	resistance to oxidizing atmospheres, high temperature strength, and creep properties	thermal processing, automotive
N06030	Alloy G-30	resists highly oxidizing environments, phosphoric acid	chemical process, phosphoric acid plants, pickling operations, nuclear waste processing
N06045	Alloy 45 TM	sulfidation, oxidation resistance	thermal processing
N06059	Alloy 59	resists highly aggressive oxidizing and reducing aqueous environments, optimum resistance to localized corrosion	chemical process, pollution control
N06075	Alloy 75	oxidation resistance	aerospace thermal processing, LBGT
N06200	Alloy C-2000	resists highly aggressive oxidizing and reducing aqueous environments, advanced general corrosion	chemical process
N06230	Alloy 230	high temperature strength, oxidation, and nitridation resistance	thermal processing

Table 7.10 Corrosion Resistant Alloys, Properties and Typical Applications (continued)

UNS No.	Common Name	General Properties	Applications
N06455	Alloy C-4	resists highly aggressive aqueous environments, resists weld heat affected zone attack	chemical process
N06600	Alloy 600	resistant to reducing environments, high temperature strength, and oxidation resistance	chemical process, heat treating, power, electronic
N06601	Alloy 601	resistance to oxidizing atmospheres, high temperature strength, and creep properties	thermal processing, automotive, chemical
N06601	Alloy 601GC	as above plus high temperature stability	thermal processing
N06617	Alloy 617	high temperature strength and creep properties. Oxidation resistance	thermal processing, aerospace
N06625	Alloy 625	resistant to many aqueous environments, high temperature strength	aerospace, chemical process, marine, automotive
N06626	Alloy 625 LCF	resistant to low cycle fatigue and many aqueous environments, high temperature strength	bellows, aerospace, automotive, chemical process
N06686	Alloy 686	resists highly aggressive oxidizing and reducing aqueous environments, optimum resistance to localized corrosion	chemical process, pollution control
N06950	Alloy G-50	hydrogen sulfide SCC resistance	oil and gas production
N06985	Alloy G-3	aqueous corrosion resistance in a variety of environments	chemical process and deep sour gas wells
N07001	WASPALOY	age hardened, high temperature strength, and oxidation resistance	aerospace
N07031	Pyromet 31	age hardened, high temperature strength, sulfidation and oxidation resistance	Automotive
N07080	Alloy 80A	age hardened, high temperature strength, sulfidation and oxidation resistance	automotive, aerospace, LBGT
N07090	Alloy 90	age hardened, high temperature strength	automotive, springs, aerospace
N07214	Alloy 214	oxidation resistance	thermal processing
N07716	Alloy 625Plus	resistant to many aqueous environments, high low temperature strength, age hardened	oil and gas production, marine
N07718	Alloy 718	age hardened, high temperature corrosion resistance	aerospace and based gas turbines, oilfield, springs
N07725	Alloy 725	resistant to many aqueous environments, high low temperature strength, age hardened	oil and gas production, marine
N07750	Alloy X-750	age hardened	automotive, spring

Table 7.10 Corrosion Resistant Alloys, Properties and Typical Applications (continued)

UNS No.	Common Name	General Properties	Applications
N07751	Alloy 751	age hardened, oxidation resistance	automotive
N07754	Alloy MA754	mechanically alloyed, high temperature strength	aerospace
N08020	Alloy 20Cb3 (Alloy 20)	resists wide range of corrosive environments and intergranular sensitization	chemical process, oil and gas recovery, acid production, radioactive waste handling
N08330	Alloy 330	resistance to oxidizing/reducing atmospheres	thermal processing
N08800	Alloy 800	resists oxidation, moderately corrosive conditions	heat exchanger tubing, power, heat treating, domestic appliances, petrochemical
N08810	Alloy 800H	high temperature strength	thermal processing petrochemical
N08811	Alloy 800HT	high temperature strength, creep and oxidation resistance	thermal processing petrochemical
N08825	Alloy 825	resists wide range of corrosive environments and intergranular sensitization	chemical process, oil and gas recovery, acid production, radioactive waste handling
N09925	Alloy 925	resists wide range of corrosive environments, high low temperature strength, age hardened	oil and gas production, marine
N10276	Alloy C-276	resists highly aggressive aqueous environments, resists localized corrosion	chemical process, pollution control, oil and gas recovery
N10665	Alloy B-2	resists highly reducing environments such as high temperature, concentrated hydrochloric acid	chemical process
N10675	Alloy B-3		
N10629	Alloy B-4		
N12160	Alloy HR-160	nitridation resistance	thermal processing
R30155	MULTIMET	oxidation resistance	thermal processing
R30188	Alloy 188	sulfidation resistance	thermal processing
R30556	Alloy 556	sulfidation, carburization, and oxidation resistance	thermal processing chemical
S35135	Alloy 684	resists hot salt corrosion, fatigue, oxidation and aqueous corrosion	automotive exhaust and other high temperature applications
---	Nickel 212	electrical	electronics
---	Nickel 240	electrical	automotive spark plugs
---	Alloy 671	Resistance to high sulfur coals and vanadium fuel oils	Superheater tubing

Table 7.10 Corrosion Resistant Alloys, Properties and Typical Applications (continued)

UNS No.	Common Name	General Properties	Applications
---	Alloy DS	resistance to oxidizing/reducing atmospheres	thermal processing, resistance heating, braking resistors
---	Alloy 803	oxidation, sulfidation resistant	thermal processing, petrochemical
---	Alloy 70	age hardened	automotive
---	Alloy 101	age hardened	automotive, aerospace
---	NIMONIC 86	high temperature oxidation resistance	aerospace, thermal processing
---	INCOCLAD 671	fuel ash corrosion resistance	thermal processing
---	NIMONIC 105	age hardened, high temperature strength, and corrosion resistance	aerospace
---	Alloy PE16	controlled thermal expansion	nuclear fuel rods

References

1.   ASM Handbook, Vol. 2: Properties and Selection: Nonferrous Alloys and Special Purpose Materials.

2.   The Nimonic Alloys, W. Betteridge and J. Heslop, Edward Arnold, 1974.

3.   "Twentieth Century Building Materials – History and Conservation" Derek H. Tielsfad, Library of Congress Cataloging in Publication Data, ISBN-0-07-032573-1, 1995.

4.   Corrosion Control in the Process Industries, C. P. Dillon, McGraw Hill, New York, 1992.

5.   Effect of Composition on Stress Corrosion Cracking of Some Alloys Containing Nickel, H. R. Copson, "Physical Metallurgy of Stress Corrosion Fracture," T. Rhodin (ed.), Interscience Publishers, Inc., New York, 1959.

6.   B. D. Craig, "Selection Guidelines for Corrosion Resistant Alloys in the Oil and Gas Industry," NiDI Technical Series No. 10 073, Toronto, Ontario, Canada, July, 1995.

7.   R. H. Moeller, et. al., "Large Diameter Cold-worked C-276 for Downhole Equipment," CORROSION/91, paper no. 30, NACE International, Houston, TX, USA, 1991.

8.   Standard Materials Requirement MR0175, "Sulfide Stress Cracking Resistant Metallic Materials for Oilfield Equipment," NACE International, Houston, TX, USA, 1998.

9.   R. B. Bhavsar and E. L. Hibner, "Evaluation of Testing Techniques for Selection of Corrosion Resistant Alloys for Sour Gas Service," CORROSION/96, paper no. 59, NACE International, Houston, TX, USA, 1996.

10.   L. M. Smith, et. al., "Material Selection for Gas Processing Plant", *Stainless Steel Europe,* pp. 21- 31, Jan./Feb. 1995.

11.   Draft #1, "Sulfide Stress Cracking and Stress Corrosion Cracking Resistant Metallic Materials for Oilfield Equipment," NACE International, Houston, TX, USA, 1997.

12.   E. L. Hibner, et. al., Effect of Alloy Content vs. PREN on the Selection of Austenitic Oil Country Tubular Goods for Sour Gas Service," CORROSIOIN/98, paper no. 98106, NACE International, Houston, TX, USA, 1998.

13.   Inco Alloys International Inc., Technical Bulletin on "Corrosion Resistant Alloys for Oil and Gas Production."

14.   Field Data from Halliburton Energy Services.

15.   Laboratory Test Data from Inco Alloys International Inc.

16.   NACE Standard Test Method MR0175-99, "Sulfide Stress Cracking Resistance Metallic Materials for Oilfield Equipment".

17.   "New Methods in I. D. Pinned Tubing for High Nickel Alloys" C. S. Tassen, H. H. Ruble, and J. C. England, The Tube and Pipe Quarterly, Vol 1, No 3, Winter 1990, p 11.

18.   "The Corrosion Resistance of Nickel-Containing Alloys in Fine Gas Desulfurization and other Scrubbing Processes." Nickel Development Institute publication #1300

19.   "Life-Cycle Cost Comparison of Alternative Alloys for FGD Components", JD Redmond, RM Darison and YM Shah, Nickel Development Institute publication #10 023.

20.   NACE International Standard RE0292-92, Item #53088, "Installation of Thin Metallic Wallpaper Lining in Air Pollution Control and Other Process Equipment."

21. "Rolled Bonded Nickel Alloy Clad Steel Plate: Proven Performance of FGD Systems" K.E. Orie et al. NACE International Corrosion 93, Paper #422

22. Industrial Waste Incineration Systems Materials Solution and Application Problems, J. J. Santolesi and D. L. Corwin, Chemical Waste Conference, Manchester, UK 1BC Technical Services Ltd. March 1990.

23. "Corrosion Resistance of Nickel and Nickel-Containing Alloys in Caustic Soda and Other Alkalies" Corrosion Engineering Bull CEB-2, International Nickel Co., Inc., 1973.

24. "The Static Corrosion of Nickel and Other Materials in Molten Caustic Soda" J. N. Gregory, N. Hodge, and J. V. Iredale, Atomic Energy Res. Estab. Haswell England, AERE Report C/M 272, March 1958.

25. "Corrosion Behavior of Steels and Nickel Alloys in High Temperature Sodium", A. W. Thorley and C. Tyzack, Proc. of the Symp. on Alkali Metal Coolants. Int. Atomic Energy Agency, Vienna 1966 pp.97-118.

26. ASTM Standard Test Method G-48, "Pitting and Crevice Corrosion Resistance of Stainless Steels and Related Alloys by the use of ferric Chloride Solution."

27. The Stress Corrosion of Metals, Hugh L. Logan, John Wiley and Sons, Inc., New York, 1966.

28. NACE International Glossary of Corrosion Related Terms, 2nd Edition, March 1995.

29. Corrosion Engineering, Mars G. Fontana and McGraw-Hill, 3rd Edition 1986.

30. ASTM Standard Practice A262, "Defecting Susceptibility to Intergranular Attack in Austenitic Stainless Steels."

31. ASTM Standard Practice G28, "Detecting Susceptibility to Intergranular Corrosion in Wrought Nickel-Rich, Chromium Bearing Alloys."

32. "High Temperature Corrosion of Engineering Alloys," G. Y. Lai, ASM International, 1990.

33. "High Performance Alloys for Thermal Processing", Inco Alloys International Inc Publication #HIAI-48-1.

34. "Sulphidation Resistance of High Temperature Alloys Tested in Reducing and Oxidizing Atmospheres," J. C. Hosier and J. A. Harris, *Industrial Heating*, June 1980, pp14-16.

35. "Assessment of the Degration of Metals and Alloys in Air-2% Cl_2 at High Temperature," FH Scott, R Prescott, P. Elliott and MJJ Al'Ahia High Temperature Technology Book (1988) pps 115-121.

36. Carburizaion", HJ Grabke, MTI Book (1988)

Chapter

8

WELD FABRICATION OF NICKEL-CONTAINING MATERIALS

Donald J. Tillack

Tillack Metallurgical Consulting, Inc.

Huntington, West Virginia

Most components used in corrosive environments are fabricated by welding. Even when forgings or castings are used, such as pump housings or valve components, welds are used to make attachment connections, for example, to piping. The heat generated during welding of these corrosion resistant materials can result in metallurgical structure changes that can alter the material's corrosion characteristics. In addition, the welded structure design and the techniques used to weld the structures can impact and alter significantly the useful life of the component.

This chapter discusses numerous classes of alloys that contain nickel and examines the techniques, procedures, and pitfalls that can occur when the materials are joined by welding. Our discussion will cover martensitic, ferritic, and austenitic stainless steels, duplex stainless steels, and nickel alloys. Many times the discussion refers to the carbon and low alloy steels for comparison, mainly because there are so many of them produced and welded compared to the nickel-containing materials.

It is not within the scope of this chapter to discuss the welding processes in any detail. Most of the commonly used welding processes, such as Gas Tungsten Arc Welding (GTAW), Gas Metal Arc Welding

(GMAW), Shielded Metal Arc Welding (SMAW), Submerged Arc Welding (SAW), Pulsed Arc Welding (PAW), etc., can be used to weld the higher alloy materials. Information on the welding processes can be obtained from American Welding Society publications.

Welding Characteristics of Austenitic Stainless Steels and Nickel Alloys

Heat penetration and weld metal viscosity are two major differences between the nickel-containing materials and carbon/low alloy steels and influence how a weld joint is designed, prepared, and welded.

Heat penetration differences are caused by the fundamental characteristics of the metals. Carbon steels have a greater ability to dissipate heat than nickel-chromium alloys, with austenitic stainless steels falling between the two, as shown in their thermal conductivity. Table 8.1 shows the differences between carbon steel (AISI 1030), austenitic and ferritic stainless steels, and nickel alloys. This influences the way in which a weld joint is prepared, especially when complete penetration is required. The nickel-based alloys vary considerably in thermal conductivity values. The nickel-chromium-iron alloys have values lower than those of either carbon steel or austenitic stainless steel, but the nickel and nickel-copper alloys have conductivity values considerably higher than those of steels.

Table 8.1 Thermal Characteristics of Selected Metals

Material	Thermal Conductivity @ 70°F (Btu/ft^2/hr/in./°F)	Thermal Expansion, 70-1000°F in./in./°F x 10^{-6}
AISI 1030	35	6.5
S30400	9.4	10.2
S31000	8.2	9.4
S31803	132	7.4
S43000	15.1	6.3
N02200	520	8.5
N04400	151	9.1
N06600	103	8.4
N10276	71	7.4
N08800	80	9.4

A key difference between carbon steel and austenitic stainless steel is their differing thermal expansion and ability to dissipate heat. A method of evaluating a metal's heat dissipation ability is to calculate its *thermal diffusivity*, by dividing its thermal conductivity by the product of its specific heat and density. The thermal diffusivity of austenitic stainless steels is significantly lower than carbon steels or ferritic/martensitic stainless steels, which restricts the migration of heat away from the weld zone.

Since austenitic stainless steels thermally expand about 1.4 times that of carbon steels, and with higher heat concentration from relatively poor thermal diffusivity, a sharper stress gradient is produced in austenitic stainless steel weldments. This leads to increased distortion when welding austenitic stainless steels, as compared to carbon steels.

The level of residual stress produced in austenitic stainless steel welds can have a significant effect on the metal's stress corrosion cracking (SCC) potential. Unfortunately, this relationship is typically not addressed in welding procedure qualification codes and standards, but should always be considered when preparing or evaluating an austenitic stainless steel welding procedure.

Since stress gradients are very difficult to measure, care must be taken when welding austenitic stainless steels to minimize their effect, especially if they are to be placed in a service environment where stress corrosion cracking is a possibility. Lowering the heat input of welding will improve the stress corrosion cracking resistance of austenitic stainless steel welds. Some common means of lower the heat input of welding are shown in the following list.

a) Use smaller diameter electrodes; particularly for maintenance repairs where large diameter electrodes are commonly used to reduce the time for repair, but at an increase cost of making a weld more susceptible to premature failure.

b) Increase the travel speed; e.g., use stringer beads rather weave beads, and if weaving is required then restrict the width of the weave, e.g., maximum of 2 times the electrode diameter.

c) Do not preheat austenitic stainless steels.

d) Do not add excess filler metal to the weld joint; large cap passes add to the heat input of welding and also result in larger weld metal volume which increases the residual welding stresses.

e) Restrict the welding interpass temperature to a maximum amount, then carefully measure and document it between each pass.

The most efficient way of minimizing the threat of stress corrosion cracking is to remove the residual stresses with an annealing heat treatment. Since this is not always possible, and may in some cases lead to structural distortion, minimizing the heat input is often the only choice.

The comparisons between recommended joint designs for carbon steel and nickel alloys are shown in Figure 8.1. This difference in penetration is extremely important when welding materials for corrosion service. If a butt joint is only welded from one side and complete penetration is not accomplished, a crevice will be present on the back side of the weld joint, which will be a site for crevice corrosion and will also be a notch for possible fatigue failure.

While there are other causes of incomplete penetration, such as narrow root gaps, and misaligned weld faces, these can be overcome for by proper fit-up and design. However, lack of penetration caused by low arc energies or thermal conductivity must be accounted for by proper design such as by specifying thinner lands.

The viscosity of the molten weld metal is very different when comparing carbon steels and nickel alloys. When welding carbon steel, the molten weld puddle flows readily into the joint, whereas a nickel alloy weld puddle is more sluggish. Therefore, the welder must manipulate the nickel alloy weld puddle, directing it to the sides of the joint, whereas the carbon steel welder has a much easier task since the weld metal flows very freely.

This difference must be accounted for when designing the weld joint by opening up the included angles on a nickel alloy butt joint to allow the welder more room for electrode manipulation during welding. If the included angle is too small, problems can occur such as lack of fusion or lack of penetration.

When welding with the SMAW process, slag entrapment at the tie-in areas can be a problem if the included angle is too small. If slag entrapment does occur, it must be removed by grinding. It cannot be removed by chipping or welding over with the next weld pass.

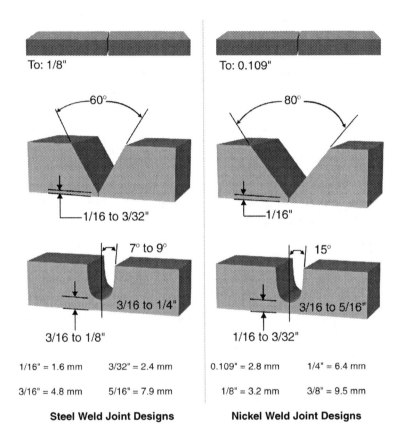

Figure 8.1 Weld joint comparisons between
carbon steels and nickel alloys.

Cleanliness in the weld zone is very important when welding higher nickel alloys because of the ability of nickel and other alloying elements to combine with elements that promote cracking, particularly sulfur and phosphorous. The weld area should be cleaned with clean solvents using clean cloths prior to welding to prevent this problem. Low melting point elements, such as tin and zinc, can form compounds that cause cracking if they are present in sufficient quantities in the weld fusion zone.

Another difference between welding carbon steels and nickel alloys is the ability to weld over the dark oxides that form during hot forming or heat treating. In carbon steel, this dark oxide melts at about the same temperature as the base material. However, the nickel and chromium oxides that form on stainless steels and nickel alloys during exposure to elevated temperatures melt at much higher temperatures, as shown in Table 8.2. Consequently, if these oxides are not removed before welding, they may not be completely absorbed into the molten weld puddle, resulting in the formation of oxide stringers throughout the welds. These oxide stringers may or may not be picked up during x-ray, and they could be the source of early fatigue failure during cyclic service. Therefore, it is recommended that all oxide be removed up to about 1 inch on both sides of the weld joint prior to welding. This is usually accomplished by grinding, but other means can be used, such as machining, blasting, or pickling.

Table 8.2 Melting Temperatures of Selected Metals and Oxides

Metal or Alloy	Melting Temp. °F (°C)	Metal Oxide	Melting Temp. °F (°C)
Iron	2798 (1537)	Fe_2O_3	2850 (1565)
Iron	2798 (1537)	Fe_3O_4	2900 (1593)
Nickel	2650 (1454)	NiO	3600 (1982)
S30400	2550-2650 (1400-1454)	Cr_2O_3	4110 (2266)

Whenever stainless steels or nickel alloys are ground, use only grinding wheels that have been used on other stainless steel or nickel alloy weldments. Do not use grinding wheels that have been used on carbon steel. This practice of using dedicated grinding equipment will minimize the problem that can occur when iron particles are carried over and embedding in higher alloy materials. If these embedded iron particles are not removed prior to putting the vessel into service, accelerated corrosion of these particles can occur. At the least, this will cause staining of the vessels; more seriously, however, these particles can be the site of pitting corrosion.

Welding Martensitic Stainless Steels

Martensitic stainless steels usually require a fairly high preheat during welding because of their high degree of hardenablity. The preheat temperature depends to a great extent on the carbon content, with carbon levels higher than 0.10% requiring a minimum of 400°F (204°C) preheat temperature. Also, with carbon levels higher than 0.20%, the interpass temperature should be maintained until welding is completed, followed by slow cooling. The post weld heat treatment temperature range will vary depending on the composition, the filler metal, and the service requirements, but is generally in the 1200-1400°F (650-760°C) range. Table 8.3 lists suggested preheat, welding heat input, and postweld heat treatment (PWHT) requirements for martensitic stainless steels.

Table 8.3 Suggested Preheat, Welding Heat Input, and Postweld Heat Treatment Requirements for Martensitic Stainless Steels

Carbon Content (%)	Approximate Preheat (°F)	Welding Heat Input	PWHT
< 0.10	60 min.	Normal	Optional
0.10 to 0.20	400-500	Normal	Cool slowly, PWHT optional
0.20 to 0.50	500-600	Normal	PWHT required
> 0.50	500-600	High	PWHT required

The postweld heat treatment is needed to lower residual welding stresses and to temper the heat-affected zone (HAZ) and weld metal. This tempering lowers the hardness and increases the toughness in these areas.

While matching filler metal compositions (listed in Table 8.4) are used to weld the martensitic stainless steels, weldability is improved if austenitic stainless steel filler metals are used. The lower yield strength and better ductility of the austenitic filler metal composition minimizes the strains on the HAZ of the martensitic stainless steel. This approach is possible if design considerations allow for the lower strength of the austenitic welds.

Table 8.4 Filler Metal Selections for Martensitic and
Ferritic Stainless Steels

Base Metals			Filler Metals
UNS No.	Alloy	Condition[a]	Electrode or Filler Rod Type
MARTENSITIC STAINLESS STEELS			
S40300	403	1 or 2	410, 410
S40300	403	3	308, 309, 310
S41600	416	1 or 2	410
S41600	416	3	308, 309, 312
S42000	420	1 or 2	420
S42000	420	3	308, 309, 310
FERRITIC STAINLESS STEELS			
S40500	405	1	405Cb, 430
S40500	405	3	308, 309, 310
S40900	409	1	409, 409Cb
S40900	409	3	308, 309, 310
S43000	430	1	430
S43000	430	3	308, 309, 310
S44600	446	1	446
S44600	446	3	308, 309, 310

a. Condition in which weldment will be placed in service:
    1 = Annealed
    2 = Hardened and stress- relieved
    3 = As-welded

The use of austenitic Ni-Cr and Ni-Cr-Fe welding products result in approximately 30% lower shrinkage stress than stainless steel due to their lower coefficient of thermal expansion. Also, nickel-based welding products allow better evolution of hydrogen since they solidify as primary austenite.

The martensitic stainless steels, similar to the low alloy steels, are subject to hydrogen-induced cracking. Appropriate measures to avoid hydrogen pick-up are necessary to prevent this from occurring. These include using proper bake-out procedures for SMAW electrodes, making sure that the weld area is completely dry, and using hydrogen free shielding gas when using GMAW and GTAW processes.

Welding Ferritic Stainless Steels

Ferritic stainless steels generally cannot be hardened by quenching, which minimizes weld cracking problems associated with martensite formation. However, S43000 (type 430), S43400 (type 434), S44200 (type 442), and S44600 (type 446) ferritic stainless steels are exceptions because of their higher levels of chromium and carbon. These steels can crack if welded under high restraint; therefore, they are usually preheated to 300°F (149°C) or higher during welding to minimize residual stresses caused by welding.

The corrosion resistance of S40500 (type 405) and S40900 (type 409) (low-chromium, aluminum/titanium stabilized ferritic stainless steels) is generally not reduced by the heat of welding. However, the higher chromium and carbon ferritic stainless steels (S43000, S43400, S44200, and S44600) will form carbides in the HAZ, which lowers the corrosion resistance of the material. These alloys are usually annealed after welding to redissolve the carbides and minimize the possibility of intergranular corrosion.

Those ferritic stainless steels that have low levels of interstitial elements [type 444, S44626 (alloy 26-1), S44700 (alloy 29-4), and S44800 (alloy 29-4-2)] do not suffer from grain boundary carbide formation during welding. However, they can be susceptible to the precipitation of intermetallic compounds during welding which can adversely affect corrosion resistance.

One of the biggest problems with welding of the ferritic stainless steels is the lowering of toughness in the HAZ due to grain growth caused by the heat of welding. Since ferritic stainless steels typically contain more than 16% chromium, this significantly decreases the size of the austenite (gamma) loop, and thereby avoids allotropic transformation upon heating. The absence of an allotropic transformation allows the room temperature ferrite grains to continue to grow, while heated by the welding process, without the ability to recrystalize and thus grain growth occurs unabated. Grain growth in ferritic stainless steels can be limited, but not eliminated, by using the same heat input restricting steps listed on page 327, a) to e).

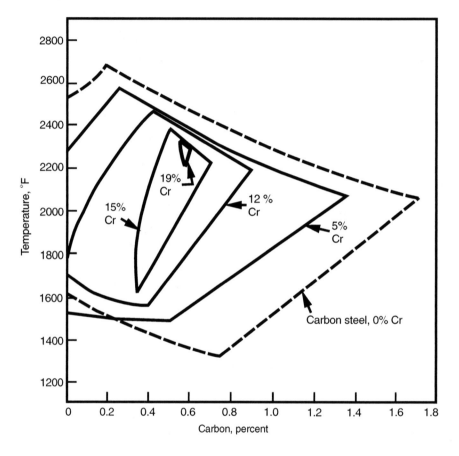

Figure 8.2 Showing the decreasing austenite (gamma) loop with increasing chromium content. Ferritic stainless steels typical contain more than 16% chromium.

Postweld heat treating of the ferritic stainless steels is done as low as possible (in the 1300-1550°F (704-843°C) range) to minimize grain coarsening. Cooling rates after postweld heat treatments should be as rapid as possible to minimize exposure in the 1000-700°F (538-370°C) range, where 885°F embrittlement problems can occur.

Typical ferritic stainless steel filler metal compositions are listed in Table 8.4.

Welding Austenitic Stainless Steels

The austenitic stainless steel family is quite extensive, with numerous variations of the standard compositions. Table 8.5 lists the selections of the filler metals used to weld austenitic stainless steels.

Table 8.5 Austenitic Stainless Steel Filler Metal Compositions

Base Metals			Filler Metals
UNS No.	Alloy	Condition[a]	Electrode or Filler Rod Type
S30400	304	1	308
S30403	304L	2 or 3	347, 308L
S30900	309	3	309
S31000	310	3	310
S31600	316	1 or 3	310
S31603	316L	2 or 3	316Cb, 316L
S31700	317	1 or 3	317
S31703	317L	2 or 3	317Cb
S32100	321	2 or 3	347
S34700	347	2 or 3	347

a. Condition in which weldment will be placed in service:
   1 = Annealed
   2 = Stress relieved
   3 = As-welded

There are two major goals when welding austenitic stainless steels: to make the welds crack-free and to ensure that the weld and weld HAZ have the same corrosion resistance properties as the base metal.

One of the major advantages of austenitic stainless steels is that they do not undergo a major metallurgical transformation during cooling. There is no quench effect, such as martensite formation that occurs with many of the lower alloyed steels. This makes the welding of the austenitic stainless steels much easier with regard to preheat, interpass temperature, heat input, and similar variables. However, some precautions should be taken when welding stainless steels, and this begins with selection of the alloy and the intended service environment.

In general, welds in austenitic stainless steels require about 20-30% less heat input than carbon steels. Excessive heat input can cause warping, loss of corrosion resistance, and weld cracking.

While austenitic stainless steels are easier to weld from many aspects, as discussed above, the austenitic nature of the microstructure also introduces a few potential problems. The biggest of these is hot cracking. A completely austenitic structure is prone to hot cracking, which occurs during solidification of the weld metal. There are several techniques to minimize and eliminate hot cracking, including weld bead contour control and ensuring the presence of small amounts of ferrite in the weld microstructure.

Weld bead contour is very influential in preventing hot cracking. During welding, the goal should be to maintain an elliptical weld puddle instead of a tear drop shaped puddle. This is illustrated in Figures 8.3a and 8.3b.[1]

 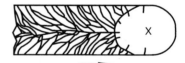

Figure 8.3a Showing weld bead shape and grain development.[1]

Figure 8.3b Exterior weld bead shape.[1]

The tear drop shaped puddle allows lower melting compounds to concentrate in the center of the weld because of the way the puddle solidifies. The stresses caused by welding are often too high for this lower strength zone, and cracking occurs down the centerline of the weld. By changing the welding parameters to obtain the elliptical shaped puddle shown, the lower melting compounds are more evenly distributed and not concentrated in a weak center line. This makes the weld more resistant to cracking.

In addition to controlling the shape of the weld bead, it is important to control the cross-sectional profile of the weld bead. Depositing a convex weld bead will help considerably in preventing weld cracking in austenitic welds by providing reinforcement in the center of the weld. The welder should always avoid depositing a concave weld bead, which is very prone to cracking because of the lack of reinforcement.

While austenite is prone to hot cracking during welding, ferrite is more resistant to hot cracking. This is used to good advantage in austenitic welds by incorporating small amounts of delta ferrite in the austenite welds, ranging from 4-10%. The percentage of delta ferrite in an austenitic weld is also known as the *Ferrite Number* (FN), e.g., 5% delta ferrite has a ferrite number of 5 (FN 5). Some welding electrode manufacturers of austenitic filler metals will include the ferrite number on the package label or report it in the mill certificate document.

Several diagrams have been developed to predict the ferrite level in a weld; these include the Schaeffler, DeLong, and WRC diagrams. The Welding Research Council (WRC) '92 diagram is the latest and most accurate diagram that includes ferrite numbers, see Figure 8.4.

It should be noted that because a weld composition is completely austenitic does not mean that there will be hot cracking – only that there is a greater tendency for hot cracking to occur. There are many completely austenitic weld compositions that are welded successfully, including S31000 stainless steel and nickel alloys. Without the presence of ferrite, however, other preventative measures, such as bead contour control, become more important.

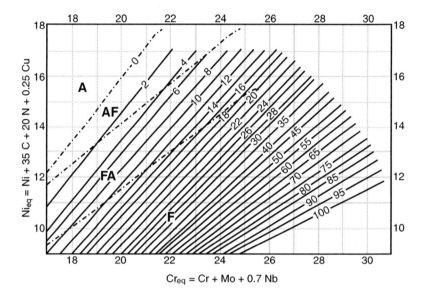

Figure 8.4 WRC '92 ferrite diagram.[2]

Effect of Thermal Treating and Exposure

There are few metals that are unaffected by thermal exposure. Most metals undergo changes in the microstructure during exposure to heat; some of these changes occur very quickly and others take time. Many of the austenitic stainless steels are very prone to changes during intermediate temperature exposure.

Sensitization and Stabilization

All of the stainless steels have two elements in common in addition to iron: chromium and carbon. One of the definitions of a stainless steel is that it contains a minimum amount of chromium (about 11% minimum), but usually the chromium level is around 18% or higher. Carbon can be present from very low levels (less that 0.01%) to fairly high levels (over 1%), depending on whether the material is in cast or wrought form and the desired strength level.

Chromium and carbon combine easily to form chromium carbides above about 800°F (427°C); generally, higher temperatures allow chromium carbides to form more easily. This is shown graphically in the time-temperature transformation (TTT) diagram for S31600 stainless steel in Figure 8.5.

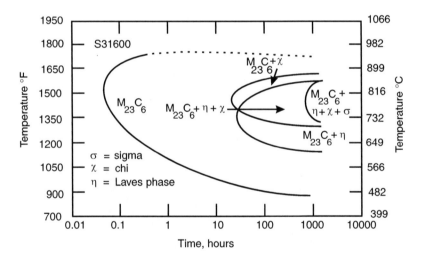

Figure 8.5 TTT diagram for S31600 stainless steel.[3]

This becomes a problem with some austenitic stainless steels during welding and during service if the service temperatures are in the chromium carbide forming range. The problem is not so much with the chromium carbides themselves, but that chromium is taken out of solution when the carbide is formed. This depletes the area around the carbide of chromium, thus lowering the corrosion resistance of the material in the region around the chromium carbide. The alloy depends on the chromium for much of its corrosion resistance; when the chromium gets tied up by the carbon as a carbide the alloy is effectively less corrosion resistant.

This chromium carbide formation depends on the carbon level and the time that the area is exposed to the intermediate temperatures required for carbide precipitation. This carbide precipitation can occur during welding in the heat-affected zone as well as during service exposure and is termed *sensitization*. The classic example of sensitization in austenitic stainless steel is shown in Figure 8.5, where a piece of S30400 stainless steel was welded and exposed to an aggressive acid-chloride corrosive environment. The sensitized area is several grains away from the fusion zone of the weld and is where the base metal had sufficient time in the 1000-1500°F (538-816°C) range for chromium carbides to precipitate, lowering the chromium level in this area. By comparison, the region immediately adjacent to the fusion zone had been exposed to higher temperatures but did not spend much time in the critical sensitizing temperature range.

Figure 8.6 Sensitized S30400 stainless steel.[4]

There are two metallurgical techniques to minimize this detrimental chromium carbide precipitation, or sensitization, in the austenitic stainless steels: lowering the carbon level or preferentially tying up the carbon with other elements, such as titanium and/or columbium.

By lowering the carbon to very low levels, the ability of chromium to form the undesirable chromium carbides is almost eliminated or at least lowered to an insignificant level.

An example of this is S30403 (type 304L), which contains a maximum of 0.03% carbon compared to S30400 which can contain up to 0.08% carbon. One disadvantage of lowering the carbon level is that the strength of the alloy is also lowered somewhat, which must be taken into consideration during the design of a component. This lower strength, however, can be offset by adding controlled amounts of nitrogen to the metal during the melting operation. In the case of the S30400 composition, the low-carbon nitrogen-added alloy is called S30453 (type 304LN), and can be supplied as a dual-certified product to meet both S30400 strength criteria and S30453 welding desirability.

The stabilization approach to controlling chromium carbide formation in austenitic stainless steels relies on the ability of some elements to form carbides more quickly than chromium. Niobium and titanium, for example, are added to S30400 as stabilizers. S34700 (type 347) uses niobium and S32100 (type 321) uses titanium as stabilizing elements.

The advantage of using stabilizers is that the carbon level can be maintained at sufficient levels to assure adequate strength. During exposure to intermediate temperatures, the stabilized grades of stainless steel form niobium carbides or titanium carbides instead of chromium carbides, thus minimizing the depletion of chromium and maintaining the corrosion resistance of the alloy.

Effect of Intermetallic Compound Formation

Another negative consequence of exposure to intermediate temperatures for many compositions is the formation of intermetallic phases that precipitate out of solution over time. Usually the time required for the formation of these phases is fairly lengthy, but some materials such as the duplex stainless steels can experience problems after relatively short (less than one hour) exposures during heat treatment cycles or possibly during welding.

The most prevalent phase of this nature that forms during exposure to intermediate temperatures is sigma phase. This is a hard, brittle compound that is high in iron and chromium and, when present in the composition, molybdenum. In addition to the brittle nature of the sigma particles, they have a plate-like structure that leads to easy fracture paths through the alloy.

Sigma formation can also rob the alloy matrix of useful alloying elements. High silicon and niobium levels also promote the formation of sigma, which results in a drastic reduction in room temperature impact strength and ductility. Its formation is to be avoided whenever possible.

Another embrittlement problem occurs with ferrite-containing materials, such as the ferritic and martensitic stainless steels, but also can occur in austenitic stainless steels that rely on low levels of ferrite for hot-cracking control in the welds. This phenomenon is termed *885 °F (or 475 °C) embrittlement*, so called because 885°F (475°C) is the optimum temperature for its formation.

If either sigma embrittlement or 885°F embrittlement has occurred, it can be taken back into solution by annealing. Rapid quenching through the temperature ranges that caused the original precipitation will eliminate the occurrence of this effect. Re-exposure to the temperatures at which 885ºF embrittlement or sigma occurs can result in re-embrittlement.

Welding 4% and 6% Molybdenum Austenitic Stainless Steels

There is one group of alloys that has a particular corrosion problem when matching filler metals are used for welding. This group is loosely called the 4% and 6% molybdenum stainless steels or *superaustenitic stainless steels.* The problem is one of microsegregation, or coring, during the cooling of the weld metal. The molybdenum in the welds tends to segregate during solidification, causing some parts of the structure to be higher and other parts lower in molybdenum. When matching filler metal compositions are used to weld these alloys, the part of the structure that is lower in molybdenum is prone to pitting attack because of the reduced pitting resistance resulting from the lower molybdenum.

The recommended filler metals to use in these situations are those that are considerably higher in molybdenum than the base metal. This provides sufficient molybdenum to compensate for the microsegregation that naturally occurs in these alloys during solidification. This problem also occurs with lower molybdenum stainless steels such as S31600 and S31700, as shown in Table 8.6.

Figure 8.7 shows the difference in molybdenum contents of base metal and weld structure for several 4% and 6% molybdenum stainless steels and various filler metals. While filler metals with molybdenum levels higher than 9% can be used, the additional corrosion benefit of these higher priced filler metals is generally not cost justified.

Table 8.6 Molybdenum Segregation Effects
in S31603 and S31703 Stainless Steels[5]

| | Composition - Weight % | | | |
| | S31603 | | S31703 | |
Location	Cr	Mo	Cr	Mo
Parent metal	16.3	2.8	18.4	3.2
Weld metal - dendrite center	14.3	1.8	14.2	2.0
Weld metal - interdendritic phase	20.1	5.7	24.0	6.6

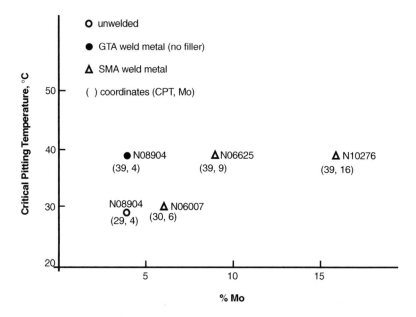

Figure 8.7 Critical pitting potential vs. molybdenum content.[6]

When using matching filler metals, pitting resistance of these welds can be improved by giving the finished weldment a high temperature anneal, but often this is not a practical solution. The anneal allows the molybdenum to diffuse, but this takes time to accomplish. If a final anneal is impractical, the use of overmatching welding products can compensate for these microsegregated, low-molybdenum conditions by realizing higher values in the as-welded condition.

Whenever a highly alloyed wrought material is used in a corrosive environment there may be a difference in the corrosion resistance of the base material and the weld metal. This difference is usually due to the dendritic nature of the weld deposit and normally there is no problem from a corrosion standpoint. In those situations where the corrosive environment is severe, however, the difference in corrosion resistance between base metal and matching weld metal might cause a problem. If so, it is advisable to use an over-alloyed filler metal composition to compensate for the lower corrosion resistance of the matching filler metal.

Table 8.7 shows how the welding method and the filler metal selection can influence corrosion resistance when N10276 (alloy C-276) is welded with a matching filler metal and an over-alloyed filler metal, using three welding processes. The GMAW-P (pulsed arc mode) matching filler metal welds had improved corrosion rates in the ASTM G 28, Practice B corrosion test, whereas the over-alloyed 686CPT welds exhibited reduced corrosion rates.

Table 8.7 Corrosion Rates for N10276 Welds
Using Different Filler Metals and Welding Processes

	Corrosion Rate in mils/year (mm/a)				
Filler	GTAW		GMAW-P		SMAW
Metal	Sheet	Plate	Sheet	Plate	Plate
N10276	71 (1.8)	56 (1.4)	361[a] (9.2)	1773[a] (45)	70 (1.8)
686CPT	80 (2.0)	74 (1.9)	63 (1.6)	162 (4.1)	72 (1.8)

a. Accelerated weld attack.

Welding Duplex Stainless Steels

The duplex stainless steels have been used since the 1930s when type 329 (S32900) stainless steel was developed. This alloy was used for many years but was hampered by weld metal corrosion problems (reference Chapter 5 on duplex stainless steels). Therefore, the alloy was used primarily as castings and forgings for such components as pump and valve bodies. With the development of the argon-oxygen decarburization (AOD) process, however, the alloy composition could be controlled more closely during melting. Controlled amounts of nitrogen could be added for additional corrosion resistance. The development and properties of the duplex stainless steels are covered in more detail in Chapter 5, but the biggest challenge with these alloys is their welding.

Most of the welding problems with the duplex stainless steels originate with the ferrite phase, which comprises roughly 50% of the structure of many of these alloys. A photomicrograph of the structure of a typical duplex stainless steel is shown in Chapter 5. As discussed above, ferrite has a much greater tendency to form the brittle intermetallic phases, such as sigma, during exposure to intermediate temperatures than does austenite. Also, ferrite is subject to 885°F embrittlement. Therefore, care must be taken when cooling through these temperature ranges, which occur after annealing and during welding. The cooling rate must be fast enough to go through these temperature ranges without causing these embrittling reactions.

However, with duplex stainless steels, the cooling rate from the annealing temperature range must be slow enough to form austenite. If the cooling rate is too fast, too much ferrite will remain in the microstructure and low toughness and corrosion resistance will result. The cooling rate must be slow enough to allow a sufficient amount of austenite to form but fast enough to avoid problems with sigma phase and 885°F embrittlement. What we have, then, is a tightrope: either too fast or too slow a cooling rate will cause problems.

This is a particular problem when using matching filler metals or when making autogenous (no added filler metal) welds because the cooling rates are too rapid to allow the austenite to form. But the problem can be solved effectively by using a filler metal that has a higher level of nickel. Nickel is an austenite former and by raising the nickel level from the nominal 5% contained in the base metal to about 9% in the filler metal, enough austenite can form during the rapid cooling of the weld to obtain the balance of austenite to ferrite.

Even if a nickel-enriched filler metal is used with the proper welding conditions and heat input, there still is a potential dilution problem when welding the root pass of butt joints. Unless there is an adequate amount of filler metal added to the weld puddle, too much ferrite can form because of too much base metal dilution in the weld. For this reason, it is a good idea to use an open root with a land, which encourages the welder to add filler metal, rather than a tight root with a feather edge.

Effect of Welding on Corrosion Resistance

Because of the sensitivity of the duplex stainless steels to cooling rates, they are much more sensitive to welding procedures, particularly heat input, than the fully austenitic stainless steels. For this reason, welding procedures and heat treatments must be taken into consideration in the design stage to obtain sound, corrosion-resistant welds. If a welded structure of duplex stainless steel can not be annealed after welding, for example, it is important for the proper weld filler metals to be specified and that proper welding conditions are used. If, however, the structure can be annealed after welding, such strict attention to these details is not necessary.

While nitrogen is intentionally added to duplex stainless steel to aid in austenite formation and for pitting resistance, it can cause problems under rapid cooling conditions because of the formation of chromium nitrides. These nitrides can be initiation sites for pitting corrosion. This is another reason for cooling rate control with these steels. When using the gas welding processes, such as GTAW and GMAW, there can be loss of nitrogen from the weld puddle by as much as 0.05%. This can be compensated for by small additions of nitrogen to the shielding gas; typically nitrogen is added in the 2-5% range. Too much nitrogen, however, can cause weld porosity problems since the solubility of nitrogen in duplex stainless steel is in the 0.20-0.35% range. Also, the nitrogen addition will cause some contamination of the tungsten electrode during GTAW welding, and will increase arc instability, weld pool turbulence, and spatter.

Precautions for Welding Duplex Stainless Steel

While duplex stainless steels are often considered to be a group of similar compositions, they can vary considerably in their response to welding and their resistance to corrosion. The 22% chromium steels are more sensitive to low heat input welding conditions than the 25% chromium superduplex stainless steels. This is primarily because of the higher nitrogen levels in the 25% chromium alloys. This increase in austenite-forming capability is particularly beneficial during the very rapid cooling encountered with low heat input welds. On the other hand, 22% chromium steels are more tolerant of high heat inputs compared to 25% chromium steels, reflecting the greater tendency of the 25% chromium steels to form intermetallic phases, such as sigma, during exposure to intermediate temperatures. The higher alloy content of the 25% chromium superduplex stainless steels makes them more susceptible to intermetallic phase formation than the lower alloyed duplex stainless steels.

The cooling times generally recommended for 22% chromium duplex stainless steel are from 1472 to 932°F (800 to 500°C) within 8 and 30 seconds, respectively, which corresponds to arc energies of 1 to 2 kJ/mm (50.8 kJ/in.) for 10 to 15 mm (0.39 to 0.59 inch) thick butt welds. A maximum interpass temperature of 300-390°F (150-200°C) is often recommended.

The preferred arc energies for the 25% chromium duplex stainless steel are 0.5-1.5 kJ/mm, which is somewhat below those recommended for 22% chromium duplex stainless steel. The lower limit is made possible because of the higher nitrogen level of 25% chromium steels, while the upper limit is restricted because of the tendency for undesirable intermetallic phase formation. These phases form more easily in thin sections because of the heat build-up associated with a low heat sink. Therefore, in thin sections, the arc energy should be at the lower end of the recommended heat input range.[7] The maximum interpass temperature for 25% Cr duplex stainless steel is 300°F (150°C).

Heat input, in kJ/in., can be determined by the formula:

$$\frac{\text{Volts x Amps x 60}}{\text{Travel Speed (in./min.) x 1000}}$$

(For kJ/mm, convert by multiplying by 0.039.)

Weld spatter can be a potential problem when welding the duplex stainless steel. Because of its sensitivity to cooling rates One problem with weld spatter is that when the molten droplet hits the base metal, it heats the base metal to fairly high temperatures in a very localized area, followed by very rapid cooling. This can cause excessive ferrite to form locally, which can lead to corrosion problems during service. The second problem with weld spatter is that if it is not removed, it can lead to crevice corrosion during service in corrosive environments.

Another negative aspect of the high ferrite content of duplex stainless steels is the possibility of hydrogen cracking. For this reason, coated electrodes used in the SMAW process must be handled and stored properly to minimize moisture pickup. This includes storing the electrodes in ovens, removing them just before welding. Welding with high moisture electrodes will very likely cause hydrogen cracking problems. Similarly, flux used with the SAW process can cause problems if it is too moist. When using the GTAW process with duplex stainless steels, hydrogen gas should not be used in the shielding gas. Hydrogen cracking has not been a problem when welding duplex stainless steels if the nominal 50% austenite - 50% ferrite balance is achieved. However, if the ferrite level exceeds about 75% there is a potential problem with hydrogen cracking from the sources mentioned above.[8]

Preheat is usually not required when welding duplex stainless steel except for those situations where high ferrite structures are being welded under high restraint conditions. Under these conditions hydrogen induced cracking is a possibility, which can be minimized by preheating to around 300°F (150°C).

Duplex stainless steels can be welded with all of the conventional welding processes with little change in procedure compared to austenitic stainless steels. However, some precautions are needed to achieve the desired resistance to corrosion.

Welding Precipitation Hardenable Austenitic Stainless Steels

The precipitation-hardening stainless steels are usually welded autogenously or with matching filler metals using the gas tungsten arc or laser beam welding methods. The weld metal responds similarly to the base metal during heat-treatment. The martensitic precipitation-hardenable stainless steel S17400 (17-4 PH) is one of the most commonly specified precipitation-hardenable stainless steels, and the matching filler metals E630-XX and ER630 are readily available. Many of the other precipitation hardenable stainless steels have weld filler metals that are either unobtainable or difficult to obtain because of low demand. In some cases, fabricators have resorted to shearing strips from sheet to use as filler metals - an acceptable practice as long as good cleanliness practices are followed. In case of poorly accessible stainless steel filler metals, the precipitation-hardened nickel alloys such as ERNiCrMo (725 NDUR®) often provide suitable properties and corrosion resistance.

Because of the diversity in microstructure and precipitation behavior, it is difficult to make general recommendations regarding filler metal selections for the precipitation-hardenable stainless steels. This is another situation where the base metal manufacturer should be consulted for specific recommendations.

The alloys hardened by aluminum, titanium, or copper precipitates usually must be annealed after welding before they are precipitation-hardened or else they will crack during the precipitation-hardening cycle. This type of cracking, called *strain-age cracking* is described more fully in the nickel alloy section following.

Most precipitation-hardenable stainless steels rely on aluminum, titanium, niobium, and/or copper. All of these elements, except niobium which precipitates more slowly, form compounds that precipitate very quickly, and this rapid precipitation rate is responsible for most of the strain-age cracking difficulties encountered.

Strain-age cracking can be eliminated or minimized by reducing the welding stresses by annealing prior to beginning the precipitation-hardening process.

Welding Nickel Alloys

As described in Chapter 7, there are numerous classes of nickel alloys with many compositions within each class. Each of these alloys has corresponding filler metals that are used during welding, and these are listed along with their designations in Table 8.8.

Table 8.8 Nickel Alloy Filler Metals

Base Metal		Filler Metal		
UNS No.	Alloy	UNS No.	GMAW/GTAW	For SMAW
N02200	200	N02061	ERNi-1	ENi-1
N04400	400	N04060	ERNiCu-7	ENiCu-7
N05500	K500	N05504	ERNiCu-8	---
N06600	600	N06082	ERNiCr-3	ENiCrFe-3
		N06076	ERNiCr-6	ENiFe-1
		N06062	ERNiCrFe-5	
---	50Ni/50Cr	N06072	ERNiCr-4	---
---	Age-Hard	N07092	ERNiCrFe-6	ENiCrFe-2
N06690	690	N06052	ERNiCrFe-7	ENiCrFe-7
N07750	X-750	N07069	ERNiCrFe-8	---
N06601	601	N06601	ERNiCrFe-11	---
N08825	825	N08065	ERNiFeCr-1	---
N07718	718	N07718	ERNiFeCr-2	---
N10001	B	N10001	ERNiMo-1	ENiMo-1
N10003	N	N10003	ERNiMo-2	---
N10004	W	N10004	ERNiMo-3	ENiMo-3
N10665	B-2	N10665	ERNiMo-7	EniMo-7
K81340	9% Ni Steel	N10008	ERNiMo-8	ENiMo-8
		N10009	ERNiMo-9	ENiMo-9
N10675	B-3	N10675	ERNiMo-10	ENiMo-10
N06007	G	N06007	ERNiCrMo-1	ENiCrMo-1
N06002	X	N06002	ERNiCrMo-2	ENiCrMo-2
N06625	625	N06625	ERNiCrMo-3	ENiCrMo-3
N10276	C-276	N10276	ERNiCrMo-4	ENiCrMo-4
N06455	C-4	N06455	ERNiCrMo-7	ENiCrMo-7
N06975	G-2	N06975	ERNiCrMo-8	---
N06007/ N06985	G/G-3	N06985	ERNiCrMo-9	ENiCrMo-9
N06022	C-22	N06022	ERNiCrMo-10	ENiCrMo-10
N06030	G-30	N06030	ERNiCrMo-11	ENiCrMo-11
N06059	59	N06059	ERNiCrMo-13	ENiCrMo-13
N06686	686	N06686	ERNiCrMo-14	ENiCrMo-14
N07725	725	N07725	ERNiCrMo-15	---
N06617	617	N06617	ERNiCrCoMo-1	ENiCrCoMo-1
N06230	230	N06231	ERNiCrWMo-1	---

In many respects the nickel alloys are somewhat easier to weld than many of the high strength alloy steels. The nickel alloys do not undergo a phase transformation, such as martensite formation, when cooled from high temperatures. They normally do not need to be preheated or post-weld heat treated since they do not rely on carbon-related phases to achieve high strength at low temperatures. A post-weld heat treatment may be helpful to relieve fabrication stresses, but it usually is not necessary with the high nickel alloys. Also, cooling rates are usually not critical, and a normal air cool is usually sufficient for most of the nickel alloys.

There are, however, some issues that need to be addressed when welding nickel alloys. These involve the necessity for cleanliness and for recognizing the characteristics of the specific alloys when designing a weld joint and preparing a welding procedure. These issues were discussed at length in the section on stainless steel welding, but bear repeating because of their importance.

Care must be taken to eliminate the presence of embrittling elements in the weld area when welding high nickel alloys. The most commonly encountered problem elements are sulfur and phosphorus, but also include lead, tin, zinc, and numerous other low-melting metal elements. These form compounds in the microstructure that cannot withstand the heat and stresses of welding and will cause cracking in the welds or heat-affected zones on high nickel alloys.

The chromium containing nickel alloys have lower thermal conductivities than steels and therefore cannot dissipate the heat of welding as fast as steels. Arc penetration is not as deep with the nickel alloys, which requires using thinner lands in the root of butt joints. Increasing the amperage will not necessarily increase the penetration of the arc. In SMAW electrodes, increasing the amperage to above the recommended range will overheat the electrode and can cause arc stability problems as well as coating breakdown and flaking. Overheating can also cause excessive puddle formation and may possibly lead to the loss of important elements, such as titanium, magnesium, manganese, and aluminum that are added to the wire or flux to control porosity.

Because the molten weld puddle is more sluggish than a steel puddle the welder needs more room to get a good weld. Therefore, the included angles in butt joints need to be opened up more than for comparable steel joints. These modified joint design recommendations are shown in Figure 8.1.

All of the nickel alloys have a single phase in the annealed condition–the austenite phase. The nickel alloy families fall into two general categories as far as welding is concerned–those that contain chromium or iron and those that do not. Chromium is a very helpful element because it helps control porosity. (It can also form undesirable phases during intermediate temperature exposure, however.) The nickel and nickel-copper alloys, which do not contain chromium, must have good gas protection, along which deoxidizing elements added to the filler metal, to avoid weld metal porosity, when welding with the GTAW and GMAW welding processes.

The shape and contour of the weld bead is very important in preventing solidification cracking in nickel alloys. This was discussed in the stainless steel section and is even more important when welding the ferrite-free nickel alloys. In summary, the molten weld puddle should have an elliptical, or rounded, shape rather than a tear-drop, or pointed, trailing edge which usually indicates a very fast travel speed. The cross-section of the weld, or the weld profile, should be convex, or reinforcing, rather than concave.

The condition of the base material will sometimes determine whether or not a weld can be made successfully. In particular, the grain size of the material is very important in choosing a welding process. The larger the grain, the more sensitivity to cracking using the high heat input welding processes. Consequently, selection of the welding process should be considered when welding large-grain size nickel-base alloys to avoid potential cracking. Table 8.9 compares the effect of grain size (and its influence, along with stress, on intergranular cracking) on the ability of several welding processes to make a successful, crack-free weld in several nickel and stainless steel alloys.

Table 8.9 Effect of Grain Size and Welding Process on Weldability[9]

Alloy	Grain Size	GMAW	EBW	GTAW	SMAW
600	Fine	X	X	X	X
	Coarse	---	---	X	X
617	Fine	X	X	X	X
	Coarse	---	---	---	X
625	Fine	X	X	X	X
	Coarse	---	---	X	X
706	Fine	---	X	X	X
	Coarse	---	---	---	X
718	Fine	---	X	X	X
	Coarse	---	---	---	X
800	Fine	X	X	X	X
	Coarse	---	X	X	X
S31600	Fine	X	X	X	X
	Coarse	---	---	X	X
S34700	Fine	X	X	X	X
	Coarse	---	---	X	X

The higher energy welding processes, such as GMAW and EB, have difficulty in several coarse grained alloys, whereas the lower energy welding process, such as GTAW and particularly SMAW, are much more forgiving. For example, even though the SMAW welding process is a much slower process, it may be necessary to use it if the base material is coarse grained and sensitive to high heat inputs during welding. The prevention of cracking when welding coarse grained material is a particular problem when high restraint is present.

Welding Precipitation Hardenable Nickel Alloys

There are numerous precipitation-hardenable nickel alloys. These materials are useful because of their very high strengths combined with their corrosion resistance. They do, however, present a challenge when it comes to welding.

The precipitation-hardenable alloys are usually welded by the GTAW process, since most applications involve relatively thin sheet or flat rolled products. There are few massive structures of precipitation-hardenable materials that require welding. Therefore, the relatively slow deposition rates of GTAW are not much of a concern. Additionally it is difficult, or almost impossible, to make a SMAW electrode that will produce the precipitation-hardenable composition in the deposited weld because of the difficulty of transferring highly oxidizable elements, such as aluminum and titanium, across the arc.

This is not a problem when the welding arc is surrounded by an inert gas, such as in the GMAW or GTAW process. However, many of the precipitation-hardenable alloys cannot endure the high heat input that is typical of the GMAW process, particularly in a coarse grained condition or when under high restraint. Therefore, most welding of precipitation-hardenable alloys is done with the GTAW process.

Almost all of the precipitation-hardenable alloys are sensitive to strain-age cracking. This cracking occurs after welding and during the precipitation-hardening thermal treatments used to increase the strengths of the alloys. The precipitation-hardening process creates stresses that may be too much for the material to withstand, with cracking as the result.

Strain-age cracking usually occurs in the HAZ of the weldment, which experiences the greatest stresses during the welding operation, but may also extend into the weld itself. Therefore, the combination of high residual stresses caused by welding and the stresses caused by the precipitation reaction can exceed the rupture strength of the base metal at the precipitation-hardening temperature, resulting in cracking. The precipitation-hardening temperature is not an effective stress-relieving temperature. This tendency to strain-age crack is dependent on the elements responsible for the precipitation-hardening reaction and the thickness of the weld joint.

Most precipitation-hardenable nickel alloys depend on aluminum, titanium, and/or niobium for the precipitation reaction. However, the precipitation reaction that relies on niobium is much more sluggish than the reaction involving aluminum or titanium, and this sluggish metallurgical reaction enables these alloys to be welded and precipitation-hardened with less possibility of cracking. N07718 (alloy 718) and N07725 (alloy 725), which are niobium-hardened, are alloys which owe a great deal of their commercial success to their ability to resist strain-age cracking.

In addition to creating problems during precipitation-hardening, the relatively high aluminum or titanium contents of the precipitation-hardenable alloys also can cause problems during the welding operation itself because of the tendency for the aluminum and titanium to oxidize easily. Oxides can form on the surface of GTAW or GMAW beads during welding. These oxides should be ground away when making multi-pass welds so that they do not create oxide inclusions in the weld or interfere with weld fusion.

Two of the biggest problems when welding the precipitation-hardenable alloys are restraint and the heat treat cycle. The alloys hardened by aluminum or titanium precipitates usually must be annealed after welding before they are precipitation-hardened or else they will crack during the precipitation-hardening cycle. The precipitation-hardenable stainless steels and nickel alloys should be welded as follows: weld in the annealed (soft) condition, anneal, precipitation-harden.

Also, strain-age cracking must also be considered when repair welds are made. Repair welds on most of the precipitation-hardenable alloys should be made on annealed material using the following sequence: anneal, repair weld, anneal, precipitation-harden.

Also, the usual precautions concerning cleanliness and dark oxide removal prior to welding, as discussed in the beginning of this chapter, should be followed when repair welding these materials.

Specialized Welding Processes or Procedures

There are some welding requirements that are somewhat specialized and require techniques that demand extra consideration. These include overlay welding, welding of clad steels, and pipe welding.

Overlay Welding

Overlay welding, or surfacing, is a specialized welding operation usually performed as a cost-saving technique to place a layer of more corrosion-resistant composition over a less costly base. The most common combination is an austenitic stainless steel or nickel alloy over a carbon steel.

Large flat surfaces, such as tube sheets, are usually overlay welded with an automated welding process such as GMAW, SAW, or ElectroSlag Surfacing (ESS). Another common overlay application is the inside surface of large process vessels that are welded in the flat position by rotating the vessels on mechanized positioners. In these situations, the objective is to deposit as much material in the least amount of time with as little dilution from the base material as possible.

As with welding the austenitic stainless steels and nickel alloys to themselves, cleanliness is an important consideration when overlay welding. Contaminants on the surface of the base material can cause cracking problems in the deposited weld if they are present in sufficient quantities.

Excessive sulfur levels are particularly a problem when welding with high nickel alloys, and occasionally this can be a problem if the base steel material has a high sulfur content, even if the surface of the material is clean.

If cracking in the first overlay layer occurs because of this situation, the deposited first layer should be machined or ground away. Then a layer of carbon steel weld metal should be applied, followed by the nickel alloy overlay. The carbon steel weld overlay acts as a buffer between the high-sulfur base material and the nickel alloy layer.

Deposition Rates During Overlaying

Since one of the objectives of overlaying is welding efficiency, the SAW, and more recently the ESS, processes are favorite choices for high volume deposition rates. The ESS process uses filler metal in the form of wide strip instead of wire, and the molten weld metal is deposited under a molten slag rather than being melted by an arc. The molten slag is usually high in fluorides that make it electroconductive (able to conduct electrical energy). By contrast, most SAW fluxes are composed of oxides that favor arc formation.

Because of the absence of an ARC, the ESS process can produce lower penetration into the base metal, thereby resulting in less dilution, particularly iron. The strip used in ESS overlaying can range from 30 mm wide x 0.5 mm thick to strip up to at least 120 mm. The ESS molten weld pool using 30mm wide strip is typically 64 mm long by 32 mm wide with a molten slag size of 127 mm long x 45 mm wide. Typical overlay techniques and deposition rates are shown in Table 8.10.

Table 8.10 Weld Overlay Techniques and Deposition Rates

Weld Process	Wire/Strip	Deposition Rates
GMAW	1.6 mm diameter wire	7 Kg/arc hour
SAW	1.6 mm diameter wire	9 Kg/arc hour
SAW	60 mm x 0.5 mm strip	14 Kg/arc hour
ESS	60 mm x 0.5 mm strip	22 Kg/arc hour

One of the most common nickel alloy filler metal overlays is Ni-Cr-Mo, such as N06625 (alloy 625), on steel. In one study comparing SAW and ESS overlaying, the ESS process showed consistently lower weld dilution than the SAW process.[10] The ESS process was also less sensitive to variations in welding parameters.

Dilution Effects During Weld Cladding

Dilution is the biggest concern when overlay welding. The objective of applying an overlay is to replace the steel surface with a more corrosion-resistant composition and, depending on the welding method selected, this usually takes anywhere from 1 to 3 layers of overlay. Welding parameters must be controlled carefully to maintain a consistent dilution of base metal into the weld layer and retain the corrosion resistance of the overlay. Side bend test are usually performed on test samples to verify that dilution is not excessive.

Dilution during cladding is heavily influenced by the welding parameters. Increasing the amperage increases the dilution by making the arc hotter and increasing the penetration, resulting in more base metal melting. Increasing the wire diameter, which incurs higher amperage, also increases dilution.

Polarity will also influence dilution. DCEN (direct current electrode negative) will give less penetration and less dilution than DCEP (direct current electrode positive), but must only be used with oscillation when welding with nickel alloys.

Varying the travel speed is very influential. A lower travel speed allows the arc to be concentrated on the molten puddle instead of on the base metal, resulting in lower penetration and lower dilution.

Changing the oscillation width and frequency also influences dilution. The wider the oscillation and the higher the oscillation frequency, the lower the penetration and dilution, generally[11]. In stringer bead overlays, a 50% overlap is critical.

There usually is not a dilution problem when overlaying a stainless steel or nickel-chromium alloy onto steel, but other combinations can be quite sensitive to dilution effects. Following are some general guidelines for the tolerance of several nickel-containing filler metals to dilution by various elements.

➢ Nickel weld metal will tolerate the addition of:
- any amount of copper
- 25-35% maximum chromium
- 40% maximum iron with coated electrodes
- 25% maximum iron with GMAW

➢ 70/30 copper nickel weld metal will tolerate the addition of:
- any amount of copper
- any amount of nickel
- 5% maximum iron
- 5% maximum chromium

➢ 70/30 nickel copper weld metal will tolerate the addition of:
- any amount of nickel or copper
- 6-8% maximum chromium
- 30% maximum iron with 70/30 Ni-Cu, SMAW electrodes
- 5% maximum iron in 70/30 Ni-Cu, GMAW if the welds are to be stress-relieved
- 15% maximum iron in 70/30 Ni-Cu, GMAW if the welds are not to be stress-relieved

➢ Ni-Cr-Fe weld metal will tolerate the addition of:
- any amount of nickel
- 15% maximum copper
- 15% maximum chromium
- approximately 40-50% of S30400
- approximately 30-40% maximum iron with SMAW electrodes
- 20-30% maximum iron with GMAW

As can be seen by the wide variation in the dilution tolerances listed above, the success of an overlay depends not only on the filler metal/base metal combination but also the welding method.

Welding of Clad Steel

Steels clad with an austenitic stainless steel or nickel alloy, in the form of plate, are often joined by welding. The cladding is either bonded to the base material by hot rolling or by explosive bonding and is sometimes an economical alternative to either overlaying or using solid alloy. In both cladding operations a true metallurgical bond is achieved between the alloy and the steel. The alloy portion of the plate is typically about 20% of the thickness of the plate. Since the alloy side of the cladding will be exposed to the corrosive environment, welding techniques must be used that will assure that no steel is exposed to the corrosive media. If the corrosive media is able to penetrate the alloy barrier, failure will occur by rapid corrosion. This makes weld design and technique even more critical.

Butt joints should be used to weld clad steels whenever possible because they offer more consistent weldability and easier inspection procedures than other types of joints. Clad steels can be welded either from the alloy side or from the steel side. Figure 8.8 illustrates two acceptable weld joint designs. Whichever approach is used, the alloy side of the joint should have a minimum of two layers of filler metal to minimize the effects of dilution from the steel.

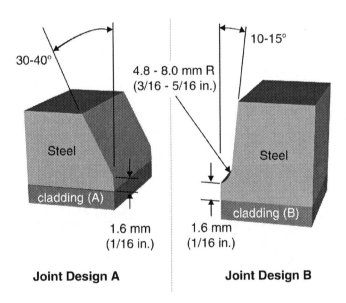

Figure 8.8 Weld joint designs for joining clad steels.[12]

Figures 8.9a and 8.9b show welding sequences that could be used when welding from either side of the clad. For relatively thin (less than ½ inch thick) clad plate, it is usually more economical and easier to use nickel alloy filler metal for the whole joint.

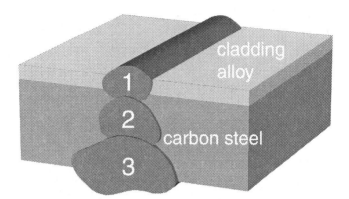

Process: GMAW-pulsed arc or SMAW
Joint Design: Square butt or bevel on carbon steel side
Sequence:
   1. Deposit bead 1 with minimum penetration
   2. Back gouge to cladding
   3. Fill as many passes as needed
Welding Filler Metals: Comparable nickel alloy filler metal

Figure 8.9a Welding sequences for joining clad steels.[12]

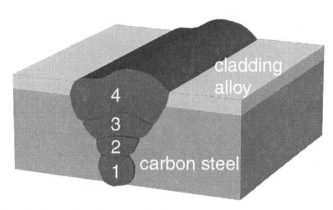

Joint Design: Bevel from the cladding alloy side
Sequence:
   1. Deposit bead 1 by GTAW or smaller diameter SMAW
   2. Deposit bead 2, penetrating into bead 1
   3. Deposit beads 3 and 4 to complete weld (SMAW or GMAW-pulsed)
   4. Welding Filler Metals: Comparable nickel alloy

Figure 8.9b Welding sequences for joining clad steels.[12]

Pipe Welding

One of the weak links in a pipe weld that is exposed to a corrosive environment is the root pass. If a complete penetration weld with a smooth contour on the ID is not achieved, accelerated corrosion will invariably occur in the crevices created by the weld.

Pipe welding can either be fairly straight-forward or it can be quite challenging. An easy pipe welding task would be joining fairly large diameter pipe that has an ID backing strip with the axis of the pipe in the horizontal position with the pipe rolled during welding so that all welding is done in the flat position. One of the most challenging pipe welding situations would be one where a fairly small diameter pipe is in the fixed vertical or 45° position with no ID backing strip nor any access to the ID after welding.

As the difficulty of making pipe welds increases, such as with fixed position small diameter pipe, the use of consumable inserts becomes more attractive. Consumable inserts are manufactured preformed shapes made of filler metal compositions that can be fitted to a beveled pipe, tackwelded in place, and fusion welded, usually with the GTAW process.

Several consumable insert designs are shown in Figure 8.10. The big advantage to them is that they allow the welder to concentrate on controlling the weld puddle without the extra burden of feeding filler metal into the joint during welding. The result is a much smoother and more consistent ID root bead contour.

From a corrosion standpoint, this technique eliminates the crevices that are present when a backing strip is used and not removed after welding. It also improves greatly the ability of a radiographic test to pick up welding defects.

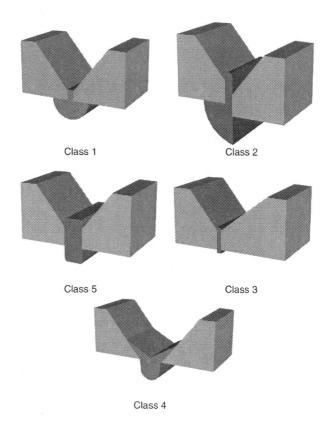

Figure 8.10 Typical consumable insert designs.

Welding of Castings

Castings offer several advantages over similar wrought compositions, including lower cost and higher strength. This is true primarily because of the nature of the casting process–the finished casting does not have to negotiate the rather severe hot and cold working operations that wrought materials must endure to reach the final product form. Lower costs are possible because of this cast-to-shape characteristic of the casting process. Higher strengths are possible because higher levels of strengthening elements, such as carbon, aluminum, titanium, and others, can be added because no forming operations are necessary after the casting is completed.

There are, however, some prices to be paid for this lower cost and higher strength–and weldability is one of them. (Others include lower ductility and lower thermal fatigue life.) Castings do not have either the weldability or repair weldability of similar compositions in the wrought form, and this difference in weldability is usually a large one.

Stainless steel castings are categorized as being either corrosion resistant or heat-resistant, with the corrosion resistant castings identified with a "C" before the identifying number and the heat-resistant stainless alloys having an "H." The stresses and heat generated during welding are the source of problems when welding either wrought or cast materials but are even more so with castings.

For the best chance of having defect-free welds in castings, the castings should be in the solution-annealed condition offering optimum ductility. For new castings, this heat treatment relieves any stresses caused by the casting process and provides homogenization of the cast dendritic structure. This homogenization treatment allows the large differences in composition in the cast structure to become more uniform through diffusion, which takes time. For repair welds in castings, the solution anneal allows time for phases that have precipitated during service to go into solution, thus increasing the ductility of the casting. This is particularly an issue for castings that have been exposed to intermediate temperatures, such as 1000-1500°F (538-816°C), during service. This temperature range will cause carbides to precipitate out of solution during service, drastically reducing ductility.

An example of what an annealing treatment can do for the ductility of a casting is shown in Figure 8.11, which contains two curves of ductility vs. annealing temperature after two and five hours annealing for the cast alloy HP. The castings had been in service for several thousand hours in the carbide-precipitation temperature range, which resulted in room temperature tensile ductility of around 4%. If these castings had been repair welded without annealing, they certainly would have cracked during welding. However, the recovery of the room temperature ductility was improved dramatically by annealing in the 2000°F (1093°C) area for either two or five hours, resulting in a room temperature ductility that is 2-3 times that of the unannealed material. This improvement of room temperature ductility would result in a successful weld repair. It should be noted that after the weld-repaired castings are returned to service carbides will precipitate again if the material is exposed to the carbide precipitation range of approximately 1000-1500°F (538-816°C).

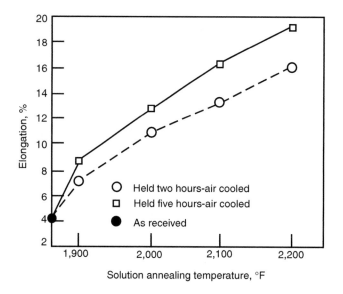

Figure 8.11 Ductility vs. annealing temperature
for HP castings.[13]

There are some procedures to follow during the welding or repair welding of castings that will help ensure crack-free welds. Low heat input can be helpful by reducing welding stresses. Light peening after welding each bead can be beneficial. Sometimes preheating can help by lowering the cooling rate, thereby reducing stresses.

In some situations, such as during butt welding of castings, *buttering* (a localized overlay) can be very helpful if the castings have very low ductility. After proper cleaning, both weld joint faces are surfaced with a layer of filler metal. Depositing filler metal on the surface of a metal is a very low-stress operation and is usually successful on castings. After buttering the weld faces, the joint can be made between the buttered surfaces. In this situation, the welder is basically joining a relatively high-ductility material (the deposited weld metal) to another relatively high-ductility material (the other deposited weld metal). The weld joint faces now consist of a layer of higher ductility filler metal which is much more likely to resist the stresses of welding. This technique can often be used to salvage a cracked casting that cannot be annealed prior to repair welding.

Maximizing Weld Corrosion Resistance Through Design

Often in corrosive situations a stainless steel or nickel alloy is specified to assure a reasonable life, usually at a considerable increase in cost compared to lower alloyed materials. The increased material costs can usually be justified because of the increased life of the component. But if the design of the fabricated structure does not take into account some basic principles, even a more expensive material will not live up to design expectations.

Following are some weld-related design or fabrication principles that can extend the life of a component in a corrosive environment:

➢ To minimize crevice corrosion
  • specify that crevices are welded shut
  • specify double-butt or double-lap welded joints
  • use consumable/removable inserts for critical single-butt joints
  • prohibit skip welding for corrosive environments; weld all crevices
  • seal weld all tubes to tubesheets
  • if crevices cannot be sealed or eliminated, open them up to allow circulation
  • remove backing rings in pipe welds after welding
➢ To minimize end grain attack
  • butter the end grain with weld metal
➢ To minimize stress corrosion cracking
  • stress-relieve susceptible material after welding or peen

➤ To minimize corrosion fatigue
  • avoid stress raisers by blending welds to the base metal
  • use butt welds instead of fillet welds
  • blend grinding marks, rough edges, and rough machine marks
➤ To minimize intergranular corrosion
  • use stabilized or low carbon welding materials
  • do not stress relieve unstabilized stainless steels in the carbide precipitation range (specify an anneal instead)
  • specify low carbon or stabilized grades of stainless steel
➤ To minimize galvanic, pitting, crevice corrosion
  • select an over matching filler metal that is cathodic to the base metal, for the intended service

Postweld Cleaning

After a component is designed, the material specified, and welding completed, there may be one very important step remaining. If the component is expected to resist a corrosive environment, and if the material selected has marginal corrosion resistance to that environment, it may be wise to inspect and clean the unit prior to putting it into service.

Accelerated corrosion can take place because markings have not been removed from the surface of stainless steel vessels. Crevice corrosion has occurred under crayon markings, for example. Scratches can disturb the protective passive film that stainless steels rely upon for their corrosion resistance, causing crevice corrosion sites. Forming equipment, such as press brakes, are made of low alloy steel. During forming operations, iron particles may become imbedded in the surface of a component which may corrode rapidly in the early stages of service. While this may do no more damage than cause an unsightly rust appearance, the iron particles may cause accelerated corrosion in the form of pits. Heat tint caused by welding is usually not a problem, but it could cause accelerated corrosion in certain environments if not removed prior to putting the component into service. Heat tint is an oxide and has a different composition compared to the underlying and adjacent base material. Therefore, it may not have the same corrosion resistance.

These examples are some of the problems that can occur when a material is specified that has marginal, but adequate, corrosion resistance to the intended environment. In that kind of situation, postweld cleaning procedures should be considered in order to avoid premature corrosion problems. An alternative solution would be to use a more corrosion-resistant (and more expensive) alloy. However, it would be much less costly to specify an adequate inspection and cleaning procedure.

Summary and Conclusions

The corrosion-resistant stainless steels and nickel alloys are very weldable. There are precautions that need to be taken, such as increased weld-area cleanliness and modifications in joint designs, but generally they are no more difficult to weld than the low alloy steels. By following the guidelines discussed in this chapter, successful welds can be made that will ensure a trouble-free design life. Thermal treatments and controls may be necessary with the duplex stainless steels, precipitation-hardenable stainless steels and nickel alloys, and the martensitic/ferritic grades of stainless steels. Some special precautions may also be necessary for castings.

References

1   Davis,G.J. and Garland, J.G., International Metals Review, 20 (1976), pg. 83.

2   "WRC 1992 Constitution Diagram for Stainless Steel Weld Metal: A Modification of the WRC 1988 Diagram", Kotecki & Siewert, Welding Journal, May 1992, pg. 171s.

3   "Corrosion of Stainless Steel", Sedriles, John Wiley & Sons (2nd Edition), 1996, pg. 18.

4   Welding Handbook, AWS, Vol.4, Seventh Edition, pg. 108.

5   "The Effect of Autogenous Welding on Chloride Pitting
        Corrosion in Austenitic Stainless Steels", Garner, Corrosion,
        Vol. 35, No. 3, March,1979, pg. 108.

6   "Corrosion of High Alloy Austenitic Stainless Steel Weldments
        in Oxidizing Environments", Garner, Materials Performance,
        Aug. 1982, pg. 10.

7   "Arc Welding Duplex Stainless Steel for Maximum Corrosion
        Resistance", Gooch & Gunn, Materials Performance, March
        1995, pg. 60.

8   "Welding Practice for the Sandvik duplex Stainless Steel SAF
        2304, 2205 and 2507," Sandvik Steel paper.

9   AWS Handbook, 8th Edition, Vol. 3, pg. 232, Table 4.8.

10   Metals Handbook, ASM International, Vol. 6, pg. 819.

11   "Electroslag and Submerged Arc Cladding with Nickel Alloy
        Strip" by Gao Devletian and Wood, International Trends in
        Welding Science and Technology, ASM International
        Conference Proceedings 1992, Gatlinburg TN, pg. 449.

12   Lukens Steel Co. brochure #803.

13   "Repair Welding High-Alloy Furnace Tubes", Avery &
        Schillmoller, Hydrocarbon Processing, Jan 1988.

Chapter

9

CORROSION RESISTANT ALLOYS FOR DOWNHOLE PRODUCTION TUBING

Liane Smith

Intetech Ltd.

The choice of materials for the production tubing in a well is a vital issue both with respect to the safety, and also the economics, of the well.

The basic material for consideration for any oil or gas production well is carbon steel and various grades are available, both standard grades listed in API5CT,[1] and also proprietary grades of individual manufacturers. The API grades are essentially low alloy steels which have rather high carbon contents in order to achieve the high yield strength necessary for downhole tubing. The proprietary grades generally cover similar materials but with increased chromium levels typically 1% or 3%. These chromium alloyed grades tend to be slightly more corrosion resistant but normally would still require the injection of corrosion inhibitor to provide sufficient resistance to corrosion.

The corrosion processes, which take place in the well, are typically due to the presence of acidic gases such as carbon dioxide and hydrogen sulfide. Carbon dioxide results in quite substantial weight loss corrosion of low alloy steels. The corrosion rates in carbon dioxide containing environments generally increase with temperature up to around 175-195°F (80-90°C). Rates of corrosion above 0.1 mm/y are generally regarded as too high to tolerate and so either corrosion inhibitors are injected to reduce the corrosion rate to below this 'manageable limit', or Corrosion Resistant Alloys (CRAs) are selected.

Around 175°F (80°C) the corrosion product film formed, which is iron carbonate, becomes a quite coherent and protective scale on the surface of the steel, so that at higher temperatures there is a tendency for corrosion rates to decrease. Nevertheless in wells where flow rates are high, this scale may be removed and this can result in very high corrosion rates. In high flow rate conditions with high shear stress at the tubing wall, corrosion inhibitors can also be less effective.

The presence of hydrogen sulfide in the produced fluids (so-called "sour" environments) can often be quite beneficial in oil and gas production systems in that it can result in much reduced corrosion rates as result of the iron sulfide scale. Most operators inject corrosion inhibitors to help to stabilise and reinforce this sulfide film. In some circumstances, in particular regimes of temperature, chloride content and pH the sulfide scale is not stable and the rate of corrosion can be very high, in the form of pitting. In such conditions more highly alloyed materials are required to resist corrosion.

Any corrosion in sour environments results in hydrogen diffusion into the steel and high strength grades of carbon and low alloy steels can be very difficult to produce with sufficient resistance to sulfide stress corrosion cracking. The choice of strength grade becomes particularly critical in sour systems and NACE MR0175[2] standard is a useful guide to the strength grades which are applicable for wells operating at different temperatures in sour conditions.

Where sour corrosion rates are expected to be too high, or where breakdown of the sulfide film is predicted with pitting attack, or where high strength downhole tubing is required, it is usual to shift from standard low alloy steel API grade tubing to different metallurgical systems, in particular to nickel base materials, which can provide the required corrosion and cracking resistance at high strength

Use of Corrosion Inhibitors

Weightloss corrosion in production environments with or without H_2S can be mitigated to some extent by the application of corrosion inhibitors in the well. Various systems exist for injecting the inhibitor, either on a continuous system (through the annulus or through a coiled tubing to the bottom of the well), or intermittently by squeezing inhibitor every few months down the well and into the surrounding formation at the bottom of the well.

Correctly selected corrosion inhibitors can be very effective in reducing the corrosion rate when they are present. However, periods of time without correctly controlled corrosion inhibitor injection can quickly result in higher corrosion rates. Thus, the use of corrosion inhibitor is generally regarded as only partially effective and it is normal to describe the effectiveness of the inhibitor by its "availability". The corrosion rate is reduced by the availability value when inhibition is carried out.

Availability values for continuous injection are generally in the range of 60-90% and for squeeze inhibitor range from 35-60% (depending upon the frequency of squeezes). The effectiveness of corrosion inhibitors is also strongly affected by the flow rates and distribution of phases flowing up the tubing. To get a fully effective corrosion inhibitor it is necessary for the water phase to be in contact with the steel surface and for the corrosion inhibitor to be carried in this phase so that there is good distribution over the steel surface.

Most wells deviate from the vertical and the change in angle up the length of the tubing results in changes in the distribution of water throughout the tubing. This results in a variation both in the corrosivity of the fluid and also in the capability of corrosion inhibitors to provide sufficient protection up the full tubing length.

The selection of carbon steels with corrosion inhibitor is technically limited as a means of preventing corrosion and also has the inevitable consequence of ongoing operating costs for purchase of chemicals and operator control. These operating costs have to be carefully considered when selecting downhole tubing.

Use of Corrosion Resistant Alloys

The choice of a corrosion resistant alloy tubing provides a number of benefits which offset the higher initial capital cost. Corrosion resistant alloy are much more resistant to corrosion than carbon steels and so no inhibitor is required. This is an immediate saving both in operating costs and in logistical support. The high corrosion resistance also means that the surface of the tube stays relatively smooth and this can be highly beneficial in improving production rates. As an example, the cost of changing from carbon steel to 13%Cr tubing in a major European gas field of 300 wells was reputedly entirely paid for by the increase in production rate because of the reduced surface roughness when compared to the originally installed carbon steel tubing. So the choice of corrosion resistant alloys may be made simply on an economic basis because of the reduced operating cost and increased production.

In some cases the corrosion rate with carbon steel may be too high and it can be shown that the tubing would not survive for a sufficient period of time before there would be a corrosion failure. Corrosion failure in tubing is detected by rise in pressure in the annulus which indicates a situation which may become unsafe. In such circumstances a workover would be carried out to replace the tubing. Replacing tubing carries a cost, not only of the new material which is installed but also the cost of the workover crew (which can be very high, particularly in remote parts of the world) and also the cost of the production which is lost over the duration of the workover. The choice of a corrosion resistant alloy tubing should be one that will last the full anticipated life time of the project and so all these potential future costs can be eliminated. In fact CRA tubing may be removed from a well at the end of its service period and inspected and potentially replaced in another well for a further period of service. So CRAs are recyclable in the production application which makes them a very sustainable materials choice.

Stainless Steels in Production Conditions

Martensitic Stainless Steels

The API 5CT standard lists 9Cr and 13Cr (UNS S42000) grades in strength levels C-75 and L-80 (minimum yield strength of 75 ksi (517 MPa) and 80 ksi (552 MPa) respectively). There are some proprietary sources of 13Cr tubing with a minimum yield strength of 95 ksi (655 MPa).

The 9Cr grade is rarely selected, mostly because the 13Cr grade is widely available and cost-effective. 13Cr tubing has been very widely applied and has provided excellent service internationally. Approximately 50% of all CRA production tubing is 13Cr, with the composition indicated in Table 9.1.

Table 9.1 Chemical Composition of CRA Grades of Downhole Tubing

CRA	C	Mn	Mo	Cr	Ni	Cu	P	S	Si
9Cr [a]	0.15	0.30-0.60	0.90-1.10	8.0-10.0	0.5	0.25	0.020	0.010	1.0
13Cr [a]	0.15-0.22	0.25-1.00	---	12.0-14.0	0.5	0.25	0.020	0.010	1.0
S13Cr	0.05	0.30-1.50	0-2.5	10.5-14.0	6.5	0.55	0.020	0.005	1.0
15Cr	0.22	1.00	1.00	14.0-16.0	2.0	0.25	0.020	0.010	1.0
22Cr	0.03	2.0	3.0	21.0-23.0	4.5-6.0	0.25	0.030	0.010	1.0
25Cr	0.03	2.0	3.0	24.0-26.0	4.5-6.0	0.25	0.030	0.010	1.0
28Cr	0.03	2.0	3.0	21.0-23.0	4.5-6.0	0.25	0.030	0.010	1.0

Single values are maximums.
a.   denotes API 5CT specification; other compositions are ranges covering typical proprietary grades.

13Cr martensitic stainless steel has excellent resistance to corrosion in oil or gas producing conditions which are free of hydrogen sulfide. Figure 9.1 illustrates the temperature boundary over a range of CO_2 partial pressures and chloride ion concentrations[3]. Above this temperature limit there is a risk of corrosion cracking or general corrosion or pitting attack with a propagation rate of more than

0.05 mm/y (2 mpy). So these temperatures represent the safe-use limits of this alloy.

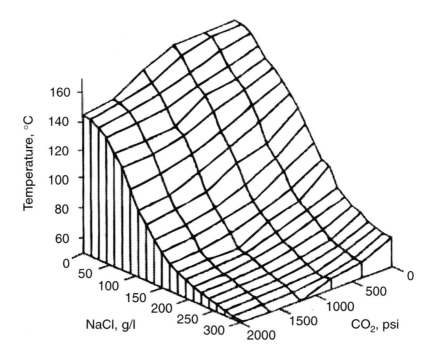

Figure 9.1 The corrosion resistance of 13Cr martensitic stainless steel in CO_2/NaCl environments in the absence of oxygen and H_2S.

Some producers make a 15Cr grade tubing but this has tended to be superseded by the development of 'Super 13Cr' (S13Cr) grades which have a much reduced carbon content, leaving more chromium in solution in the steel and improving the corrosion resistance compared to the standard 13Cr grade. These materials also show superior resistance to CO_2 corrosion, offering, in general a temperature limit which is approximately 30°C higher than the limit indicated in Figure 9.1 for standard 13Cr grades.

The proprietary S13Cr grades can also be manufactured at higher strengths than the standard 13Cr grade and are typically produced in yield strengths from 80 ksi (552 MPa) to 110 ksi (759 MPa). Deeper wells or higher pressure wells will tend to require S13Cr for the beneficial mechanical properties as well as the improved corrosion resistance.

In production environments containing no H_2S, or only traces, the 13Cr and S13Cr materials are capable of withstanding the majority of conditions practically encountered. They do not need corrosion inhibitors to provide additional protection.

In slightly sour conditions the limits of 13Cr and Super 13Cr steels have been investigated and generally show reasonable resistance to general corrosion in very low levels of H_2S[4]. However, in sour environments these materials tend to pit if the pH drops below 4 and this can then cause failure (particularly of sealing areas).

At higher H_2S levels there is a risk of sulfide stress corrosion cracking depending upon the pH of the environment. The limits of H_2S and pH to avoid cracking have been established in Figure 9.2 with the pH = 4 limit representing the pitting corrosion limit. The pH in this figure is evaluated at room temperature, as the material is most sensitive to cracking at low temperature. This is a safe approach to evaluating the material for service since there would be no risk of cracking even if the tubing was removed from the well after being saturated with hydrogen due to exposure to sour service conditions. Individual manufacturers may be able to show slightly better cracking resistance in Super 13Cr grades compared to the standard 13Cr grades, but Figure 9.2 summarises the typical room temperature properties of both types of 13Cr steel[5].

It is expected that both these materials would be able to withstand higher concentrations of H_2S without cracking when the temperature is raised. This has some benefit for wells which are completed with 13Cr or Super 13Cr when fields are not producing H_2S, but which gradually become 'sour' during production. In such cases the tubing is already at the well temperature, and particularly when the temperature is high, the tubing can be left in situ without immediate risk of cracking even when the H_2S level rises above the limits indicated in Figure 9.2.

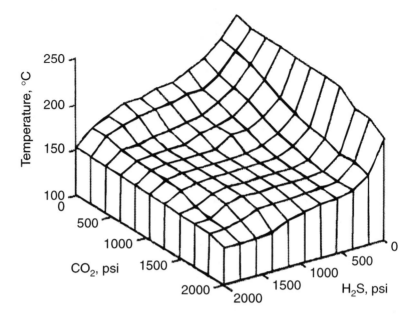

Figure 9.2 Ranges of pH and partial pressure of H_2S within which sulfide stress cracking is likely to arise in 13Cr and Super 13Cr tubing materials. (The pH 4 limit represents a limit below which there may be localised pitting corrosion).

This experience is similar to the noted reduced sensitivity to cracking of high strength carbon steels at higher well temperatures, e.g. above 175°F (80°C). Actual limits of H_2S tolerable at higher temperature would have to be defined through laboratory testing as there is relatively little published data on high temperature cracking resistance. Since such tubing would be saturated with hydrogen at the well temperature care would have to be taken when removing the tubing as it might crack when it cools to surface temperature. Whilst this approach would not be a safe basis for selecting tubing material for a new well completion, it does allow existing wells to continue to produce when the conditions become more aggressive than normal design guidelines would allow.

Duplex and Austenitic Stainless Steels

For more aggressive well conditions than martensitic stainless steel tubing can tolerate the next class of materials which are used are the duplex stainless steels. The range of compositions of typical grades are listed in Table 9.1. The duplex stainless steels can also be produced in higher strengths than the martensitic stainless steels because they can be cold worked. The high strength condition may be required for deeper wells.

The 22Cr duplex stainless steels are used in the solution annealed condition at a yield strength grade of 65 ksi (450 MPa) or, more typically, cold worked to yield strengths of 95 ksi (655 MPa), 110 ksi (760 MPa) and 140 ksi (965 MPa). The yield strength of solution annealed 25Cr duplex stainless steel is higher, 80 ksi (550 MPa), and it can be cold worked to 110 ksi (760 MPa) and 125 ksi (860 MPa).

In production conditions without H_2S the duplex stainless steels are suitable for use at temperatures up to well beyond any normal production conditions. Above about 375°F (190°C) [22Cr] or 410°F (210°C) [25Cr] some localised corrosion may start to be noted.

In sour conditions the limiting level of H_2S depends strongly upon the chloride content of the environment and also on the pH. General guidelines suggest that in gas systems with low chloride content (less than 1000 ppm) and low pH (around 3.7-4.2) both types of duplex can tolerate about 1bar of H_2S. In oil producing systems with typically 150,000 ppm of chloride ions and higher pH (around 5 or higher) the 22Cr can tolerate about 0.02 bar (0.3 psi) and the 25Cr can tolerate up to 0.14 bar (2 psi) H_2S.

Whilst these H_2S limits are higher than the martensitic stainless steels, the duplex stainless steels are not an appropriate choice for wells with significant H_2S content and for such conditions higher alloyed materials are required. The only stainless steel which is used at any significant level of H_2S in production tubing is Alloy 28. The typical composition is given in Table 9.1.

This steel is cold worked to a yield strength of 110 ksi (760 MPa) and is used in H_2S producing wells without corrosion cracking or general corrosion (<0.05 mm/y (2 mpy)) within the limits indicated by Figure 9.3. pH 4 represents the limit below which general corrosion may be excessive.

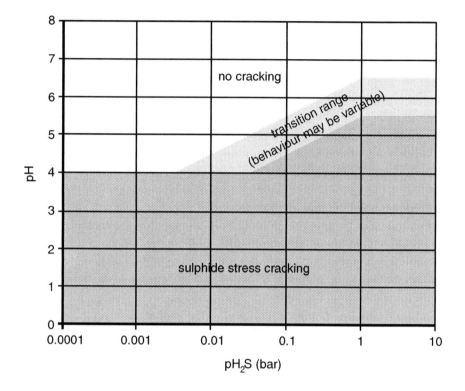

Figure 9.3 The corrosion resistance of Alloy 28 in H_2S/CO_2 environments in the absence of elemental sulfur.

For any well conditions which cannot be safely produced using the stainless steels indicated above, nickel alloy production tubing is required.

References

1.   API5CT - Specification for Tubing and Casing

2.   NACE MR0175 Standard Materials Requirements - Sulfide Stress Cracking Resistant - Metallic Materials for Oilfield Equipment

3.   Craig, B.D. "Corrosion Resistant Alloys in the Oil and Gas Industry - Selection Guidelines", NiDI Technical Series No 10073, Second edition, 2000

4.   Asahi H et al, "Corrosion and Mechanical Properties of Weldable Martensitic Stainless Line Pipes", Supermartensitic Stainless Steels '99, Brussels, Belgium, May 27-28, 1999

5.   "The Electronic Corrosion Engineer", software produced by Intetech Ltd 2001

Appendix

1

ABBREVIATED TERMS

AAI	Arco Alaska Inc.
AISI	American Iron and Steel Institute
ANSI	American National Standards Institute
API	American Petroleum Institute
ASME	American Society of Mechanical Engineers
ASTM	American Society for Testing and Materials
BS	British Standard
CII	Lined pipe product produced by NSC
CITHP	Closed in tubing head pressure
CLI	Creusot Loire Industrie
CPT	Critical pitting temperature
CRA	Corrosion-resistant alloy
CRC	CRC-Evans Automatic Welding
CTOD	Crack tip opening displacement
DHF	Dai-Ichi High Frequency
DIN	Deutsche Institut für Normung
DR	Radius of bend expressed as multiple of pipe diameter
DWTT	Drop weight tear test
EFC	European Federation of Corrosion
ENP	Electroless nickel plating
ESW	Electroslag welding
FAC	Flow accelerated corrosion
FCAW	Flux cored arc welding
FGD	Flue Gas Desulphurisation
GMAW	Gas metal arc welding
GTAW	Gas tungsten arc welding
HAZ	Heat-Affected Zone

HIP	Hot isostatic pressing
HSLA	High strength low alloy
ID	Internal diameter
JSW	Japan Steel Works
LIDB	Liquid interface diffusion bonding
LNG	Liquefied natural gas
NACE	National Association of Corrosion Engineers (now NACE International)
NDE	Non-destructive examination
NKK	NKK Corporation
NSC	Nippon Steel Corporation
OCTG	Oil Country Tubular Goods
OD	Outside diameter
OOR	Out-of-roundness
PASSO	Processo Arcos Saipem Saldatura Orbitale
PGMAW	Pulsed gas metal arc welding
PGTAW	Pulsed gas tungsten arc welding
PQR	Procedure Qualification Record
PTA	Pure Terephthalic Acid
PWHT	Post welding heat treatment
PWR	Pressurised Water Reactor
QT	Quenched and tempered
RT	Radiographic testing
SAW	Submerged arc welding
SCC	Stress corrosion cracking
SDH	Side drilled hole
SMAW	Shielded metal arc welding
SMYS	Specified minimum yield strength
SSC	Sulfide stress corrosion cracking
SSCV	Semi-submersible crane vessel
SWC	Stepwise cracking
TDS	Total dissolved solids
TMCP	Thermo-mechanical control process
UNS	Unified Numbering System
UO	U-ing, O-ing, (pipe forming)
UOE	U-ing, O-ing, Expansion, (pipe forming)
UT	Ultrasonic testing
V-A	Voest-Alpine

Appendix

2

TRADE NAMES

3M Company

Nextel

Allegheny Ludlum Corp. (ATI Properties Inc.)

AL-6X	AL-825
AL-6XN	AL-4750
AL-6XN PLUS	ALFA IV
AL 29-4	E-BRITE
AL 29-4C	E-BRITE 26-1
AL 29-4-2	Sealmet
AL-36	203 EZ
AL-42	JS 700
AL-52	

Allied-Signal Inc.

FVS-0611	FVS-1212
FVS-0812	Metglas

Allison Gas Turbine, Division of General Motors Corp.

AF-56

Aluminum Company of America (Alcoa)

CU78	CZ42
CW67	

Allvac Metals Company A Teledyne Company

René 41

Armco Inc.

Nitronic	15-5 PH
PH 13-8 Mo	17-4 PH
PH 15-7 Mo	18 SR
12 SR	21-6-9

Avesta Sheffield AB

254 SMO	353 MA
Monit	SAF 2507
44LN	PRODEC
153 MA	2205 Code Plus Two
253 MA	654 SMO

British Steel Corp.

Esshete

Canon-Muskegon Corp.

CM 247 LC	CMSX

Carpenter Technology Corp.

Custom 450	883 PLUS
Custom 455	20Cb-3
Gall-Tough™	20Mo-4
Glass Sealing "49"	20Mo-6
Invar "36"	Carpenter Project 70
Kovar	Extendo-Die™
Low Expansion "42"	Pyromet
TrimRite	Pyromet 718
7-Mo PLUS	Pyromet Alloy 350
18-18 PLUS	Pyromet Alloy 355
NeutroSorb PLUS	Pyrowear Alloy 53
Custom Age 625 PLUS	

Creusot-Loire Industrie

URANUS   CLI 168 HE
CLI 170 HE

Crucible Materials Corp.

SEA-CURE

Deloro Stellite, Inc.

Stellite

DMV Stainless

DMV

Duraloy Technologies, Inc.

22H   MO-RE
Super 22H   Supertherm

E.I. Dupont De Nemours & Company, Inc.

Kapton   Tefzell
Teflon

Fansteel Inc

Tantology   Tribocor

Haynes International Inc.

Hastelloy   Haynes

Imphy, S.A.

Elinvar   Invar

Krupp Thyssen Nirosta GmbH

NIROSTA

Krupp VDM

Nicrofer   Cronifer

Lockheed Missile and Space Co.

Transage 134	Transage 175

Martin Marietta Corp.

MAR-M	Weldalite

Meighs Limited

Ferralium

Pfizer Hospital Products Group, Inc.

Vitallium

Pratt & Whitney Aircraft

PWA 1484

Rolled Alloys

RA 85H	RA 333
RA 253 MA	

Rolls Royce, Inc.

RR 2000	SRR 99

Sandvik AB

SANICRO	Sandvik SAF 2507
Sandvik SAF 2304	Sandvik SAF 2205

SCM Metal Products, Inc.

GlidCop

SINTERMETALLWERK KREBSOGE GmbH

DISPAL

Snecma/Onera

AM1

Special Metals Corp.

625LCF	NILO
686CPT	MAXORB
800HT	CORRONEL
INCO-WELD	INCOCLAD
INCO-CORED	INCOTHERM
INCOFLUX	INCOLOY
NI-ROD	NIMONIC
NI-SPAN-C	UDIMET
FERRY	MONEL
DURANICKEL	INCONEL
RESISTOHM	

SPS Technologies, Inc.

MP35N

Standard Pressed Steel Co.

MP (Multiphase)

Sumitomo

DP3™	DP3W™

Thompson Company

Refrasil

United Technologies, Inc.

Waspaloy

Universal Cyclops Steel Corp.

Unitemp

Uss, Division of USX Corp.

Cryogenic Tenelon	Tenelon

Vallourec

290 Mo™	29Cr-3Mo™

Vallourec Mannesmann Tubes	
VM 22	VM 25

Vallourec Welded
VLX

Vereingte Deutsche Metallwerks (VDM)	
Cronifer	1925HMo™

Wier Materials Ltd.
ZERON 100™

Westinghouse Electric Corp.
Discaloy

Appendix

3

HARDNESS CONVERSION NUMBERS

APPROXIMATE HARDNESS CONVERSION NUMBERS FOR NONAUSTENITIC STEELS[a, b]

Rockwell C 150 kgf Diamond HRC	Vickers HV	Brinell 3000 kgf 10 mm ball HB	Knoop 500 gf HK	Rockwell A 60 kgf Diamond HRA	Rockwell Superficial Hardness			Approximate Tensile Strength ksi (MPa)
					15 kgf Diamond HR15N	30 kgf Diamond HR30N	45 kgf Diamond HR45N	
68	940	---	920	85.6	93.2	84.4	75.4	---
67	900	---	895	85.0	92.9	83.6	74.2	---
66	865	---	870	84.5	92.5	82.8	73.3	---
65	832	739[d]	846	83.9	92.2	81.9	72.0	---
64	800	722[d]	822	83.4	91.8	81.1	71.0	---
63	772	706[d]	799	82.8	91.4	80.1	69.9	---
62	746	688[d]	776	82.3	91.1	79.3	68.8	---
61	720	670[d]	754	81.8	90.7	78.4	67.7	---
60	697	654[d]	732	81.2	90.2	77.5	66.6	---
59	674	634[d]	710	80.7	89.8	76.6	65.5	351 (2420)
58	653	615	690	80.1	89.3	75.7	64.3	338 (2330)
57	633	595	670	79.6	88.9	74.8	63.2	325 (2240)
56	613	577	650	79.0	88.3	73.9	62.0	313 (2160)
55	595	560	630	78.5	87.9	73.0	60.9	301 (2070)
54	577	543	612	78.0	87.4	72.0	59.8	292 (2010)
53	560	525	594	77.4	86.9	71.2	58.6	283 (1950)
52	544	512	576	76.8	86.4	70.2	57.4	273 (1880)
51	528	496	558	76.3	85.9	69.4	56.1	264 (1820)
50	513	482	542	75.9	85.5	68.5	55.0	255 (1760)
49	498	468	526	75.2	85.0	67.6	53.8	246 (1700)
48	484	455	510	74.7	84.5	66.7	52.5	238 (1640)
47	471	442	495	74.1	83.9	65.8	51.4	229 (1580)

APPROXIMATE HARDNESS CONVERSION NUMBERS FOR NONAUSTENITIC STEELS[a, b] (Continued)

Rockwell C 150 kgf Diamond HRC	Vickers HV	Brinell 3000 kgf 10 mm ball HB	Knoop 500 gf HK	Rockwell A 60 kgf Diamond HRA	Rockwell Superficial Hardness			Approximate Tensile Strength ksi (MPa)
					15 kgf Diamond HR15N	30 kgf Diamond HR30N	45 kgf Diamond HR45N	
46	458	432	480	73.6	83.5	64.8	50.3	221 (1520)
45	446	421	466	73.1	83.0	64.0	49.0	215 (1480)
44	434	409	452	72.5	82.5	63.1	47.8	208 (1430)
43	423	400	438	72.0	82.0	62.2	46.7	201 (1390)
42	412	390	426	71.5	81.5	61.3	45.5	194 (1340)
41	402	381	414	70.9	80.9	60.4	44.3	188 (1300)
40	392	371	402	70.4	80.4	59.5	43.1	182 (1250)
39	382	362	391	69.9	79.9	58.6	41.9	177 (1220)
38	372	353	380	69.4	79.4	57.7	40.8	171 (1180)
37	363	344	370	68.9	78.8	56.8	39.6	166 (1140)
36	354	336	360	68.4	78.3	55.9	38.4	161 (1110)
35	345	327	351	67.9	77.7	55.0	37.2	156 (1080)
34	336	319	342	67.4	77.2	54.2	36.1	152 (1050)
33	327	311	334	66.8	76.6	53.3	34.9	149 (1030)
32	318	301	326	66.3	76.1	52.1	33.7	146 (1010)
31	310	294	318	65.8	75.6	51.3	32.5	141 (970)
30	302	286	311	65.3	75.0	50.4	31.3	138 (950)
29	294	279	304	64.6	74.5	49.5	30.1	135 (930)
28	286	271	297	64.3	73.9	48.6	28.9	131 (900)
27	279	264	290	63.8	73.3	47.7	27.8	128 (880)
26	272	258	284	63.3	72.8	46.8	26.7	125 (860)
25	266	253	278	62.8	72.2	45.9	25.5	123 (850)
24	260	247	272	62.4	71.6	45.0	24.3	119 (820)

APPROXIMATE HARDNESS CONVERSION NUMBERS FOR NONAUSTENITIC STEELS[a, b] (Continued)

Rockwell C 150 kgf Diamond HRC	Vickers HV	Brinell 3000 kgf 10 mm ball HB	Knoop 500 gf HK	Rockwell A 60 kgf Diamond HRA	Rockwell Superficial Hardness			Approximate Tensile Strength ksi (MPa)
					15 kgf Diamond HR15N	30 kgf Diamond HR30N	45 kgf Diamond HR45N	
23	254	243	266	62.0	71.0	44.0	23.1	117 (810)
22	248	237	261	61.5	70.5	43.2	22.0	115 (790)
21	243	231	256	61.0	69.9	42.3	20.7	112 (770)
20	238	226	251	60.5	69.4	41.5	19.6	110 (760)

a. This table gives the approximate interrelationships of hardness values and approximate tensile strength of steels. It is possible that steels of various compositions and processing histories will deviate in hardness-tensile strength relationship from the data presented in this table. The data in this table should not be used for austenitic stainless steels, but have been shown to be applicable for ferritic and martensitic stainless steels. Where more precise conversions are required, they should be developed specially for each steel composition, heat treatment, and part.

b. All relative hardness values in this table are averages of tests on various metals whose different properties prevent establishment of exact mathematical conversions. These values are consistent with ASTM A 370-91 for nonaustenitic steels. It is recommended that ASTM standards A 370, E 140, E 10, E 18, E 92, E 110 and E 384, involving hardness tests on metals, be reviewed prior to interpreting hardness conversion values.

c. Carbide ball, 10mm.

d. This Brinell hardness value is outside the recommended range for hardness testing in accordance with ASTM E 10.

APPROXIMATE HARDNESS CONVERSION NUMBERS FOR NONAUSTENITIC STEELS [a, b]

Rockwell B 100 kgf 1/16" ball HRB	Vickers HV	Brinell 3000 kgf 10 mm HB	Knoop 500 gf HK	Rockwell A 60 kgf Diamond HRA	Rockwell Superficial Hardness			Approximate Tensile Strength ksi (MPa)
					15 kgf 1/16" ball HR15T	30 kgf 1/16" ball HR30T	45 kgf 1/16" ball HR45T	
100	240	240	251	61.5	93.1	83.1	72.9	116 (800)
99	234	234	246	60.9	92.8	82.5	71.9	114 (785)
98	228	228	241	60.2	92.5	81.8	70.9	109 (750)
97	222	222	236	59.5	92.1	81.1	69.9	104 (715)
96	216	216	231	58.9	91.8	80.4	68.9	102 (705)
95	210	210	226	58.3	91.5	79.8	67.9	100 (690)
94	205	205	221	57.6	91.2	79.1	66.9	98 (675)
93	200	200	216	57.0	90.8	78.4	65.9	94 (650)
92	195	195	211	56.4	90.5	77.8	64.8	92 (635)
91	190	190	206	55.8	90.2	77.1	63.8	90 (620)
90	185	185	201	55.2	89.9	76.4	62.8	89 (615)
89	180	180	196	54.6	89.5	75.8	61.8	88 (605)
88	176	176	192	54.0	89.2	75.1	60.8	86 (590)
87	172	172	188	53.4	88.9	74.4	59.8	84 (580)
86	169	169	184	52.8	88.6	73.8	58.8	83 (570)
85	165	165	180	52.3	88.2	73.1	57.8	82 (565)
84	162	162	176	51.7	87.9	72.4	56.8	81 (560)
83	159	159	173	51.1	87.6	71.8	55.8	80 (550)
82	156	156	170	50.6	87.3	71.1	54.8	77 (530)
81	153	153	167	50.0	86.9	70.4	53.8	73 (505)
80	150	150	164	49.5	86.6	69.7	52.8	72 (495)
79	147	147	161	48.9	86.3	69.1	51.8	70 (485)

APPROXIMATE HARDNESS CONVERSION NUMBERS FOR NONAUSTENITIC STEELS [a, b] (Continued)

Rockwell B 100 kgf 1/16" ball HRB	Vickers HV	Brinell 3000 kgf 10 mm HB	Knoop 500 gf HK	Rockwell A 60 kgf Diamond HRA	Rockwell Superficial Hardness			Approximate Tensile Strength ksi (MPa)
					15 kgf 1/16" ball HR15T	30 kgf 1/16" ball HR30T	45 kgf 1/16" ball HR45T	
78	144	144	158	48.4	86.0	68.4	50.8	69 (475)
77	141	141	155	47.9	85.6	67.7	49.8	68 (470)
76	139	139	152	47.3	85.3	67.1	48.8	67 (460)
75	137	137	150	46.8	85.0	66.4	47.8	66 (455)
74	135	135	147	46.3	84.7	65.7	46.8	65 (450)
73	132	132	145	45.8	84.3	65.1	45.8	64 (440)
72	130	130	143	45.3	84.0	64.4	44.8	63 (435)
71	127	127	141	44.8	83.7	63.7	43.8	62 (425)
70	125	125	139	44.3	83.4	63.1	42.8	61 (420)
69	123	123	137	43.8	83.0	62.4	41.8	60 (415)
68	121	121	135	43.3	82.7	61.7	40.8	59 (405)
67	119	119	133	42.8	82.4	61.0	39.8	58 (400)
66	117	117	131	42.3	82.1	60.4	38.7	57 (395)
65	116	116	129	41.8	81.8	59.7	37.7	56 (385)
64	114	114	127	41.4	81.4	59.0	36.7	---
63	112	112	125	40.9	81.1	58.4	35.7	---
62	110	110	124	40.4	80.8	57.7	34.7	---
61	108	108	122	40.0	80.5	57.0	33.7	---
60	107	107	120	39.5	80.1	56.4	32.7	---
59	106	106	118	39.0	79.8	55.7	31.7	---
58	104	104	117	38.6	79.5	55.0	30.7	---
57	103	103	115	38.1	79.2	54.4	29.7	---
56	101	101	114	37.7	78.8	53.7	28.7	---

APPROXIMATE HARDNESS CONVERSION NUMBERS FOR NONAUSTENITIC STEELS[a,b] (Continued)

Rockwell B 100 kgf 1/16" ball HRB	Vickers HV	Brinell 3000 kgf 10 mm HB	Knoop 500 gf HK	Rockwell A 60 kgf Diamond HRA	Rockwell Superficial Hardness			Approximate Tensile Strength ksi (MPa)
					15 kgf 1/16" ball HR15T	30 kgf 1/16" ball HR30T	45 kgf 1/16" ball HR45T	
55	100	100	112	37.2	78.5	53.0	27.7	---
54	---	---	111	36.8	78.2	52.4	26.7	---
53	---	---	110	36.3	77.9	51.7	25.7	---
52	---	---	109	35.9	77.5	51.0	24.7	---
51	---	---	108	35.5	77.2	50.3	23.7	---
50	---	---	107	35.0	76.9	49.7	22.7	---
49	---	---	106	34.6	76.6	49.0	21.7	---
48	---	---	105	34.1	76.2	48.3	20.7	---
47	---	---	104	33.7	75.9	47.7	19.7	---
46	---	---	103	33.3	75.6	47.0	18.7	---
45	---	---	102	32.9	75.3	46.3	17.7	---
44	---	---	101	32.4	74.9	45.7	16.7	---
43	---	---	100	32.0	74.6	45.0	15.7	---
42	---	---	99	31.6	74.3	44.3	14.7	---
41	---	---	98	31.2	74.0	43.7	13.6	---
40	---	---	97	30.7	73.6	43.0	12.6	---
39	---	---	96	30.3	73.3	42.3	11.6	---
38	---	---	95	29.9	73.0	41.6	10.6	---
37	---	94	29.5	78.0	41.0	9.6		---
36	---	---	93	29.1	72.3	40.3	8.6	---
35	---	---	92	28.7	72.0	39.6	7.6	---
34	---	---	91	28.2	71.7	39.0	6.6	---
33	---	---	90	27.8	71.4	38.3	5.6	---

CASTI Handbook of Stainless Steels & Nickel Alloys – Second Edition

APPROXIMATE HARDNESS CONVERSION NUMBERS FOR NONAUSTENITIC STEELS [a, b] (Continued)				Rockwell A	Rockwell Superficial Hardness			Approximate
Rockwell B 100 kgf 1/16" ball HRB	Vickers HV	Brinell 3000 kgf 10 mm HB	Knoop 500 gf HK	60 kgf Diamond HRA	15 kgf 1/16" ball HR15T	30 kgf 1/16" ball HR30T	45 kgf 1/16" ball HR45T	Tensile Strength ksi (MPa)
32	---	---	89	27.4	71.0	37.6	4.6	---
31	---	---	88	27.0	70.7	37.0	3.6	---
30	---	---	87	26.6	70.4	36.3	2.6	---

a. This table gives the approximate interrelationships of hardness values and approximate tensile strength of steels. It is possible that steels of various compositions and processing histories will deviate in hardness-tensile strength relationship from the data presented in this table. The data in this table should not be used for austenitic stainless steels, but have been shown to be applicable for ferritic and martensitic stainless steels. Where more precise conversions are required, they should be developed specially for each steel composition, heat treatment, and part.

b. All relative hardness values in this table are averages of tests on various metals whose different properties prevent establishment of exact mathematical conversions. These values are consistent with ASTM A 370-91 for nonaustenitic steels. It is recommended that ASTM standards A 370, E 140, E 10, E 18, E 92, E 110 and E 384, involving hardness tests on metals, be reviewed prior to interpreting hardness conversion values.

APPROXIMATE HARDNESS NUMBERS FOR AUSTENITIC STEELS[a]

Rockwell C 150 kgf, Diamond HRC	Rockwell A 60 kgf, Diamond HRA	Rockwell Superficial Hardness		
		15 kgf, Diamond HR15N	30 kgf, Diamond HR30N	45 kgf, Diamond HR45N
48	74.4	84.1	66.2	52.1
47	73.9	83.6	65.3	50.9
46	73.4	83.1	64.5	49.8
45	72.9	82.6	63.6	48.7
44	72.4	82.1	62.7	47.5
43	71.9	81.6	61.8	46.4
42	71.4	81.0	61.0	45.2
41	70.9	80.5	60.1	44.1
40	70.4	80.0	59.2	43.0
39	69.9	79.5	58.4	41.8
38	69.3	79.0	57.5	40.7
37	68.8	78.5	56.6	39.6
36	68.3	78.0	55.7	38.4
35	67.8	77.5	54.9	37.3
34	67.3	77.0	54.0	36.1

APPROXIMATE HARDNESS NUMBERS FOR AUSTENITIC STEELS[a] (Continued)

Rockwell C 150 kgf, Diamond HRC	Rockwell A 60 kgf, Diamond HRA	Rockwell Superficial Hardness		
		15 kgf, Diamond HR15N	30 kgf, Diamond HR30N	45 kgf, Diamond HR45N
33	66.8	76.5	53.1	35.0
32	66.3	75.9	52.3	33.9
31	65.8	75.4	51.4	32.7
30	65.3	74.9	50.5	31.6
29	64.8	74.4	49.6	30.4
28	64.3	73.9	48.8	29.3
27	63.8	73.4	47.9	28.2
26	63.3	72.9	47.0	27.0
25	62.8	72.4	46.2	25.9
24	62.3	71.9	45.3	24.8
23	61.8	71.3	44.4	23.6
22	61.3	70.8	43.5	22.5
21	60.8	70.3	42.7	21.3
20	60.3	69.8	41.8	20.2

a. All relative hardness values in this table are averages of tests on various metals whose different properties prevent establishment of exact mathematical conversions. These values are consistent with ASTM A 370-91 for austenitic steels. It is recommended that ASTM standards A 370, E 140, E 10, E 18, E 92, E 110 and E 384, involving hardness tests on metals, be reviewed prior to interpreting hardness conversion values.

Appendix

4

UNIT CONVERSIONS

METRIC CONVERSION FACTORS

To Convert From	To	Multiply By	To Convert From	To	Multiply By
Angle			**Mass per unit time**		
degree	rad	1.745 329 E -02	lb/h	kg/s	1.259 979 E - 04
Area			lb/min	kg/s	7.559 873 E - 03
in.²	mm²	6.451 600 E + 02	lb/s	kg/s	4.535 924 E - 01
in.²	cm²	6.451 600 E + 00	**Mass per unit volume (includes density)**		
in.²	m²	6.451 600 E - 04	g/cm³	kg/m³	1.000 000 E + 03
ft²	m²	9.290 304 E - 02	lb/ft³	g/cm³	1.601 846 E - 02
Bending moment or torque			lb/ft³	kg/m³	1.601 846 E + 01
lbf - in.	N - m	1.129 848 E - 01	lb/in.³	g/cm³	2.767 990 E + 01
lbf - ft	N - m	1.355 818 E + 00	lb/in.³	kg/m³	2.767 990 E + 04
kgf - m	N - m	9.806 650 E + 00	**Power**		
ozf - in.	N - m	7.061 552 E - 03	Btu/s	kW	1.055 056 E + 00
Bending moment or torque per unit length			Btu/min	kW	1.758 426 E - 02
lbf - in./in.	N - m/m	4.448 222 E + 00	Btu/h	W	2.928 751 E - 01
lbf - ft/in.	N - m/m	5.337 866 E + 01	erg/s	W	1.000 000 E - 07
Corrosion rate			ft - lbf/s	W	1.355 818 E + 00
mils/yr	mm/yr	2.540 000 E - 02	ft - lbf/min	W	2.259 697 E - 02
mils/yr	μ/yr	2.540 000 E + 01	ft - lbf/h	W	3.766 161 E - 04
Current density			hp (550 ft - lbf/s)	kW	7.456 999 E - 01
A/in.²	A/cm²	1.550 003 E - 01	hp (electric)	kW	7.460 000 E - 01
A/in.²	A/mm²	1.550 003 E - 03	**Power density**		
A/ft²	A/m²	1.076 400 E + 01	W/in.²	W/m²	1.550 003 E + 03

METRIC CONVERSION FACTORS (Continued)

To Convert From	To	Multiply By	To Convert From	To	Multiply By
Electricity and magnetism			**Pressure (fluid)**		
gauss	T	1.000 000 E - 04	atm (standard)	Pa	1.013 250 E + 05
maxwell	μWb	1.000 000 E - 02	bar	Pa	1.000 000 E + 05
mho	S	1.000 000 E + 00	in. Hg (32°F)	Pa	3.386 380 E + 03
Oersted	A/m	7.957 700 E + 01	in. Hg (60°F)	Pa	3.376 850 E + 03
Ω - cm	Ω - m	1.000 000 E - 02	lbf/in.^{2} (psi)	Pa	6.894 757 E + 03
Ω circular - mil/ft	μΩ - m	1.662 426 E - 03	torr (mm Hg, 0°C)	Pa	1.333 220 E + 02
Energy (impact other)			**Specific heat**		
ft•lbf	J	1.355 818 E + 00	Btu/lb - °F	J/kg - °K	4.186 800 E + 03
Btu (thermochemical)	J	1.054 350 E + 03	cal/g - °C	J/kg - °K	4.186 800 E + 03
cal (thermochemical)	J	4.184 000 E + 00	**Stress (force per unit area)**		
kW - h	J	3.600 000 E + 06	tonf/in.^{2} (tsi)	MPa	1.378 951 E + 01
W - h	J	3.600 000 E + 03	kgf/mm^2	MPa	9.806 650 E + 00
Flow rate			ksi	MPa	6.894 757 E + 00
ft^3/h	L/min	4.719 475 E - 01	lbf/in.^{2} (psi)	MPa	6.894 757 E - 03
ft^3/min	L/min	2.831 000 E + 01	N/mm^2	Pa	1.000 000 E + 00
gal/h	L/min	6.309 020 E - 02	**Temperature**		
gal/min	L/min	3.785 412 E + 00	°F	°C	5/9 (F - 32)
Force			°R	°K	5/9
lbf	N	4.448 222 E + 00	**Temperature interval**		
kip (1000 lbf)	N	4.448 222 E + 03	°F	°C	5/9
tonf	kN	8.896 443 E + 00	**Thermal conductivity**		
kgf	N	9.806 650 E + 00	Btu - in./s - ft^2 - °F	W/m - °K	5.192 204 E + 02
lbf/in.	N/m	1.751 268 E + 02	cal/cm - s - °C	W/m - °K	4.184 000 E + 02

METRIC CONVERSION FACTORS (Continued)

To Convert From	To	Multiply By	To Convert From	To	Multiply By
Force per unit length					
lbf/ft	N/m	1.459 390 E + 01	Btu/ft - h - °F	W/m - °K	1.730 735 E + 00
			Btu - in./h. ft^2 - °F	W/m - °K	1.442 279 E - 01
Fracture toughness			**Thermal expansion**		
ksi in.$^{1/2}$	MPa m $^{1/2}$	1.098 800 E + 00	in./in. - °C	m/m - °K	1.000 000 E + 00
Heat content			in./in. - °F	m/m - °K	1.800 000 E + 00
Btu/lb	kJ/kg	2.326 000 E + 00	**Velocity**		
cal/g	kJ/kg	4.186 800 E + 00	ft/h	m/s	8.466 667 E - 05
Heat input			ft/min	m/s	5.080 000 E - 03
J/in.	J/m	3.937 008 E + 01	ft/s	m/s	3.048 000 E - 01
kJ/in.	kJ/mm	3.937 008 E - 02	in./s	m/s	2.540 000 E - 02
Length			km/h	m/s	2.777 778 E - 01
A	nm	1.000 000 E - 01	mph	km/h	1.609 344 E + 00
μin.	μm	2.540 000 E - 02	**Velocity of rotation**		
mil	μm	2.540 000 E + 01	rev/min (rpm)	rad/s	1.047 164 E - 01
in.	mm	2.540 000 E + 01	rev/s	rad/s	6.283 185 E + 00
in.	cm	2.540 000 E + 00	**Viscosity**		
ft	m	3.048 000 E - 01	poise	Pa - s	1.000 000 E - 01
yd	m	9.144 000 E - 01	stokes	m^2/s	1.000 000 E - 04
mile	km	1.609 300 E + 00	ft^2/s	m^2/s	9.290 304 E - 02
Mass			in.^{2}/s	mm^2/s	6.451 600 E + 02
oz	kg	2.834 952 E - 02	**Volume**		
lb	kg	4.535 924 E - 01	in.^{3}	m^3	1.638 706 E - 05
ton (short 2000 lb)	kg	9.071 847 E + 02	ft^3	m^3	2.831 685 E - 02
ton (short 2000 lb)	kg x 10^3	9.071 847 E - 01	fluid oz	m^3	2.957 353 E - 05

METRIC CONVERSION FACTORS (Continued)

To Convert From	To	Multiply By	To Convert From	To	Multiply By
Mass per unit area			**Volume per unit time**		
ton (long 2240 lb)	kg	1.016 047 E + 03	gal (U.S. liquid)	m^3	3.785 412 E - 03
$kg \times 10^3$ = 1 metric ton			ft^3/min	m^3/s	4.719 474 E - 04
oz/in.^{2}	kg/m^2	4.395 000 E + 01	ft^3/s	m^3/s	2.831 685 E - 02
oz/ft^2	kg/m^2	3.051 517 E - 01	in.^{3}/min	m^3/s	2.731 177 E - 07
oz/yd^2	kg/m^2	3.390 575 E - 02	**Wavelength**		
lb/ft^2	kg/m^2	4.882 428 E + 00	A	nm	1.000 000 E - 01

SI PREFIXES

Prefix	Symbol	Exponential Expression	Multiplication Factor
exa	E	10^{18}	1 000 000 000 000 000 000
peta	P	10^{15}	1 000 000 000 000 000
tera	T	10^{12}	1 000 000 000 000
giga	G	10^{9}	1 000 000 000
mega	M	10^{6}	1 000 000
kilo	k	10^{3}	1 000
hecto	h	10^{2}	100
deka	da	10^{1}	10
Base Unit	---	10^{0}	1
deci	d	10^{-1}	0.1
centi	c	10^{-2}	0.01
milli	m	10^{-3}	0.001
micro	μ	10^{-6}	0.000 001
nano	n	10^{-9}	0.000 000 001
pico	p	10^{-12}	0.000 000 000 001
femto	f	10^{-15}	0.000 000 000 000 001
atto	a	10^{-18}	0.000 000 000 000 000 001

THE GREEK ALPHABET

A, α - Alpha	I, ι - Iota
B, β - Beta	K, κ - Kappa
Γ, γ - Gamma	Λ, λ - Lambda
Δ, δ - Delta	M, μ - Mu
E, ε - Epsilon	N, ν - Nu
Z, ζ - Zeta	Ξ, ξ - Xi
H, η - Eta	O, o - Omicron
Θ, θ - Theta	Π, π - Pi
P, ρ - Rho	
Σ, σ - Sigma	
T, τ - Tau	
Y, υ - Upsilon	
Φ, φ - Phi	
X, χ - Chi	
Ψ, ψ - Psi	
Ω, ω - Omega	

CASTI Handbook of Stainless Steels & Nickel Alloys – Second Edition

Appendix

5

PIPE DIMENSIONS

DIMENSIONS OF WELDED AND SEAMLESS PIPE[a]

Nominal Pipe Size (in.)	Outside Diameter (in.)	Nominal Wall Thickness (in.)						
		Schedule 5S	Schedule 10S	Schedule 10	Schedule 20	Schedule 30	Schedule Standard	Schedule 40
1/8	0.405	---	0.049	---	---	---	0.068	0.068
1/4	0.540	---	0.065	---	---	---	0.088	0.088
3/8	0.675	---	0.065	---	---	---	0.091	0.091
1/2	0.840	0.065	0.083	---	---	---	0.109	0.109
3/4	1.050	0.065	0.083	---	---	---	0.113	0.113
1	1.315	0.065	0.109	---	---	---	0.133	0.133
1 1/4	1.660	0.065	0.109	---	---	---	0.140	0.140
1 1/2	1.900	0.065	0.109	---	---	---	0.145	0.145
2	2.375	0.065	0.109	---	---	---	0.154	0.154
2 1/2	2.875	0.083	0.120	---	---	---	0.203	0.203
3	3.5	0.083	0.120	---	---	---	0.216	0.216
3 1/2	4.0	0.083	0.120	---	---	---	0.226	0.226
4	4.5	0.083	0.120	---	---	---	0.237	0.237
5	5.563	0.109	0.134	---	---	---	0.258	0.258
6	6.625	0.109	0.134	---	---	---	0.280	0.280
8	8.625	0.109	0.148	---	0.250	0.277	0.322	0.322
10	10.75	0.134	0.165	---	0.250	0.307	0.365	0.365
12	12.75	0.156	0.180	---	0.250	0.330	0.375	0.406
14 O.D.	14.0	0.156	0.188	0.250	0.312	0.375	0.375	0.438
16 O.D.	16.0	0.165	0.188	0.250	0.312	0.375	0.375	0.500
18 O.D.	18.0	0.165	0.188	0.250	0.312	0.438	0.375	0.562
20 O.D.	20.0	0.188	0.218	0.250	0.375	0.500	0.375	0.594
22 O.D.	22.0	0.188	0.218	0.250	0.375	0.500	0.375	---

DIMENSIONS OF WELDED AND SEAMLESS PIPE[a] (Continued)

Nominal Pipe Size (in.)	Outside Diameter (in.)	Nominal Wall Thickness (in.)							
		Schedule 5S	Schedule 10S	Schedule 10	Schedule 20	Schedule 30	Schedule Standard	Schedule 40	
24 O.D.	24.0	0.218	0.250	0.250	0.375	0.562	0.375	0.688	
26 O.D.	26.0	---	---	0.312	0.500	---	0.375	---	
28 O.D.	28.0	---	---	0.312	0.500	0.625	0.375	---	
30 O.D.	30.0	0.250	0.312	0.312	0.500	0.625	0.375	---	
32 O.D.	32.0	---	---	0.312	0.500	0.625	0.375	0.688	
34 O.D.	34.0	---	---	0.312	0.500	0.625	0.375	0.688	
36 O.D.	36.0	---	---	0.312	0.500	0.625	0.375	0.750	
42 O.D.	42.0	---	---	---	---	---	0.375	---	

a. See next page for heavier wall thicknesses

DIMENSIONS OF WELDED AND SEAMLESS PIPE

Nominal Pipe Size (in.)	Outside Diameter (in.)	Nominal Wall Thickness (in.)							
		Schedule 60	Extra Strong	Schedule 80	Schedule 100	Schedule 120	Schedule 140	Schedule 160	XX Strong
1/8	0.405	---	0.095	0.095	---	---	---	---	---
1/4	0.540	---	0.119	0.119	---	---	---	---	---
3/8	0.675	---	0.126	0.126	---	---	---	---	---
1/2	0.840	---	0.147	0.147	---	---	---	0.188	0.294
3/4	1.050	---	0.154	0.154	---	---	---	0.219	0.308
1	1.315	---	0.179	0.179	---	---	---	0.250	0.358
1 1/4	1.660	---	0.191	0.191	---	---	---	0.250	0.382
1 1/2	1.900	---	0.200	0.200	---	---	---	0.281	0.400
2	2.375	---	0.218	0.218	---	---	---	0.344	0.436
2 1/2	2.875	---	0.276	0.276	---	---	---	0.375	0.552
3	3.5	---	0.300	0.300	---	---	---	0.438	0.600
3 1/2	4.0	---	0.318	0.318	---	---	---	---	---
4	4.5	---	0.337	0.337	---	0.438	---	0.531	0.674
5	5.563	---	0.375	0.375	---	0.500	---	0.625	0.750
6	6.625	---	0.432	0.432	---	0.562	---	0.719	0.864
8	8.625	0.406	0.500	0.500	0.594	0.719	0.812	0.906	0.875
10	10.75	0.500	0.500	0.594	0.719	0.844	1.000	1.125	1.000
12	12.75	0.562	0.500	0.688	0.844	1.000	1.125	1.312	1.000
14 O.D.	14.0	0.594	0.500	0.750	0.938	1.094	1.250	1.406	---
16 O.D.	16.0	0.656	0.500	0.844	1.031	1.219	1.438	1.594	---
18 O.D.	18.0	0.750	0.500	0.938	1.156	1.375	1.562	1.781	---
20 O.D.	20.0	0.812	0.500	1.031	1.281	1.500	1.750	1.969	---
22 O.D.	22.0	0.875	0.500	1.125	1.375	1.625	1.875	2.125	---

DIMENSIONS OF WELDED AND SEAMLESS PIPE (Continued)										
Nominal Pipe Size (in.)	Outside Diameter (in.)	Nominal Wall Thickness (in.)								
		Schedule 60	Extra Strong	Schedule 80	Schedule 100	Schedule 120	Schedule 140	Schedule 160	XX Strong	
24 O.D.	24.0	0.969	0.500	1.218	1.531	1.812	2.062	2.344	---	
26 O.D.	26.0	---	0.500	---	---	---	---	---	---	
28 O.D.	28.0	---	0.500	---	---	---	---	---	---	
30 O.D.	30.0	---	0.500	---	---	---	---	---	---	
32 O.D.	32.0	---	0.500	---	---	---	---	---	---	
34 O.D.	34.0	---	0.500	---	---	---	---	---	---	
36 O.D.	36.0	---	0.500	---	---	---	---	---	---	
42 O.D.	42.0	---	0.500	---	---	---	---	---	---	

Appendix

6

STAINLESS STEEL WELDING FILLER METALS

AWS A5.4 Chemical Composition Requirements for Undiluted Weld Metal

CHEMICAL COMPOSITION REQUIREMENTS FOR UNDILUTED WELD METAL		
AWS A5.4 Classification[c]	UNS Number[d]	Weight Percent [a,b]
E209-XX[e]	W32210	C 0.06 Cr 20.5-24.0 Ni 9.5-12.0 Mo 1.5-3.0 Mn 4.0-7.0 Si 0.90 P 0.04 S 0.03 N 0.10-0.30 Cu 0.75
E219-XX	W32310	C 0.06 Cr 19.0-21.5 Ni 5.5-7.0 Mo 0.75 Mn 8.0-10.0 Si 1.00 P 0.04 S 0.03 N 0.10-0.30 Cu 0.75
E240-XX	W32410	C 0.06 Cr 17.0-19.0 Ni 4.0-6.0 Mo 0.75 Mn 10.5-13.5 Si 1.00 P 0.04 S 0.03 N 0.10-0.30 Cu 0.75
E307-XX	W30710	C 0.04-0.14 Cr 18.0-21.5 Ni 9.0-10.7 Mo 0.5-1.5 Mn 3.30-4.75 Si 0.90 P 0.04 S 0.03 Cu 0.75
E308-XX	W30810	C 0.08 Cr 18.0-21.0 Ni 9.0-11.0 Mo 0.75 Mn 0.5-2.5 Si 0.90 P 0.04 S 0.03 Cu 0.75
E308H-XX	W30810	C 0.04-0.08 Cr 18.0-21.0 Ni 9.0-11.0 Mo 0.75 Mn 0.5-2.5 Si 0.90 P 0.04 S 0.03 Cu 0.75
E308L-XX	W30813	C 0.04 Cr 18.0-21.0 Ni 9.0-11.0 Mo 0.75 Mn 0.5-2.5 Si 0.90 P 0.04 S 0.03 Cu 0.75
E308Mo-XX	W30820	C 0.08 Cr 18.0-21.0 Ni 9.0-12.0 Mo 2.0-3.0 Mn 0.5-2.5 Si 0.90 P 0.04 S 0.03 Cu 0.75
E308MoL-XX	W30823	C 0.04 Cr 18.0-21.0 Ni 9.0-12.0 Mo 2.0-3.0 Mn 0.5-2.5 Si 0.90 P 0.04 S 0.03 Cu 0.75
E309-XX	W30910	C 0.15 Cr 22.0-25.0 Ni 12.0-14.0 Mo 0.75 Mn 0.5-2.5 Si 0.90 P 0.04 S 0.03 Cu 0.75
E309L-XX	W30913	C 0.04 Cr 22.0-25.0 Ni 12.0-14.0 Mo 0.75 Mn 0.5-2.5 Si 0.90 P 0.04 S 0.03 Cu 0.75
E309Cb-XX	W30917	C 0.12 Cr 22.0-25.0 Ni 12.0-14.0 Mo 0.75 Cb (Nb) plus Ta 0.70-1.00 Mn 0.5-2.5 Si 0.90 P 0.04 S 0.03 Cu 0.75
E309Mo-XX	W30920	C 0.12 Cr 22.0-25.0 Ni 12.0-14.0 Mo 2.0-3.0 Mn 0.5-2.5 Si 0.90 P 0.04 S 0.03 Cu 0.75
E309-MoL-XX	W30923	C 0.04 Cr 22.0-25.0 Ni 12.0-14.0 Mo 2.0-3.0 Mn 0.5-2.5 Si 0.90 P 0.04 S 0.03 Cu 0.75
E310-XX	W31010	C 0.08-0.20 Cr 25.0-28.0 Ni 20.0-22.5 Mo 0.75 Mn 1.0-2.5 Si 0.75 P 0.03 S 0.03 Cu 0.75
E310H-XX	W31015	C 0.35-0.45 Cr 25.0-28.0 Ni 20.0-22.5 Mo 0.75 Mn 1.0-2.5 Si 0.75 P 0.03 S 0.03 Cu 0.75
E310Cb-XX	W31017	C 0.12 Cr 25.0-28.0 Ni 20.0-22.0 Mo 0.75 Cb (Nb) plus Ta 0.70-1.00 Mn 1.0-2.5 Si 0.75 P 0.03 S 0.03 Cu 0.75
E310Mo-XX	W31020	C 0.12 Cr 25.0-28.0 Ni 20.0-22.0 Mo 2.0-3.0 Mn 1.0-2.5 Si 0.75 P 0.03 S 0.03 Cu 0.75
E312-XX	W31310	C 0.15 Cr 28.0-32.0 Ni 8.0-10.5 Mo 0.75 Mn 0.5-2.5 Si 0.90 P 0.04 S 0.03 Cu 0.75
E316-XX	W31610	C 0.08 Cr 17.0-20.0 Ni 11.0-14.0 Mo 2.0-3.0 Mn 0.5-2.5 Si 0.90 P 0.04 S 0.03 Cu 0.75
E316H-XX	W31610	C 0.04-0.08 Cr 17.0-20.0 Ni 11.0-14.0 Mo 2.0-3.0 Mn 0.5-2.5 Si 0.90 P 0.04 S 0.03 Cu 0.75
E316L-XX	W31613	C 0.04 Cr 17.0-20.0 Ni 11.0-14.0 Mo 2.0-3.0 Mn 0.5-2.5 Si 0.90 P 0.04 S 0.03 Cu 0.75
E317-XX	W31710	C 0.08 Cr 18.0-21.0 Ni 12.0-14.0 Mo 3.0-4.0 Mn 0.5-2.5 Si 0.90 P 0.04 S 0.03 Cu 0.75

CHEMICAL COMPOSITION REQUIREMENTS FOR UNDILUTED WELD METAL (Continued)

AWS A5.4 Classification[c]	UNS Number[d]	Weight Percent [a,b]
E317L-XX	W31713	C 0.04 Cr 18.0-21.0 Ni 12.0-14.0 Mo 3.0-4.0 Mn 0.5-2.5 Si 0.90 P 0.04 S 0.03 Cu 0.75
E318-XX	W31910	C 0.08 Cr 17.0-20.0 Ni 11.0-14.0 Mo 2.0-3.0 Cb (Nb) plus Ta 6 X C, min to 1.00 max Mn 0.5-2.5 Si 0.90 P 0.04 S 0.03 Cu 0.75
E320-XX	W88021	C 0.07 Cr 19.0-21.0 Ni 32.0-36.0 Mo 2.0-3.0 Cb (Nb) plus Ta 8 X C, min to 1.00 max Mn 0.5-2.5 Si 0.60 P 0.04 S 0.03 Cu 3.0-4.0
E320LR-XX	W88022	C 0.03 Cr 19.0-21.0 Ni 32.0-36.0 Mo 2.0-3.0 Cb (Nb) plus Ta 8 X C, min to 0.40 max Mn 1.50-2.50 Si 0.30 P 0.020 S 0.015 Cu 3.0-4.0
E330-XX	W88331	C 0.18-0.25 Cr 14.0-17.0 Ni 33.0-37.0 Mo 0.75 Mn 1.0-2.5 Si 0.90 P 0.04 S 0.03 Cu 0.75
E330H-XX	W88335	C 0.35-0.45 Cr 14.0-17.0 Ni 33.0-37.0 Mo 0.75 Mn 1.0-2.5 Si 0.90 P 0.04 S 0.03 Cu 0.75
E347-XX	W34710	C 0.08 Cr 18.0-21.0 Ni 9.0-11.0 Mo 0.75 Cb (Nb) plus Ta 8 X C, min to 1.00 max Mn 0.5-2.5 Si 0.90 P 0.04 S 0.03 Cu 0.75
E349-XX[e,f,g]	W34910	C 0.13 Cr 18.0-21.0 Ni 8.0-10.0 Mo 0.35-0.65 Cb (Nb) plus Ta 0.75-1.20 Mn 0.5-2.5 Si 0.90 P 0.04 S 0.03 Cu 0.75
E383-XX	W88028	C 0.03 Cr 26.5-29.0 Ni 30.0-33.0 Mo 3.2-4.2 Mn 0.5-2.5 Si 0.90 P 0.02 S 0.02 Cu 0.6-1.5
E385-XX	W88904	C 0.03 Cr 19.5-21.5 Ni 24.0-26.0 Mo 4.2-5.2 Mn 1.0-2.5 Si 0.75 P 0.03 S 0.02 Cu 1.2-2.0
E410-XX	W41010	C 0.12 Cr 11.0-13.5 Ni 0.7 Mo 0.75 Mn 1.0 Si 0.90 P 0.04 S 0.03 Cu 0.75
E410NiMo-XX	W41016	C 0.06 Cr 11.0-12.5 Ni 4.0-5.0 Mo 0.40-0.70 Mn 1.0 Si 0.90 P 0.04 S 0.03 Cu 0.75
E430-XX	W43010	C 0.10 Cr 15.0-18.0 Ni 0.6 Mo 0.75 Mn 1.0 Si 0.90 P 0.04 S 0.03 Cu 0.75
E502-XX[h]	W50210	C 0.10 Cr 4.0-6.0 Ni 0.4 Mo 0.45-0.65 Mn 1.0 Si 0.90 P 0.04 S 0.03 Cu 0.75
E505-XX[h]	W50410	C 0.10 Cr 8.0-10.5 Ni 0.4 Mo 0.85-1.20 Mn 1.0 Si 0.90 P 0.04 S 0.03 Cu 0.75
E630-XX	W37410	C 0.05 Cr 16.00-16.75 Ni 4.5-5.0 Mo 0.75 Cb (Nb) plus Ta 0.15-0.30 Mn 0.25-0.75 Si 0.75 P 0.04 S 0.03 Cu 3.25-4.00
E16-8-2-XX	W36810	C 0.10 Cr 14.5-16.5 Ni 7.5-9.5 Mo 1.0-2.0 Mn 0.5-2.5 Si 0.60 P 0.03 S 0.03 Cu 0.75
E7Cr-XX[h]	W50310	C 0.10 Cr 6.0-8.0 Ni 0.04 Mo 0.45-0.65 Mn 1.0 Si 0.90 P 0.04 S 0.03 Cu 0.75
E2209-XX	W39209	C 0.04 Cr 21.5-23.5 Ni 8.5-10.5 Mo 2.5-3.5 Mn 0.5-2.0 Si 0.90 P 0.04 S 0.03 N 0.08-0.20 Cu 0.75
E2553-XX	W39553	C 0.06 Cr 24.0-27.0 Ni 6.5-8.5 Mo 2.9-3.9 Mn 0.5-1.5 Si 1.0 P 0.04 S 0.03 N 0.10-0.25 Cu 1.5-2.5

a. Analysis shall be made for the elements for which specific values are shown in the table. If, however, the presence of other elements is indicated in the course of routine analysis, further analysis shall be made to determine that the total of these other elements, except iron, is not present in excess of 0.50 percent.

b. Single values are maximum percentages.

CHEMICAL COMPOSITION REQUIREMENTS FOR UNDILUTED WELD METAL (Continued)

c. Classification suffix -XX may be -15, -16, -17, -25, or -26. See AWS A5.4 Section A8 of the Appendix for an explanation.

d. SAE/ASTM Unified Number System for Metals and Alloys.

e. Vanadium shall be 0.10 to 0.30 percent.

f. Titanium shall be 0.15 percent max.

g. Tungsten shall be from 1.25 to 1.75 percent.

h. This grade also will appear in the next revision of AWS A5.5, *Specification for Low Alloy Steel Electrodes for Shielded Metal Arc Welding*. It will be deleted from AWS A5.4 at the first revision of A5.4 following publication of the revised A5.5

TYPE OF WELDING CURRENT AND POSITION OF WELDING

AWS A5.4 Classification[a]	Welding Current[b]	Welding Position[c]
EXXX(X)-15	dcep	All[d]
EXXX(X)-25	dcep	H, F
EXXX(X)-16	dcep or ac	All[d]
EXXX(X)-17	dcep or ac	All[d]
EXXX(X)-26	dcep or ac	H, F

a. See AWS A5.4 Section A8, Classification as to Useability, for explanation of positions.

b. dcep = Direct current electrode positive (reverse polarity); ac = Alternating current.

c. The abbreviations F and H indicate welding positions (AWS A5.4 Figure 3) as follows:

    F = Flat

    H = Horizontal

d. Electrodes 3/16 in. (4.8 mm) and larger are not recommended for welding all positions.

ALL-WELD-METAL MECHANICAL PROPERTY REQUIREMENTS				
	Tensile Strength, min			Heat
AWS A5.4 Classification	ksi	MPa	% Elongation, min.	Treatment
E209-XX	100	690	15	None
E219-XX	90	620	15	None
E240-XX	100	690	15	None
E307-XX	85	590	30	None
E308-XX	80	550	35	None
E308H-XX	80	550	35	None
E308L-XX	75	520	35	None
E308Mo-XX	80	550	35	None
E308MoL-XX	75	520	35	None
E309-XX	80	550	30	None
E309Cb-XX	80	550	30	None
E309Mo-XX	80	550	30	None
E309MoL-XX	75	520	30	None
E310-XX	80	550	30	None
E310H-XX	90	620	10	None
E310Cb-XX	80	550	25	None
E310Mo-XX	80	550	30	None
E312-XX	95	660	22	None
E316-XX	75	520	30	None
E316H-XX	75	520	30	None
E316L-XX	70	490	30	None
E317-XX	80	550	30	None
E317L-XX	75	520	30	None
E318-XX	80	550	25	None
E320-XX	80	550	30	None
E320LR-XX	75	520	30	None

ALL-WELD-METAL MECHANICAL PROPERTY REQUIREMENTS (Contined)				Heat
	Tensile Strength, min			
AWS A5.4 Classification	ksi	MPa	% Elongation, min.	Treatment
E330-XX	75	520	25	None
E330H-XX	90	620	10	None
E347-XX	75	520	30	None
E349-XX	100	690	25	None
E383-XX	75	520	30	None
E385-XX	75	520	30	None
E410-XX	75	450	20	a
E410NiMo-XX	110	760	15	c
E430-XX	65	450	20	d
E502-XX	60	420	20	b
E505-XX	60	420	20	b
E630-XX	135	930	7	e
E16-8-2-XX	80	550	35	None
E7Cr-XX	60	420	20	b
E2209-XX	100	690	20	None
E2553-XX	110	760	15	None

a. Heat to 1350 to 1400°F (730 to 760°C), hold for one hour, furnace cool at a rate of 100°F (60°C) per hour to 600°F (315°C) and air cool to ambient.
b. Heat to 1550 to 1600°F (840 to 870°C), hold for two hours, furnace cool at a rate not exceeding 100°F (55°C) per hour to 1100°F (595°C) and air cool to ambient.
c. Heat to 1100 to 1150°F (595 to 620°C), hold for one hour, and air cool to ambient.
d. Heat to 1400 to 1450°F (760 to 790°C), hold for two hours, furnace cool at a rate not exceeding 100°F (55°C) per hour to 1100°F (595°C) and air cool to ambient.
e. Heat to 1875 to 1925°F (1025 to 1050°C), hold for one hour, and air cool to ambient, and then precipitation harden at 1135 to 1165°F (610 to 630°C), hold for four hours, and air cool to ambient.

STANDARD SIZES AND LENGTHS

AWS A5.4 Electrode Size, (Diameter of Core Wire)[a]		Standard Lengths[b,c]	
in.	mm	in.	mm
1/16	1.6	9	230
5/64	2.0	9	230
3/32	2.4	9, 12, 14[d]	230, 305, 350[d]
1/8	3.2	14, 18[d]	350, 460[d]
5/32	4.0	14, 18[d]	350, 460[d]
3/16	4.8	14, 18[d]	350, 460[d]
7/32	5.6	14, 18	350, 460
1/4	6.4	14, 18	350, 460

a. Tolerance on the diameter shall be ± 0.002 in. (± 0.05 mm).
b. Tolerance on length shall be ± 1/4 in. (± 6.4 mm).
c. Other sizes and lengths shall be as agreed upon between purchaser and supplier.
d. These lengths are intended only for the EXXX-25 and EXXX-26 types.

AWS A5.4 CLASSIFICATION SYSTEM

The system of classification is similar to that used in other filler metal specifications. The letter "E" at the beginning of each number indicates an electrode. The first three digits designate the classification as to its composition. (Occasionally, a number of digits other than three is used and letters may follow the digits to indicate a specific composition.) The last two digits designate the classification as to usability with respect to position of welding and type of current as described in AWS A5.4 Appendix A8. The smaller sizes of EXXX(X)-15, EXXX(X)-16, or EXXX(X)-17 electrodes [up to and including 5/32 in. (4.0 mm)] included in this specification are used in all welding positions.

The mechanical tests measure strength and ductility, qualities which are often of lesser importance than the corrosion and heat resisting properties. These mechanical test requirements, however, provide an assurance of freedom from weld metal flaws, such as check cracks and serious dendritic segregations which, if present, may cause failure in service.

It is recognized that for certain applications, supplementary tests may be required. In such cases, additional tests to determine specific properties, such as corrosion resistance, scale resistance, or strength at elevated temperatures may be required as agreed upon between supplier and purchaser.

AWS A5.9 Bare Stainless Steel Welding Electrodes & Rods

CHEMICAL COMPOSITION REQUIREMENTS		
AWS A5.9 Classification[c,d]	UNS Number[e]	Composition, Weight Percent[a,b]
ER209	S20980	C 0.05 Cr 20.5-24.0 Ni 9.5-12.0 Mo 1.5-3.0 Mn 4.0-7.0 Si 0.90 P 0.03 S 0.03 N 0.10-0.30 Cu 0.75 Other Elements V 0.10-0.30
ER218	S21880	C 0.10 Cr 16.0-18.0 Ni 8.0-9.0 Mo 0.75 Mn 7.0-9.0 Si 3.5-4.5 P 0.03 S 0.03 N 0.08-0.18 Cu 0.75
ER219	S21980	C 0.05 Cr 19.0-21.5 Ni 5.5-7.0 Mo 0.75 Mn 8.0-10.0 Si 1.00 P 0.03 S 0.03 N 0.10-0.30 Cu 0.75
ER240	S24080	C 0.05 Cr 17.0-19.0 Ni 4.0-6.0 Mo 0.75 Mn 10.5-13.5 Si 1.00 P 0.03 S 0.03 N 0.10-0.30 Cu 0.75
ER307	S30780	C 0.04-0.14 Cr 19.5-22.0 Ni 8.0-10.7 Mo 0.5-1.5 Mn 3.3-4.75 Si 0.30-0.65 P 0.03 S 0.03 Cu 0.75
ER308	S30880	C 0.08 Cr 19.5-22.0 Ni 9.0-11.0 Mo 0.75 Mn 1.0-2.5 Si 0.30-0.65 P 0.03 S 0.03 Cu 0.75
ER308H	S30880	C 0.04-0.08 Cr 19.5-22.0 Ni 9.0-11.0 Mo 0.50 Mn 1.0-2.5 Si 0.30-0.65 P 0.03 S 0.03 Cu 0.75
ER308L	S30883	C 0.03 Cr 19.5-22.0 Ni 9.0-11.0 Mo 0.75 Mn 1.0-2.5 Si 0.30-0.65 P 0.03 S 0.03 Cu 0.75
ER308Mo	S30882	C 0.08 Cr 18.0-21.0 Ni 9.0-12.0 Mo 2.0-3.0 Mn 1.0-2.5 Si 0.30-0.65 P 0.03 S 0.03 Cu 0.75
ER308LMo	S30886	C 0.04 Cr 18.0-21.0 Ni 9.0-12.0 Mo 2.0-3.0 Mn 1.0-2.5 Si 0.30-0.65 P 0.03 S 0.03 Cu 0.75
ER308Si	S30881	C 0.08 Cr 19.5-22.0 Ni 9.0-11.0 Mo 0.75 Mn 1.0-2.5 Si 0.65-1.00 P 0.03 S 0.03 Cu 0.75
ER308LSi	S30888	C 0.03 Cr 19.5-22.0 Ni 9.0-11.0 Mo 0.75 Mn 1.0-2.5 Si 0.65-1.00 P 0.03 S 0.03 Cu 0.75
ER309	S30980	C 0.12 Cr 23.0-25.0 Ni 12.0-14.0 Mo 0.75 Mn 1.0-2.5 Si 0.30-0.65 P 0.03 S 0.03 Cu 0.75
ER309L	S30983	C 0.03 Cr 23.0-25.0 Ni 12.0-14.0 Mo 0.75 Mn 1.0-2.5 Si 0.30-0.65 P 0.03 S 0.03 Cu 0.75
ER309Mo	S30982	C 0.12 Cr 23.0-25.0 Ni 12.0-14.0 Mo 2.0-3.0 Mn 1.0-2.5 Si 0.30-0.65 P 0.03 S 0.03 Cu 0.75
ER309LMo	S30986	C 0.03 Cr 23.0-25.0 Ni 12.0-14.0 Mo 2.0-3.0 Mn 1.0-2.5 Si 0.30-0.65 P 0.03 S 0.03 Cu 0.75
ER309Si	S30981	C 0.12 Cr 23.0-25.0 Ni 12.0-14.0 Mo 0.75 Mn 1.0-2.5 Si 0.65-1.00 P 0.03 S 0.03 Cu 0.75
ER309LSi	S30988	C 0.03 Cr 23.0-25.0 Ni 12.0-14.0 Mo 0.75 Mn 1.0-2.5 Si 0.65-1.00 P 0.03 S 0.03 Cu 0.75
ER310	S31080	C 0.08-0.15 Cr 25.0-28.0 Ni 20.0-22.5 Mo 0.75 Mn 1.0-2.5 Si 0.30-0.65 P 0.03 S 0.03 Cu 0.75
ER312	S31380	C 0.15 Cr 28.0-32.0 Ni 8.0-10.5 Mo 0.75 Mn 1.0-2.5 Si 0.30-0.65 P 0.03 S 0.03 Cu 0.75
ER316	S31680	C 0.08 Cr 18.0-20.0 Ni 11.0-14.0 Mo 2.0-3.0 Mn 1.0-2.5 Si 0.30-0.65 P 0.03 S 0.03 Cu 0.75
ER316H	S31680	C 0.04-0.08 Cr 18.0-20.0 Ni 11.0-14.0 Mo 2.0-3.0 Mn 1.0-2.5 Si 0.30-0.65 P 0.03 S 0.03 Cu 0.75

CHEMICAL COMPOSITION REQUIREMENTS (Continued)

AWS A5.9 Classification[c,d]	UNS Number[e]	Composition, Weight Percent[a,b]
ER316L	S31683	C 0.03 Cr 18.0-20.0 Ni 11.0-14.0 Mo 2.0-3.0 Mn1.0-2.5 Si 0.30-0.65 P 0.03 S 0.03 Cu 0.75
ER316Si	S31681	C 0.08 Cr 18.0-20.0 Ni 11.0-14.0 Mo 2.0-3.0 Mn1.0-2.5 Si 0.65-1.00 P 0.03 S 0.03 Cu 0.75
ER316LSi	S31688	C 0.03 Cr 18.0-20.0 Ni 11.0-14.0 Mo 2.0-3.0 Mn1.0-2.5 Si 0.65-1.00 P 0.03 S 0.03 Cu 0.75
ER317	S31780	C 0.08 Cr 18.5-20.5 Ni 13.0-15.0 Mo 3.0-4.0 Mn1.0-2.5 Si 0.30-0.65 P 0.03 S 0.03 Cu 0.75
ER317L	S31783	C 0.03 Cr 18.5-20.5 Ni 13.0-15.0 Mo 3.0-4.0 Mn1.0-2.5 Si 0.30-0.65 P 0.03 S 0.03 Cu 0.75
ER318	S31980	C 0.08 Cr 18.0-20.0 Ni 11.0-14.0 Mo 2.0-3.0 Mn1.0-2.5 Si 0.30-0.65 P 0.03 S 0.03 Cu 0.75 Cb[g] 8 X C min/1.0 max
ER320	N08021	C 0.07 Cr 19.0-21.0 Ni 32.0-36.0 Mo 2.0-3.0 Mn 2.5 Si 0.60 P 0.03 S 0.03 Cu 3.0-4.0 Cb[g] 8 X C min/1.0 max
ER320LR	N08022	C 0.025 Cr 19.0-21.0 Ni 32.0-36.0 Mo 2.0-3.0 Mn 1.5 - 2.0 Si 0.15 P 0.015 S 0.02 Cu 3.0-4.0 Cb[g] 8 X C min/0.40 max
ER321	S32180	C 0.08 Cr 18.5-20.5 Ni 9.0-10.5 Mo 0.75 Mn 1.0-2.5 Si 0.30-0.65 P 0.03 S 0.03 Cu 0.75 Ti 9 X C min/1.0 max
ER330	N08331	C 0.18-0.25 Cr 15.0-17.0 Ni 34.0-37.0 Mo 0.75 Mn 1.0-2.5 Si 0.30-0.65 P 0.03 S 0.03 Cu 0.75
ER347	S34780	C 0.08 Cr 19.0-21.5 Ni 9.0-11.0 Mo 0.75 Mn 1.0-2.5 Si 0.30-0.65 P 0.03 S 0.03 Cu 0.75 Cb[g] 10 X C min/1.0 max
ER347Si	S34788	C 0.08 Cr 19.0-21.5 Ni 9.0-11.0 Mo 0.75 Mn 1.0-2.5 Si 0.65-1.00 P 0.03 S 0.03 Cu 0.75 Cb[g] 10 X C min/1.0 max
ER383	N08028	C 0.025 Cr 26.5-28.5 Ni 30.0-33.0 Mo 3.2-4.2 Mn 1.0-2.5 Si 0.50 P 0.02 S 0.03 Cu 0.70-1.5
ER385	N08904	C 0.025 Cr 19.5-21.5 Ni 24.0-26.0 Mo 4.2-5.2 Mn 1.0-2.5 Si 0.50 P 0.02 S 0.03 Cu 1.2-2.0
ER409	S40900	C 0.08 Cr 10.5-13.5 Ni 0.6 Mo 0.50 Mn 0.08 Si 0.8 P 0.03 S 0.03 Cu 0.75 Other Elements Cb[g] 10 X C min/1.5 max
ER409Cb	S40940	C 0.08 Cr 10.5-13.5 Ni 0.6 Mo 0.50 Mn 0.08 Si 1.0 P 0.04 S 0.03 Cu 0.75 Other Elements Cb[g] 10 X C min/0.75 max
ER410	S41080	C 0.12 Cr 11.5-13.5 Ni 0.6 Mo 0.75 Mn 0.6 Si 0.5 P 0.03 S 0.03 Cu 0.75
ER410NiMo	S41086	C 0.06 Cr 11.0-12.5 Ni 4.0-5.0 Mo 0.4-0.7 Mn 0.6 Si 0.5 P 0.03 S 0.03 Cu 0.75
ER420	S42080	C 0.25-0.40 Cr 12.0-14.0 Ni 0.6 Mo 0.75 Mn 0.6 Si 0.5 P 0.03 S 0.03 Cu 0.75
ER430	S43080	C 0.10 Cr 15.5-17.0 Ni 0.6 Mo 0.75 Mn 0.6 Si 0.5 P 0.03 S 0.03 Cu 0.75
ER446LMo	S44687	C 0.015 Cr 25.0-27.5 Ni[f] Mo 0.75-1.50 Mn 0.4 Si 0.4 P 0.02 S 0.02 0.015 Cu[f]
ER502[h]	S50280	C 0.10 Cr 4.6-6.0 Ni 0.6 Mo 0.45-0.65 Mn 0.6 Si 0.5 P 0.03 S 0.03 Cu 0.75
ER505[h]	S50480	C 0.10 Cr 8.0-10.5 Ni 0.5 Mo 0.8-1.2 Mn 0.6 Si 0.5 P 0.03 S 0.03 Cu 0.75
ER630	S17480	C 0.05 Cr 16.0-16.75 Ni 4.5-5.0 Mn 0.25-0.75 P 0.03 S 0.03 Cu 3.25-4.00 Cb[g] 0.15-0.30
ER19-10H	S30480	C 0.04-0.08 Cr 18.5-20.0 Ni 9.0-11.0 Mo 0.25 Mn 1.0-2.0 Si 0.30-0.65 P 0.03 S 0.03 Cu 0.75 Cb[g] 0.05, Ti 0.5

CHEMICAL COMPOSITION REQUIREMENTS (Continued)

AWS A5.9 Classification[c,d]	UNS Number[e]	Composition, Weight Percent[a,b]
ER16-8-2	S16880	C 0.10 Cr 14.5-16.5 Ni 7.5-9.5 Mo 1.0-2.0 Si 0.30-0.65 P 0.03 S 0.03 Cu 0.75
ER2209	S39209	C 0.03 Cr 21.5-23.5 Ni 7.5-9.5 Mo 2.5-3.5 Mn 0.50-2.0 Si 0.90 P 0.03 S 0.03 N 0.08-0.20 Cu 0.75
ER2553	S39553	C 0.04 Cr 24.0-27.0 Ni 4.5-6.5 Mn 2.9-3.9 Mn 1.5 Si 1.0 P 0.04 S 0.03 N 0.10-0.25 Cu 1.5-2.5
ER3556	R30556	C 0.05-0.15 Cr 21.0-23.0 Ni 19.0-22.5 Mo 2.5-4.0 Mn 0.50-2.00 Si 0.20-0.80 P 0.04 S 0.015 N 0.10-0.30 Co 16.0-21.0, W 2.0-3.5, Cb 0.30, Ta 0.30-1.25, Al 0.10-0.50, Zr 0.001-0.10, La 0.005-0.10, B 0.02

a. Analysis shall be made for the elements for which specific values are shown in this table. If the presence of other elements is indicated in the course of this work, the amount of those elements shall be determined to ensure that their total, excluding iron, does not exceed 0.50 percent.

b. Single values shown are maximum percentages.

c. In the designator for composite, stranded, and strip electrodes, the "R" shall be deleted. A designator "C" shall be used for composite and stranded electrodes and a designator "Q" shall be used for strip electrodes. For example, ERXXX designates a solid wire and EQXXX designates a strip electrode of the same general analysis, and the same UNS number. However, ECXXX designates a composite metal cored or stranded electrode and may not have the same UNS number. Consult ASTM/SAE Uniform Numbering System for the proper UNS Number.

d. For special applications, electrodes and rods may be purchased with less than the specified silicon content.

e. SAE/ASTM Unified Numbering System for Metals and Alloys.

f. Nickel + copper equals 0.5 percent maximum.

g. Cb (Nb) may be reported as Cb (Nb) + Ta.

h. These classifications also will be included in the next revision of ANSI/AWS A5.28, *Specification for Low Alloy Steel Filler Metals for Gas Shielded Metal Arc Welding*. They will be deleted from ANSI/AWS A5.9 in the first revision following publication of the revised ANSI/AWS A5.28 document.

TENSILE REQUIREMENTS FOR ALL-WELD-METAL

AWS A5.9 Classification	Tensile Strength, min		% Elongation, min	Heat Treatment
	ksi	MPa		
E209-XX	100	690	15	None
E219-XX	90	620	15	None
E240-XX	100	690	15	None
E307-XX	85	590	30	None
E308-XX	80	550	35	None
E308H-XX	80	550	35	None
E308L-XX	75	520	35	None
E308Mo-XX	80	550	35	None
E308MoL-XX	75	520	35	None
E309-XX	80	550	30	None
E309L-XX	75	520	30	None
E309Cb-XX	80	550	30	None
E309Mo-XX	80	550	30	None
E309MoL-XX	75	520	30	None
E310-XX	80	550	30	None
E310H-XX	90	620	10	None
E310Cb-XX	80	550	25	None
E310Mo-XX	80	550	30	None
E312-XX	95	660	22	None
E316-XX	75	520	30	None
E316H-XX	75	520	30	None
E316L-XX	70	490	30	None
E317-XX	80	550	30	None
E317L-XX	75	520	30	None
E318-XX	80	550	25	None
E320-XX	80	550	30	None

TENSILE REQUIREMENTS FOR ALL-WELD-METAL (Continued)

AWS A5.9 Classification	Tensile Strength, min		% Elongation, min	Heat Treatment
	ksi	MPa		
E320LR-XX	75	520	30	None
E330-XX	75	520	25	None
E330H-XX	90	620	10	None
E347-XX	75	520	30	None
E349-XX	100	690	25	None
E383-XX	75	520	30	None
E385-XX	75	520	30	None
E410-XX	75	450	20	a
E410NiMo-XX	110	760	15	c
E430-XX	65	450	20	d
E502-XX	60	420	20	b
E505-XX	60	420	20	b
E630-XX	135	930	7	e
E16-8-2-XX	80	550	35	None
E7Cr-XX	60	420	20	b
E2209-XX	100	690	20	None
E2553-XX	110	760	15	None

a. Heat to 1350 to 1400°F (730 to 760°C), hold for one hour, furnace cool at a rate of 100°F (55°C) per hour to 600°F (315°C) and air cool to ambient.
b. Heat to 1550 to 1600°F (840 to 870°C), hold for two hours, furnace cool at a rate not exceeding 100°F (55°C) per hour to 1100°F (595°C) and air cool to ambient.
c. Heat to 1100 to 1150°F (595 to 620°C), hold for one hour, and air cool to ambient.
d. Heat to 1400 to 1450°F (760 to 790°C), hold for two hours, furnace cool at a rate not exceeding 100°F (55°C) per hour to 1100°F (595°C) and air cool to ambient.
e. Heat to 1875 to 1925°F (1025 to 1050°C), hold for one hour, and air cool to ambient, and then precipitation harden at 1135 to 1165°F (610 to 630°C), hold for four hours, and air cool to ambient.

AWS A5.9 CLASSIFICATION SYSTEM

The chemical composition of the filler metal is identified by a series of numbers and, in some cases, chemical symbols, the letters L, H, and LR, or both. Chemical symbols are used to designate modifications of basic alloy types, e.g., ER308Mo. The letter "H" denotes carbon content restricted to the upper part of the range that is specified for the standard grade of the specific filler metal. The letter "L" denotes carbon content in the lower part of the range that is specified for the corresponding standard grade of filler metal. The letters "LR" denote low residuals (see AWS A5.9 appendix A8.30).

The first two designators may be "ER" for solid wires that may be used as electrodes or rods; or they may be "EC" for composite cored or stranded wires; or they may be "EQ" for strip electrodes.

The three digit number such as 308 in ER308 designates the chemical composition of the filler metal.

AWS A5.22 Stainless Steel Electrodes for Flux Cored Arc Welding and Stainless Steel Flux Cored Rods for Gas Tungsten Arc Welding

AWS A5.22	UNS	CHEMICAL COMPOSITION REQUIREMENTS FOR UNDILUTED WELD METAL
E307TX-X	W30731	C 0.13 Cr 18.0-20.5 Ni 9.0-10.5 Mo 0.5-1.5 Mn 3.30-4.75 Si 1.0 P 0.04 S 0.03 Cu 0.5
E308TX-X	W30831	C 0.08 Cr 18.0-21.0 Ni 9.0-11.0 Mo 0.5 Mn 0.5-2.5 Si 1.0 P 0.04 S 0.03 Cu 0.5
E308LTX-X	W30835	C 0.04 Cr 18.0-21.0 Ni 9.0-11.0 Mo 0.5 Mn 0.5-2.5 Si 1.0 P 0.04 S 0.03 Cu 0.5
E308HTX-X	W30831	C 0.04-0.08 Cr 18.0-21.0 Ni 9.0-11.0 Mo 0.5 Mn0.5-2.5 Si 1.0 P 0.04 S 0.03 Cu 0.5
E308MoTX-X	W30832	C 0.08 Cr 18.0-21.0 Ni 9.0-11.0 Mo 2.0-3.0 Mn 0.5-2.5 Si 1.0 P 0.04 S 0.03 Cu 0.5
E308LMoTX-X	W30838	C 0.04 Cr 18.0-21.0 Ni 9.0-12.0 Mo 2.0-3.0 Mn 0.5-2.5 Si 1.0 P 0.04 S 0.03 Cu 0.5
E309TX-X	W30931	C 0.10 Cr 22.0-25.0 Ni 12.0-14.0 Mo 0.5 Mn 0.5-2.5 Si 1.0 P 0.04 S 0.03 Cu 0.5
E309LCbTX-X	W30932	C 0.04 Cr 22.0-25.0 Ni 12.0-14.0 Mo 0.5 Cb(Nb) + Ta 0.70-1.00 Mn 0.5-2.5 Si 1.0 P 0.04 S 0.03 Cu 0.5
E309LTX-X	W30935	C 0.04 Cr 22.0-25.0 Ni 12.0-14.0 Mo 0.5 Mn 0.5-2.5 Si 1.0 P 0.04 S 0.03 Cu 0.5
E309MoTX-X	W30939	C 0.12 Cr 21.0-25.0 Ni 12.0-16.0 Mo 2.0-3.0 Mn 0.5-2.5 Si 1.0 P 0.04 S 0.03 Cu 0.5
E309LMoTX-X	W30938	C 0.04 Cr 21.0-25.0 Ni 12.0-16.0 Mo 2.0-3.0 Mn 0.5-2.5 Si 1.0 P 0.04 S 0.03 Cu 0.5
E309LNiMoTX-X	W30936	C 0.04 Cr 20.5-23.5 Ni 15.0-17.0 Mo 2.5-3.5 Mn 0.5-2.5 Si 1.0 P 0.04 S 0.03 Cu 0.5
E310TX-X	W31031	C 0.20 Cr 25.0-28.0 Ni 20.0-22.5 Mo 0.5 Mn 1.0-2.5 Si 1.0 P 0.03 S 0.03 Cu 0.5
E312TX-X	W31331	C 0.15 Cr 28.0-32.0 Ni 8.0-10.5 Mo 0.5 Mn 0.5-2.5 Si 1.0 P 0.04 S 0.03 Cu 0.5
E316TX-X	W31631	C 0.08 Cr 17.0-20.0 Ni 11.0-14.0 Mo 2.0-3.0 Mn 0.5-2.5 Si 1.0 P 0.04 S 0.03 Cu 0.5
E316LTX-X	W31635	C 0.04 Cr 17.0-20.0 Ni 11.0-14.0 Mo 2.0-3.0 Mn 0.5-2.5 Si 1.0 P 0.04 S 0.03 Cu 0.5
E317LTX-X	W31735	C 0.04 Cr 18.0-21.0 Ni 12.0-14.0 Mo 3.0-4.0 Mn 0.5-2.5 Si 1.0 P 0.04 S 0.03 Cu 0.5
E347TX-X	W34731	C 0.08 Cr 18.0-21.0 Ni 9.0-11.0 Mo 0.5 Cb(Nb) + Ta 8 x C min to 1.0 max Mn 0.5-2.5 Si 1.0 P 0.04 S 0.03 Cu 0.5
E409TX-X[e]	W40931	C 0.10 Cr 10.5-13.5 Ni 0.60 Mo 0.60 Mn 0.80 Si 1.0 P 0.04 S 0.03 Cu 0.5
E410TX-X	W41031	C 0.12 Cr 11.0-13.5 Ni 0.60 Mo 0.60 Mn 0.5 Mn 1.2 Si 1.0 P 0.04 S 0.03 Cu 0.5
E410NiMoTX-X	W41036	C 0.06 Cr 11.0-12.5 Ni 4.0-5.0 Mo 0.40-0.70 Mn 1.0 Si 1.0 P 0.04 S 0.03 Cu 0.5
E410NiTiTX-X[e]	W41038	C 0.04 Cr 11.0-12.0 Ni 3.6-4.5 Mo 0.5 Mn 0.70 Si 0.50 P 0.03 S 0.03 Cu 0.5
E430TX-X	W43031	C 0.10 Cr 15.0-18.0 Ni 0.60 Mo 0.5 Mn 1.2 Si 1.0 P 0.04 S 0.03 Cu 0.5

ALL-WELD-METAL CHEMICAL COMPOSITION REQUIREMENTS (Continued)		
AWS A5.22	UNS	
E502TX-X[f]	W50231	C 0.10 Cr 4.0-6.0 Ni 0.40 Mo 0.45-0.65 Mn 1.2 Si 1.0 P 0.04 S 0.03 Cu 0.5
E505TX-X[f]	W50431	C 0.10 Cr 8.0-10.5 Ni 0.40 Mo 0.85-1.20 Mn 1.2 Si 1.0 P 0.04 S 0.03 Cu 0.5
E307T0-3	W30733	C 0.13 Cr 19.5-22.0 Ni 9.0-10.5 Mo 0.5-1.5 Mn 3.30-4.75 Si 1.0 P 0.04 S 0.03 Cu 0.5
E308T0-3	W30833	C 0.08 Cr 19.5-22.0 Ni 9.0-11.0 Mo 0.5 Mn 0.5-2.5 Si 1.0 P 0.04 S 0.03 Cu 0.5
E308LT0-3	W30837	C 0.03 Cr 19.5-22.0 Ni 9.0-11.0 Mo 0.5 Mn 0.5-2.5 Si 1.0 P 0.04 S 0.03 Cu 0.5
E308HT0-3	W30833	C 0.04-0.08 Cr 19.5-22.0 Ni 9.0-11.0 Mo 0.5 Mn 0.5-2.5 Si 1.0 P 0.04 S 0.03 Cu 0.5
E308MoT0-3	W30839	C 0.08 Cr 18.0-21.0 Ni 9.0-11.0 Mo 2.0-3.0 Mn 0.5-2.5 Si 1.0 P 0.04 S 0.03 Cu 0.5
E308LMoT0-3	W30838	C 0.03 Cr 18.0-21.0 Ni 9.0-12.0 Mo 2.0-3.0 Mn 0.5-2.5 Si 1.0 P 0.04 S 0.03 Cu 0.5
E308HMoT0-3	W30830	C 0.07-0.12 Cr 19.0-21.5 Ni 9.0-10.7 Mo 1.8-2.4 Mn 1.25-2.25 Si 0.25-0.80 P 0.04 S 0.03 Cu 0.5
E309T0-3	W30933	C 0.10 Cr 23.0-25.5 Ni 12.0-14.0 Mo 0.5 Mn 0.5-2.5 Si 1.0 P 0.04 S 0.03 Cu 0.5
E309LT0-3	W30937	C 0.03 Cr 23.0-25.5 Ni 12.0-14.0 Mo 0.5 Mn 0.5-2.5 Si 1.0 P 0.04 S 0.03 Cu 0.5
E309LCbT0-3	W30934	C 0.03 Cr 23.0-25 5 Ni 12.0-14.0 Mo 0.5 Cb(Nb) + Ta 0.70-1.00 Mn 0.5-2.5 Si 1.0 P 0.04 S 0.03 Cu 0.5
E309MoT0-3	W30939	C 0.12 Cr 21.0-25.0 Ni 12.0-16.0 Mo 2.0-3.0 Mn 0.5-2.5 Si 1.0 P 0.04 S 0.03 Cu 0.5
E309LMoT0-3	W30938	C 0.04 Cr 21.0-25.0 Ni 12.0-16.0 Mo 2.0-3.0 Mn 0.5-2.5 Si 1.0 P 0.04 S 0.03 Cu 0.5
E310T0-3	W31031	C 0.20 Cr 25.0-28.0 Ni 20.0-22.5 Mo 0.5 Mn 1.0-2.5 Si 1.0 P 0.03 S 0.03 Cu 0.5
E312T0-3	W31231	C 0.15 Cr 28.0-32.0 Ni 8.0-10.5 Mo 0.5 Mn 0.5-2.5 Si 1.0 P 0.04 S 0.03 Cu 0.5
E316T0-3	W31633	C 0.08 Cr 18.0-20.5 Ni 11.0-14.0 Mo 2.0-3.0 Mn 0.5-2.5 Si 1.0 P 0.04 S 0.03 Cu 0.5
E316LT0-3	W31637	C 0.03 Cr 18.0-20.5 Ni 11.0-14.0 Mo 2.0-3.0 Mn 0.5-2.5 Si 1.0 P 0.04 S 0.03 Cu 0.5
E316LKT0-3[g]	W31630	C 0.04 Cr 17.0-22.0 Ni 11.0-14.0 Mo 2.0-3.0 Mn 0.5-2.5 Si 1.0 P 0.04 S 0.03 Cu 0.5
E317LT0-3	W31737	C 0.03 Cr 18.5-21.0 Ni 13.0-15.0 Mo 3.0-4.0 Mn 0.5-2.5 Si 1.0 P 0.04 S 0.03 Cu 0.5
E347T0-3	W34733	C 0.08 Cr 19.0-21.5 Ni 9.0-11.0 Mo 0.5 Cb(Nb) + Ta 8 x C min to 1.0 max Mn 0.5-2.5 Si 1.0 P 0.04 S 0.03 Cu 0.5
E409T0-3[e]	W40931	C 0.10 Cr 10.5-13.5 Ni 0.60 Mo 0.5 Mn 0.80 Si 1.0 P 0.04 S 0.03 Cu 0.5
E410T0-3	W41031	C 0.12 Cr 11.0-13.5 Ni 0.60 Mo 0.5 Mn 1.0 Si 1.0 P 0.04 S 0.03 Cu 0.5
E410NiMoT0-3	W41036	C 0.06 Cr 11.0-12.5 Ni 4.0-5.0 Mo 0.40-0.70 Mn 1.0 Si 1.0 P 0.04 S 0.03 Cu 0.5
E410NiTiT0-3[e]	W41038	C 0.04 Cr 11.0-12.0 Ni 3.6-4.5 Mo 0.5 Mn 0.70 Si 0.50 P 0.03 S 0.03 Cu 0.5
E430T0-3	W43031	C 0.10 Cr 15.0-18.0 Ni 0.60 Mo 0.5 Mn 1.0 Si 1.0 P 0.04 S 0.03 Cu 0.5

ALL-WELD-METAL CHEMICAL COMPOSITION REQUIREMENTS (Continued)

AWS A5.22	UNS	
E2209T0-X	W39239	C 0.04 Cr 21.0-24.0 Ni 7.5-10.0 Mo 2.5-4.0 Mn 0.5-2.0 Si 1.0 P 0.04 S 0.03 N 0.08-2.0 Cu 0.5
E2553T0-X	W39533	C 0.04 Cr 24.0-27.0 Ni 8.5-10.5 Mo 2.9-3.9 Mn 0.5-1.5 Si 0.75 P 0.04 S 0.03 N 0.10-0.20 Cu 1.5-2.5
EXXXTX-G[h]		Not Specified
R308LT1-5	W30835	C 0.03 Cr 18.0-21.0 Ni 9.0-11.0 Mn 0.5 Mo 0.5 Mn 0.5-2.5 Si 1.2 P 0.04 S 0.03 Cu 0.5
R309LT1-5	W30935	C 0.03 Cr 22.0-25.0 Ni 12.0-14.0 Mo 0.5 Mn 0.5-2.5 Si 1.2 P 0.04 S 0.03 Cu 0.5
R316LT1-5	W31635	C 0.03 Cr 17.0-20.0 Ni 11.0-14.0 Mo 2.0-3.0 Mn 0.5-2.5 Si 1.2 P 0.04 S 0.03 Cu 0.5
R347T1-5	W34731	C 0.08 Cr 18.0-21.0 Ni 9.0-11.0 Mo 0.5 Cb(Nb) + Ta 8 x C min to 1.0 max Mn 0.5-2.5 Si 1.2 P 0.04 S 0.03 Cu 0.5

a. The weld metal shall be analyzed for the specific elements in this table. If the presence of other elements is indicated in the course of this work, the amount of those elements shall be determined to ensure that their total (excluding iron) does not exceed 0.50%.

b. Single values shown are maximum.

c. In this table, the "X" following the "T" refers to the position of welding (1 for all-position operation or 0 for flat or horizontal operation) and the "X" following the dash refers to the shielding medium (–1, –4, or –5) as shown in the AWS A5.22 Classification column, Table 2. For information concerning the "G", see AWS A5.22 paragraphs A2.3.7 and A2.3.8 of the Annex. In A5.22-80, the position of welding was not included in the classification. Accordingly, electrodes classified herein as either EXXXT0-1 or EXXXT1-1 would both have been classified EXXXT-1 and so forth.

d. ASTM /SAE Unified Number System for Metals and Alloys

e. Titanium–10 x C min., 1.5% max.

f. See footnote 3 on page 1 of AWS A 5.22.

g. This alloy is designed for cryogenic applications.

h. See paragraphs A2.3.7 and A2.3.8 of AWS A5.22.

TENSION TEST REQUIREMENTS

AWS A5.22 Classification[a]	Tensile Strength, minimum		Elongation Percent, Min.	Postweld Heat Treatment
	ksi	MPa		
E307TX-X	85	590	30	None
E308TX-X	80	550	35	None
E308LTX-X	75	520	35	None
E308HTX-X	80	550	35	None

TENSION TEST REQUIREMENTS (Continued)

AWS A5.22 Classification[a]	Tensile Strength, minimum		Elongation Percent, Min.	Postweld Heat Treatment
	ksi	MPa		
E308MoTX-X	80	550	35	None
E308LMoTX-X	75	520	35	None
E309TX-X	80	550	30	None
E309LCbTX-X	75	520	30	None
E309LTX-X	75	520	30	None
E309MoTX-X	80	550	25	None
E309LMoTX-X	75	520	25	None
E309LNiMoTX-X	75	520	25	None
E310TX-X	80	550	30	None
E312TX-X	95	660	22	None
E316TX-X	75	520	30	None
E316LTX-X	70	485	30	None
E317LTX-X	75	520	20	None
E347TX-X	75	520	30	None
E409TX-X	65	450	15	None
E410TX-X	75	520	20	b
E410NiMoTX-X	110	760	15	c
E410NiTiTX-X	110	760	15	c
E430TX-X	65	450	20	d
E502TX-X	60	415	20	e
E505TX-X	60	415	20	e
E308HMoT0-3	80	550	30	None
E316LKT0-3	70	485	30	None
E2209TX-X	100	690	20	None
E2553TX-X	110	760	15	None
EXXXTX-G	Not specified			

TENSION TEST REQUIREMENTS (Continued)

AWS A5.22 Classification[a]	Tensile Strength, minimum		Elongation Percent, Min.	Postweld Heat Treatment
	ksi	MPa		
R308LT1-5	75	520	35	None
R309LT1-5	75	520	30	None
R316LT1-5	70	485	30	None
R347T1-5	75	520	30	None

a. In this table, the "X" following the "T" refers to the position of welding (1 for all-position or 0 for flat or horizontal operation) and the "X" following the dash refers to the shielding medium (–1,–3, or –4) as shown in the AWS A5.22 Classification.

b. The weld test assembly (or the blank from it, from which the tensile test specimen is to be machined) shall be heated to a temperature between 1350 and 1400°F (732 and 760°C), held for 1 hour, then furnace cooled to 600°F (316°C) at a rate not to exceed 100°F (55°C) per hour, then cooled in air to room temperature.

c. The weld test assembly (or the blank from it, from which the tensile test specimen is to be machined) shall be heated to a temperature between 1100 and 1150°F (593 and 621°C), held for 1 hour, then cooled in air to room temperature.

d. The weld test assembly (or the blank from it, from which the tensile test specimen is to be machined) shall be heated to a temperature between 1400 and 1450°F (760 and 788°C), held for 4 hours, then furnace cooled to 1100°F (593°C) at a rate not to exceed 100°F (55°C) per hour, then cooled in air to room temperature.

e. The weld test assembly (or the blank from it, from which the tensile test specimen is to be machined) shall be heated to a temperature between 1550 and 1600°F (840 and 870°C), held for 2 hours, then furnace cooled to 1100°F (593°C) at a rate not to exceed 100°F (55°C) per hour, then cooled in air to room temperature.

CLASSIFICATION SYSTEM, SHIELDING MEDIUM, POLARITY, AND WELDING PROCESS

AWS A5.22 Designations[a,b]	External Shielding Medium[c]	Welding Polarity	Welding Process
EXXXTX-1	CO_2	DCEP	FCAW
EXXXTX-3	none (self-shielded)	DCEP	FCAW
EXXXTX-4	75-80% Argon/remainder CO_2	DCEP	FCAW
RXXXT1-5	100% Argon	DCEN	GTAW
EXXXTX-G	Not Specified[d]	Not Specified[d]	FCAW
RXXXT1-G	Not Specified[d]	Not Specified[d]	GTAW

| CLASSIFICATION SYSTEM, SHIELDING MEDIUM, POLARITY, AND WELDING PROCESS (Continued) |

a. See AWS A5.22 paragraph 1.1 and its footnote 1 regarding the elimination of the EXXXT-2 classifications that existed in the previous revision of the document.

b. The letters "XXX" stand for the designation of the chemical composition (see Table 1 of AWS A5.22). The "X" after the "T" designates the position of operation. A "0" indicates flat or horizontal operation; a "1" indicates all-position operation.

c. The requirement for the use of specified external shielding medium shall not be construed to restrict the use of any other medium for which the electrodes are found suitable, for any application other than the classification tests.

d. See AWS A5.22 Annex A2.3.7 to A2.3.9 for additional information.

AWS A5.22 CLASSIFICATION SYSTEM

The system used for identifying the electrode and rod classifications in this specification follows the standard pattern used in other AWS filler metal specifications. An example of the method of classification follows;

EXXXTX-X

E   Indicates a welding electrode.

XXX Designates a composition of the weld metal.

T   Designates a flux-cored welding electrode or rod.

X   Designates recommended position of welding:0 = flat and horizontal; 1 = all position.

X   Designates the external shielding medium to be employed during welding specified for classification (see AWS A5.22, Table 2).

RXXXT1-5

R   Indicates a welding rod.

XXX Designates composition of the weld metal.

T   Designates a flux-cored welding electrode or rod.

1   Designates recommended position of welding: 1 = all position.

5   Designates the external shielding gas to be employed during welding. Type of shielding is 100% Argon.

The shielding designations, denoting shielding from the core materials as well as from any externally applied gas, are provided in Table 2 of AWS A5.22. This does not exclude the use of alternate gas mixtures as agreed upon between purchaser and supplier. The use of alternate gas mixtures may have an effect on welding characteristics, deposit composition, and mechanical properties of the weld, such that classification requirements may not be met.

Appendix

7

NICKEL AND NICKEL ALLOY
WELDING FILLER METALS

AWS A5.11 Chemical Composition Requirements for Undiluted Weld Metal

AWS A5.11 Classification	UNS Number[c]	Composition, Weight Percent[a,b]
ENi-1	W82141	C 0.10 Mn 0.75 Fe 0.75 P 0.03 S 0.02 Si 1.25 Cu 0.25 Ni[d] 92.0 min. Al 1.0 Ti 1.0 to 4.0 Other Elements Total 0.50
ENi-Cu-7	W84190	C 0.15 Mn 4.00 Fe 2.5 P 0.02 S 0.015 Si 1.5 Cu Rem Ni[d] 62.0–69.0 Al 0.75 Ti 1.0 Other Elements Total 0.50
ENiCrFe-1	W86132	C 0.08 Mn 3.5 Fe 11.0 P 0.03 S 0.015 Si 0.75 Cu 0.50 Ni[d] 62.0 min. Cr 13.0 to 17.0 Cb Plus Ta 1.5 to 4.0[f] Other Elements Total 0.50
ENiCrFe-2	W86133	C 0.10 Mn 1.0 to 3.5 Fe 12.0 P 0.03 S 0.02 Si 0.75 Cu 0.50 Ni[d] 62.0 min. Co[e] Cr 13.0 to 17.0 Cb Plus Ta 0.5 to 3.0[f] Mo 0.50 to 2.50 Other Elements Total 0.50
ENiCrFe-3	W86182	C 0.10 Mn 5.0 to 9.5 Fe 10.0 P 0.03 S 0.015 Si 1.0 Cu 0.50 Ni[d] 59.0 min. Co[e] Ti 1.0 Cr 13.0 to 17.0 Cb Plus Ta 1.0 to 2.5[f] Other Elements Total 0.50
ENiCrFe-4	W86134	C 0.20 Mn 1.0 to 3.5 Fe 12.0 P 0.03 S 0.02 Si 1.0 Cu 0.50 Ni[d] 60.0 min. Cr 13.0 to 17.0 Cb Plus Ta 1.0 to 3.5 Mo 1.0 to 3.5 Other Elements Total 0.50
ENiMo-1	W80001	C 0.07 Mn 1.0 Fe 4.0 to 7.0 P 0.04 S 0.03 Si 1.0 Cu 0.50 Ni[d] Rem Co 2.5 Cr 1.0 Mo 26.0 to 30.0 V 0.60 W 1.0 Other Elements Total 0.50
ENiMo-3	W80004	C 0.12 Mn 1.0 Fe 4.0 to 7.0 P 0.04 S 0.03 Si 1.0 Cu 0.50 Ni[d] Rem Co 2.5 Cr 2.5 to 5.5 Mo 23.0 to 27.0 V 0.60 W 1.0 Other Elements Total 0.50
ENiMo-7	W80665	C 0.02 Mn 1.75 Fe 2.0 P 0.04 S 0.03 Si 0.2 Cu 0.50 Ni[d] Rem Co 1.0 Cr 1.0 Mo 26.0 to 30.0 W 1.0 Other Elements Total 0.50
ENiCrCoMo-1	W86117	C 0.05 to 0.15 Mn 0.30 to 2.5 Fe 5.0 P 0.03 S 0.015 Si 0.75 Cu 0.50 Ni[d] Rem Co 9.0 to 15.0 Cr 21.0 to 26.0 Cb Plus Ta 1.0 Mo 8.0 to 10.0 Other Elements Total 0.50
ENiCrMo-1	W86007	C 0.05 Mn 1.0 to 2.0 Fe 18.0 to 21.0 P 0.04 S 0.03 Si 1.0 Cu 1.5 to 2.5 Ni[d] Rem Co 2.5 Cr 21.0 to 23.5 Cb Plus Ta 1.75 to 2.50 Mo 5.5 to 7.5 W 1.0 Other Elements Total 0.50
ENiCrMo-2	W86002	C 0.05 to 0.15 Mn 1.0 Fe 17.0 to 20.0 P 0.04 S 0.03 Si 1.0 Cu 0.50 Ni[d] Rem Co 0.50 to 2.50 Cr 20.5 to 23.0 Mo 8.0 to 10.0 W 0.20 to 1.0 Other Elements Total 0.50

CHEMICAL COMPOSITION REQUIREMENTS FOR UNDILUTED WELD METAL (Continued)

AWS A5.11 Classification	UNS Number[c]	Composition, Weight Percent[a,b]
ENiCrMo-3	W86112	C 0.10 Mn 1.0 Fe 7.0 P 0.03 S 0.02 Si 0.75 Cu 0.50 Ni[d] 55.0 min. Co[e] Cr 20.0 to 23.0 Cb Plus Ta 3.15 to 4.15 Mo 8.0 to 10.0 Other Elements Total 0.50
ENiCrMo-4	W80276	C 0.02 Mn 1.0 Fe 4.0 to 7.0 P 0.04 S 0.03 Si 0.2 Cu 0.50 Ni[d] Rem Co 2.5 Cr 14.5 to 16.5 Mo 15.0 to 17.0 V 0.35 W 3.0 to 4.5 Other Elements Total 0.50
ENiCrMo-5	W80002	C 0.10 Mn 1.0 Fe 4.0 to 7.0 P 0.04 S 0.03 Si 1.0 Cu 0.50 Ni[d] Rem Co 2.5 Cr 14.5 to 16.5 Mo 15.0 to 17.0 V 0.35 W 3.0 to 4.5 Other Elements Total 0.50
ENiCrMo-6	W86620	C 0.10 Mn 2.0 to 4.0 Fe 10.0 P 0.03 S 0.02 Si 1.0 Cu 0.50 Ni[d] 55.0 min. Cr 12.0 to 17.0 Cb Plus Ta 0.5 to 2.0 Mo 5.0 to 9.0 W 1.0 to 2.0 Other Elements Total 0.50
ENiCrMo-7	W86455	C 0.015 Mn 1.5 Fe 3.0 P 0.04 S 0.03 Si 0.2 Cu 0.50 Ni[d] Rem Co 2.0 Ti 0.70 Cr 14.0 to 18.0 Mo 14.0 to 17.0 W 0.5 Other Elements Total 0.50
ENiCrMo-9	W86985	C 0.02 Mn 1.0 Fe 18.0 to 21.0 P 0.04 S 0.03 Si 1.0 Cu 1.5 to 2.5 Ni[d] Rem Co 5.0 Cr 21.0 to 23.5 Cb Plus Ta 0.5 Mo 6.0 to 8.0 W 1.5 Other Elements Total 0.50
ENiCrMo-10	W86022	C 0.02 Mn 1.0 Fe 2.0 to 6.0 P 0.03 S 0.015 Si 0.2 Cu 0.50 Ni[d] Rem Co 2.5 Cr 20.0 to 22.5 Mo 12.5 to 14.5 V 0.35 W 2.5 to 3.5 Other Elements Total 0.50
ENiCrMo-11	W86030	C 0.03 Mn 1.5 Fe 13.0 to 17.0 P 0.04 S 0.02 Si 1.0 Cu 1.0 to 2.4 Ni[d] Rem Co 5.0 Cr 28.0 to 31.5 Cb Plus Ta 0.3 to 1.5 Mo 4.0 to 6.0 W 1.5 to 4.0 Other Elements Total 0.50
ENiCrMo-12	W86040	C 0.03 Mn 2.2 Fe 5.0 P 0.03 S 0.02 Si 0.7 Cu 0.50 Ni[d] Rem Cr 20.5 to 22.5 Cb Plus Ta 1.0 to 2.8 Mo 8.8 to 10.0 Other Elements Total 0.50

a. The weld metal shall be analyzed for the specific elements for which values are shown in this table. If the presence of other elements is indicated in the course of the work, the amount of those elements shall be determined to ensure that their total does not exceed the limit specified for "Other Elements Total".

b. Single values are maximum, except where otherwise specified.

c. SAE/ASTM Unified Numbering System for Metals and Alloys.

d. Includes incidental cobalt.

e. Cobalt - 0.12 maximum, when specified.

f. Tantalum - 0.30 maximum, when specified.

ALL-WELD-METAL TENSION TEST REQUIREMENTS

AWS A5.11 Classification	Tensile Strength, min.		% Elongation[a], min.
	psi	MPa	
ENi-1	60,000	410	20
ENiCu-7	70,000	480	30
ENiCrFe-1, ENiCrFe-2, ENiCrFe-3	80,000	550	30
ENiCrFe-4	95,000	650	20
ENiMo-1, ENiMo-3	100,000	690	25
ENiMo-7	110,000	760	25
ENiCrCoMo-1	90,000	620	25
ENiCrMo-1	90,000	620	20
ENiCrMo-2	95,000	650	20
ENiCrMo-3	110,000	760	30
ENiCrMo-4	100,000	690	25
ENiCrMo-5	100,000	690	25
ENiCrMo-6	90,000	620	35
ENiCrMo-7	100,000	690	25
ENiCrMo-9	90,000	620	25
ENiCrMo-10	100,000	690	25
ENiCrMo-11	85,000	585	25
ENiCrMo-12	95,000	650	35

a. The elongation shall be determined from a gage length equal to 4 times the gage diameter.

AWS A5.11 CLASSIFICATION SYSTEM

The system for identifying the electrode classifications in this specification follows the standard pattern used in other AWS filler metal specifications. The letter "E" at the beginning of the classification designation stands for electrode.

Since the electrodes are classified according to the chemical composition of the weld metal they deposit, the chemical symbol "Ni" appears right after the "E", as a means of identifying the electrodes as nickel-base alloys. The other symbols (Cr, Cu, Fe, Mo, and Co) in the designations are intended to group the electrodes according to their principal alloying elements. The individual designations are made up of these symbols and a number at the end of the designation (ENiMo-1 and ENiMo-3, for example). These numbers separate one composition from another, within a group, and are not repeated within that group.

AWS A5.14 Nickel & Nickel Alloy Bare Welding Electrodes & Rods

CHEMICAL COMPOSITION REQUIREMENTS FOR NICKEL AND NICKEL ALLOY ELECTRODES & RODS		
AWS A5.14 Classification	UNS Number[c]	Composition, Weight Percent[a,b]
ERNi-1	N02061	C 0.15 Mn 1.0 Fe 1.0 P 0.03 S 0.015 Si 0.75 Cu 0.25 Ni[d] 93.0 min Al 1.5 Ti 2.0 to 3.5 Other Elements Total 0.50
ERNiCu-7	N04060	C 0.15 Mn 4.0 Fe 2.5 P 0.02 S 0.015 Si 1.25 Cu Rem Ni[d] 62.0 to 69.0 Al 1.25 Ti 1.5 to 3.0 Other Elements Total 0.50
ERNiCr-3	N06082	C 0.10 Mn 2.5 to 3.5 Fe 3.0 P 0.03 S 0.015 Si 0.50 Cu 0.50 Ni[d] 67.0 min Co[e] Ti 0.75 Cr 18.0 to 22.0 Cb plus Ta 2.0 to 3.0[f] Other Elements Total 0.50
ERNiCrFe-5	N06062	C 0.08 Mn 1.0 Fe 6.0 to 10.0 P 0.03 S 0.015 Si 0.35 Cu 0.50 Ni[d] 70.0 min Co[e] Cr 14.0 to 17.0 Cb plus Ta 1.5 to 3.0[f] Other Elements Total 0.50
ERNiCrFe-6	N07092	C 0.08 Mn 2.0 to 2.7 Fe 8.0 P 0.03 S 0.015 Si 0.35 Cu 0.50 Ni[d] 67.0 min Ti 2.5 to 3.5 Cr 14.0 to 17.0 Other Elements Total 0.50
ERNiFeCr-1	N08065	C 0.05 Mn 1.0 Fe 22.0 P 0.03 S 0.03 Si 0.50 Cu 1.50 to 3.0 Ni[d] 38.0 to 46.0 Al 0.20 Ti 0.60 to 1.2 Cr 19.5 to 23.5 Mo 2.5 to 3.5 Other Elements Total 0.50
ERNiFeCr-2[g]	N07718	C 0.08 Mn 0.35 Fe Rem P 0.015 S 0.015 Si 0.35 Cu 0.30 Ni[d] 50.0 to 55.0 Al 0.20 to 0.80 Ti 0.65 to 1.15 Cr 17.0 to 21.0 Cb plus Ta 4.75 to 5.50 Mo 2.80 to 3.30 Other Elements Total 0.50
ERNiMo-1	N10001	C 0.08 Mn 1.0 Fe 4.0 to 7.0 P 0.025 S 0.03 Si 1.0 Cu 0.50 Ni[d] Rem Co 2.5 Cr 1.0 Mo 26.0 to 30.0 V 0.20 to 0.40 W 1.0 Other Elements Total 0.50
ERNiMo-2	N10003	C 0.04 to 0.08 Mn 1.0 Fe 5.0 P 0.015 S 0.02 Si 1.0 Cu 0.50 Ni[d] Rem Co 0.20 Cr 6.0 to 8.0 Mo 15.0 to 18.0 V 0.50 W 0.50
ERNiMo-3	N10004	C 0.12 Mn 1.0 Fe 4.0 to 7.0 P 0.04 S 0.03 Si 1.0 Cu 0.50 Ni[d] Rem Co 2.5 Cr 4.0 to 6.0 Mo 23.0 to 26.0 V 0.60 W 1.0 Other Elements Total 0.50
ERNiMo-7	N10665	C 0.02 Mn 1.0 Fe 2.0 P 0.04 S 0.03 Si 0.10 Cu 0.50 Ni[d] Rem Co 1.0 Cr 1.0 Mo 26.0 to 30.0 W 1.0 Other Elements Total 0.50

CHEMICAL COMPOSITION REQUIREMENTS FOR NICKEL AND NICKEL ALLOY ELECTRODES & RODS (Continued)

AWS A5.14 Classification	UNS Number[c]	Composition, Weight Percent[a,b]
ERNiCrMo-1	N06007	C 0.05 Mn 1.0 to 2.0 Fe 18.0 to 21.0 P 0.04 S 0.03 Si 1.0 Cu 1.5 to 2.5 Ni[d] Rem Co 2.5 Cr 21.0 to 23.5 Cb plus Ta 1.75 to 2.50 Mo 5.5 to 7.5 W 1.0 Other Elements Total 0.50
ERNiCrMo-2	N06002	C 0.05 to 0.15 Mn 1.0 Fe 17.0 to 20.0 P 0.04 S 0.03 Si 1.0 Cu 0.50 Ni Rem Co 0.50 to 2.5 Cr 20.5 to 23.0 Mo 8.10 to 10.0 W 0.20 to 1.0 Other Elements Total 0.50
ERNiCrMo-3	N06625	C 0.10 Mn 0.50 Fe 5.0 P 0.02 S 0.015 Si 0.50 Cu 0.50 Ni 58.0 min Al 0.40 Ti 0.40 Cr 20.0 to 23.0 Cb plus Ta 3.15 to 4.15 Mo 8.10 to 10.0
ERNiCrMo-4	N10276	C 0.02 Mn 1.0 Fe 4.0 to 7.0 P. 0.04 S 0.03 Si 0.08 Cu 0.50 Ni[d] Rem Co 2.5 Cr 14.5 to 16.5 Mo 15.0 to 17.0 V 0.35 W 3.0 to 4.5 Other Elements Total 0.50
ERNiCrMo-7	N06455	C 0.015 Mn 1.0 Fe 3.0 P 0.04 S 0.03 Si 0.08 Cu 0.50 Ni[d] Rem Co 2.0 Ti 0.70 Cr 14.0 to 18.0 Mo 14.0 to 18.0 W 0.50 Other Elements Total 0.50
ERNiCrMo-8	N06975	C 0.03 Mn 1.0 Fe Rem P 0.03 S 0.03 Si 1.0 Cu 0.7 to 1.20 Ni[d] 47.0 to 52.0 Ti 0.70 to 1.50 Cr 23.0 to 26.0 Mo 5.0 to 7.0 Other Elements Total 0.50
ERNiCrMo-9	N06985	C 0.015 Mn 1.0 Fe 18.0 to 21.0 P 0.04 S 0.03 Si 1.0 Cu 1.5 to 2.5 Ni[d] Rem Co 5.0 Cr 21.05 to 23.5 Cb plus Ta 0.50 Mo 6.0 to 8.0 W 1.5 Other Elements Total 0.50
ERNiCrMo-10	N06022	C 0.015 Mn 0.50 Fe 2.0 to 6.0 P 0.20 Si 0.010 Si 0.08 Cu 0.50 Ni Rem Co 2.5 Cr 20.0 to 22.5 Mo 12.5 to 14.5 V 0.35 W 2.5 to 4.5 Other Elements Total 0.50
ERNiCrMo-11	N06030	C 0.30 Mn 1.5 Fe 13.0 to 17.0 P 0.04 S 0.02 Si 0.80 Cu 1.0 to 2.4 Ni Rem Co 5.0 Cr 28.0 to 31.5 Cb + Ta 0.30 to 1.50 Mo 4.0 to 6.0 W 1.5 to 4.0 Other Elements Total 0.50
ERNiCrCoMo-1	N06617	C 0.05 to 0.15 Mn 1.0 Fe 3.0 P 0.03 S 0.015 Si 1.0 Cu 0.50 Ni Rem Co 10.0 to 15.0 Al 0.80 to 1.50 Ti 0.60 Cr 20.0 to 24.0 Mo 8.0 to 10.0 Other Elements Total 0.50

a. The filler metal shall be analyzed for the specific elements for which values are shown in this table. If the presence of other elements is indicated in the course of this work, the amount of those elements shall be determined to ensure that their total does not exceed the limit specified for "Other Elements Total" in this table. "Rem" stands for Remainder.

b. Single values are maximum, except where otherwise specified.

c. SAE/ASTM Unified Numbering System for Metals and Alloys.

d. Includes incidental cobalt

e. Cobalt - 0.12 maximum, when specified.

f. Tantalum - 0.30 maximum, when specified.

g. Boron is 0.006 percent maximum.

COMPARISON OF CLASSIFICATIONS

Classification in A5.14		Military Designation[b, c]	Corresponding Classification in AWS A5 11-88[c]
Present Classification	Previous Classification[a, c]		
ERNi-1	ERNi-1	EN61 & RN61	ENi-1
ERNiCu-7	ERNiCu-7	EN60 & RN60	ENiCu-7
ERNiCr-3	ERNiCr-3	EN82 & RN82	ENiCrFe-3
ERNiCrFe-5	ERNiCrFe-5	EN62 & RN62	ENiCrFe-1
ERNiCrFe-6	ERNiCrFe-6	EN6A & RN6A	-
ERNiFeCr-1	ERNiFeCr-1	-	-
ERNiFeCr-2	Not Classified	-	-
ERNiMo-1	ERNiMo-1	-	ENiMo-1
ERNiMo-2	ERNiMo-2	-	-
ERNiMo-3	ERNiMo-3	-	ENiMo-3
ERNiMo-7	ERNiMo-7	-	ENiMo-7
ERNiCrMo-1	ERNiCrMo-1	-	ENiCrMo-1
ERNiCrMo-2	ERNiCrMo-2	-	ENiCrMo-2
ERNiCrMo-3	ERNiCrMo-3	EN625 & RN625	ENiCrMo-3
ERNiCrMo-4	ERNiCrMo-4	-	ENiCrMo-4
ERNiCrMo-7	ERNiCrMo-7	-	ENiCrMo-7
ERNiCrMo-8	ERNiCrMo-8	-	-
ERNiCrMo-9	ERNiCrMo-9	-	ENiCrMo-9
ERNiCrMo-10	Not Classified	-	ENiCrMo-10
ERNiCrMo-11	Not Classified	-	ENiCrMo-11
ERNiCrCoMo-1	Not Classified	-	ENiCrCoMo-1

a. AWS A5.14-83
b. Mil-E-21562.
c. Specifications are not exact duplicates. Information is supplied only for general comparison.

TYPICAL WELD METAL TENSILE STRENGTHS[a]		
AWS A5.14 Classification	**psi**	**MPa**
ERNi-1	55,000	380
ERNiCu-7	70,000	480
ERNiCr-3, ERNiCrFe-5, ERNiCrFe-6, ERNiFeCr-1	80,000	550
ERNiFeCr-2	165,000[b]	1,138[b]
ERNiMo-1, ERNiMo-2, ERNiMo-3	100,000	690
ERNiMo-7	110,000	760
ERNiCrMo-1, ERNiCrMo-8, ERNiCrMo-9, ERNiCrMo-11	85,000	590
ERNiCrCoMo-1	90,000	620
ERNiCrMo-2	95,000	660
ERNiCrMo-3	110,000	760
ERNiCrMo-4, ERNiCrMo-7, ERNiCrMo-10	100,000	690

a. Tensile strength in as-welded condition unless otherwise specified.
b. Age hardened condition: heat to 1,325°F (718°C), hold at temperature for eight hours, furnace cool to 1,150°F (620°C) at 100°F (55°C)/hour and then air cool.

AWS A5.14 CLASSIFICATION SYSTEM

The system for classifying the filler metals in this specification follows the standard pattern used in other AWS filler metal specifications. The letter "ER" at the beginning of each classification designation stands for electrode and rod, indicating that the filler metal may be used either way.

Since the filler metals are classified according to their chemical composition, the chemical symbol "Ni" appears right after the "ER" as a means of identifying the filler metals as nickel-base alloys. The other symbols (Cr, Cu, Fe, and Mo) in the designations are intended to group the filler metals according to their principal alloying elements. The individual designations are made up of these symbols and a number at the end of the designation (ERNiMo-1 and ERNiMo-2, for example). These numbers separate one composition from another within a group and are not repeated within that group.

Subject Index

A

Acid,
   Acetic, 144, 236, 261, 278
   Alkalies, 190
   Carbonic, 189
   Formic, 236, 261, 278
   Hydrochloric, 8, 185-186, 236, 265-266, 276-277, 299, 312, 347
   Hydrofluoric, 64, 185, 236, 277, 312, 337
   Inorganic, 189-190, 236
   Napthenic, 312
   Nitric, 2, 7, 54, 58, 144, 187-188, 214, 230, 236, 265, 277-278, 300, 343
   Organic, 4, 12, 55, 140, 144, 230, 236, 261, 278-279, 300
   Phosphoric, 14, 189, 230, 235-236, 245, 262, 273, 298, 300, 345
   Polythionic, 312
   Sulfuric, 2, 11-12, 24, 54, 144, 186-187, 233-235, 244-245, 263-264, 274, 288, 298, 313, 335
   Sulfurous, 189

Aircraft, Aerospace and Missile (Components), 56, 102, 104, 164, 288, 318, 345-348

AOD (Argon-Oxygen Decarburization), 3, 16, 132, 245, 246, 373

Appliances, Domestic, 142, 164, 288, 321, 347

Architectural, 7, 140, 141, 143, 161, 164, 232, 298

Automotive
   Ferritic Stainless Steels
      Applications, 142, 149
      Corrosion, 149
      Emission Controls, 164

B

Bearing, 105
   Surface, 112, 318
   Housing, 315

Black Liquor, 272

C

Carburization, 24, 31, 64, 65, 69, 70, 73, 76-77, 81, 313, 322, 325, 330, 335, 341-342, 347

Castings
   ACI Alloy Designation System, 39-41
   Data
      Chemical Compositions
         Corrosion Resistant Alloys,
            ACI Alloys, 33-35
            Nickel Alloys, 62
            Stainless Steels
               Austenitic Stainless Steels, 57
               Ferritic Stainless Steels, 53
               Martensitic Stainless Steels, 55
               Precipitation Hardening Stainless Steels, 56
               Superaustenitic Stainless Steels, 62
         Heat Resistant Alloys,
            ACI Alloys, 36-37, 65
            ASTM A 494 Alloys, 38
            HP-Modified Alloys, 77, 78, 79
      Heat Treatment, 52
      Metallurgy, General, 41-45
         Ferrite, 46-50
            Determining Ferrite in Welds, 51
            Intergranular Attack, 47
            Stress Corrosion Cracking, 47
               Schaefler Diagram, 51
         Volumetric Change, 45

H

Halogenation, 300, 325, 340-341

Heat Exchanger, 164, 182-183, 236, 271-273, 277-278, 298, 347

Heater, *see also Tubing, Heater*
    Hot Water, 142, 232, 298

Heating,
    Electrical Elements, 321
    Resistance Heating, 345, 348

Heat Treating Furnace Parts, 69, 325-326

Hydrogen Embrittlement, 241, 303, 304, 306, 318, 343

Hydrogen Sulfide (H_2S, Sour Gas), 111, 239, 241, 278, 303, 304, 337, 346, 400, 401, 405-406, 407-408
    see also Natural Gas Production

M

Martensitic Stainless Steels
    Corrosion, Oxidation Resistance, Typical Applications, 102-105
    Data
        Chemical Compositions, 92
        Cross References, American Standards, 100-101
        Mechanical Properties
            Annealed Bars, Nominal Properties, 94
            ASTM A 276 (Bars and Shapes), 93-94
            Quenched and Tempered Bars, Nominal Properties, 95-96
            Wrought ASTM Standards, 97-98
        Physical Properties, 99
    Downhole Production Tubing Applications, 403
    General Description
        S40300 (Type 403), 90
        S41000 (Type 410), 88

N

R

S

Stainless Steels (Continued)
    Precipitation-Hardening, *see Precipitation-Hardening Stainless Steels*
    Superaustenitic, *see Superaustenitic Stainless Steels*

Sulfidation, 322, 324, 325, 335, 338-340, 345-348, 352

Superaustenitic Stainless Steels, General, 243-244
    Applications, 269
        Flue Gas Desulfurization, 275
        General, 279
        Hydrochloric Acid, 276-277
        Hydrofluoric Acid, 277
        Microbiological, 273
        Nitric, 277-278
        Organic Acids, 278
        Phosphoric Acid, 273
        Pulp and Paper, 272-273
        Seawater, 272
        Sour Gas Environments, 278
        Sulfuric Acid, 274
        Urea Synthesis, 278
        Water, 271-272
    ASTM Specifications, 250-251
    Compositions, 247
    Corrosion Properties
        Acids,
            Hydrochloric, 265-266
            Nitric, 265
            Organic, 261
            Phosphoric, 262
            Sulfuric, 263-264
        General Corrosion, 257, 261
        Intergranular Corrosion, 257-258
        Pitting and Crevice Corrosion, 258-260
            ASTM G 48, 256
        Sour Gas Environment, 266
        Stress Corrosion Cracking, 260-261

T

Alloy Index

ASTM Standards

Nickel Alloys

UNS No.	Common Name	Trade Name	Page No.
N02061	AWS ERNi-1, AWS ENi-1	---	379
N02100	ACI CZ-100	---	38, 63
N02200	alloy 200	Nickel 200	291, 294, 298, 300, 301, 318, 328, 337, 345, 354, 379
N02201	alloy 201	Nickel 201	291, 294, 327, 328, 329, 345
N02211	alloy 211	Nickel 211	291, 345
N03301	---	Duranickel alloy 301	329
N04060	AWS ERNiCu-7, AWS ENiCu-7	---	379
N04400	alloy 400	Monel alloy 400	2, 287, 291, 294, 298, 299, 300, 301, 305, 306, 312, 318, 319, 329, 330, 340, 345, 354, 379
N04405	---	Monel alloy R-504	298, 305
N05500	alloy 500	Monel K-500	4, 288, 291, 294, 298, 302, 304, 305, 306, 345, 379
N05504	AWS ERNiCu-8	---	379
N06002	alloy X, AWS ERNiCrMo-2, AWS ENiCrMo-2	Hastelloy X	291, 305, 313, 317, 318, 320, 339, 345, 379
N06003	80Ni-20Cr	Brightray alloys C and S	291, 321, 345
N06004	65Ni-15Cr	Brightray alloy B	291, 321, 345
N06006	ACI HX	---	37
N06007	alloy G, AWS ERNiCrMo-1, AWS ENiCrMo-1	Hastelloy G	14, 305, 372, 379
N06022	C-22, AWS ERNiCrMo-10, AWS ENiCrMo-10	Hastelloy C-22	10, 274, 275, 291, 294, 300, 319, 336, 345, 379

Nickel Alloys (Continued)

UNS No.	Common Name	Trade Name	Page No.
N06025	---	Nicrofer 6025HT	291, 325, 334, 338, 341, 345
N06030	alloy G-30, AWS ERNiCrMo-11, AWS ENiCrMo-11	Hastelloy G-30	14, 291, 294, 300, 345, 379
N06040	ACI CY-40	---	38
N06045	alloy 45TM	Nicrofer 603GT Nicrofer 45TM	289, 291, 324, 339, 345
N06050	ACI HX50	---	37, 291, 294
N06052	AWS ERNiCrFe-7, AWS ENiCrFe-7	---	379
N06059	alloy 59, AWS ERNiCrMo-13, AWS ENiCrMo-13	Nicrofer 5923hMo	11, 285, 289, 291, 294, 298, 300, 319, 336, 345, 379
N06062	AWS ERNiCrFe-5	---	379
N06072	AWS ERNiCr-4	Inconel Filler Metal 72,	325, 379
N06075	alloy 75	Nimonic alloy 75	291, 314, 316, 329, 330, 345
N06076	AWS ERNiCr-6, AWS ENiFe-1	---	379
N06082	AWS ERNiCr-3, AWS ENiCrFe-3	---	379
N06200	alloy C-2000	Hastelloy C-2000	11, 289, 300, 345
N06230	alloy 230	Haynes 230, RA230	291, 302, 314, 315, 324, 326, 343, 345, 379
N06231	AWS ERNiCrWMo-1	---	379
N06250	---	SM2050	305
N06255	---	SM 2250	305
N06263	---	Hastelloy X	317
N06455	alloy C-4, AWS ERNiCrMo-7, AWS ENiCrMo-7	Hastelloy C-4	10, 291, 295, 300, 346, 379

Nickel Alloys (Continued)

UNS No.	Common Name	Trade Name	Page No.
N06600	alloy 600	Inconel alloy 600	288, 291, 295, 302, 305, 314, 318, 320, 324, 325, 326, 327, 329, 330, 333, 339, 340, 341, 343, 346, 354, 379
N06601	alloy 601, alloy 601GC, AWS ERNiCrFe-11	Inconel 601, Inconel 601GC	289, 291, 295, 319, 324, 325, 326, 333, 334, 338, 339, 340, 343, 346, 379
N06617	alloy 617, AWS ERNiCrCoMo-1, AWS ENiCrCoMo-1	Inconel 617	288, 292, 295, 302, 314, 315, 324, 326, 339, 341, 343, 346, 379
N06625	alloy 625, AWS ERNiCrMo-3, AWS ENiCrMo-3	Inconel 625	11, 259, 288, 289, 292, 295, 298, 300, 305, 306, 307, 308, 312, 314, 317, 319, 324, 334, 346, 372, 379, 386
N06626	alloy 625LCF	Inconel 625LCF	292, 295, 312, 315, 334, 346
N06686	alloy 686, AWS ERNiCrMo-14, AWS ENiCrMo-14	Inconel 686	289, 292, 295, 298, 300, 319, 336, 346, 379
N06690	alloy 690	Inconel 690	288, 292, 295, 298, 320, 325, 379
N06950	alloy G-50	Hastelloy G-50	294, 303, 305, 306, 346
N06975	alloy G-2, AWS ERNiCrMo-8	Hastelloy G-2	305, 379

Nickel Alloys (Continued)

UNS No.	Common Name	Trade Name	Page No.
N06985	alloy G-3, AWS ERNiCrMo-9, AWS ENiCrMo-9	Hastelloy G-3	14, 292, 295, 298, 300, 302, 303, 305, 306, 307, 309, 346, 379
N07001	---	Waspaloy	292, 316, 317, 318, 346
N07031	---	Pyromet 31	292, 346
N07069	AWS ERNiCrFe-8		379
N07080	alloys 80 and 80A	Nimonic 80 and 80A	288, 292, 315, 316, 320, 330, 333, 346
N07090	alloy 90	Nimonic 90	292, 315, 317, 333, 346
N07092	AWS ERNiCrFe-6, AWS ENiCrFe-2	---	379
N07214	alloy 214	---	289, 292, 324, 326, 338, 340, 341, 343, 346
N07263	---	Nimonic 263	315
N07500	---	UDIMET 500	292
N07716	---	Carpenter 625PLUS	289, 292, 295, 298, 304, 305, 306, 346
N07718	alloy 718, AWS ERNiFeCr-2	Inconel 718	11, 288, 292, 304, 305, 306, 307, 310, 315, 317, 318, 320, 346, 379, 383
N07719	alloy 718SPF	Inconel 718SPF	292, 315
N07725	alloy 725, AWS ERNiCrMo-15	Inconel 725	289, 292, 295, 298, 304, 305, 306, 307, 311, 346, 379, 383
N07750	alloy X-750	Inconel X-750	292, 304, 305, 306, 316, 320, 334, 346, 379
N07751	alloy 751	Inconel 751	293, 319, 333, 347

Nickel Alloys (Continued)

UNS No.	Common Name	Trade Name	Page No.
N07754	---	Inconel alloy MA754	293, 316, 318, 347
N08001	ACI HW	---	37
N08004	ACI HU	---	37
N08005	ACI HU50	---	37
N08006	ACI HW50	---	37
N08007 (formerly J95150)	ACI CN-7M	---	13, 21, 57
N08020	alloy 20	Carpenter 20Cb-3	13, 232, 233, 244, 245, 248, 250, 252, 257, 259, 263, 264, 265, 269, 271, 277, 278, 288, 293, 296, 324, 347
N08024	---	Carpenter 20Mo-4	247, 248, 250, 252, 259, 278
N08026	---	Carpenter 20Mo-6	21, 248, 250, 252, 259
N08028	alloy 28	Sanicro 28	20, 245, 248, 250, 252, 259, 260, 262, 263, 264, 265, 266, 271, 273, 277, 278, 306, 307, 309
N08031	alloy 31	Nicrofer, 3127hMo, VDM alloy 31	21, 245, 248, 251, 252, 259, 262, 263, 264, 265, 266, 273, 274, 277, 285
N08050	ACI HT50	---	37
N08065	AWS ERNiFeCr-1	---	379
N08151	CT15C	---	69
N08320	alloy 20 modified	---	248, 251, 252, 257, 259
N08330	alloy 330	---	293, 302, 326, 338, 339, 347

Nickel Alloys (Continued)

UNS No.	Common Name	Trade Name	Page No.
N08366	---	AL-6X	20, 245, 246, 248, 251, 252, 259, 273
N08367	---	AL-6XN	21, 246, 248, 251, 252, 254, 256, 257, 259, 261, 262, 263, 264, 265, 266, 267, 270, 276, 278
N08603	ACI HT30	---	37
N08604	ACI HL	---	36
N08605	ACI HT	---	37
N08613	ACI HL30	---	36
N08614	ACI HL40	---	36
N08700	---	JS700	248, 251, 252, 257, 259, 262, 273
N08705	ACI HP	---	37
N08800	alloy 800	Inconel 800	243, 288, 293, 296, 298, 302, 305, 312, 313, 314, 319, 320, 326, 339, 340, 347, 354
N08801	alloy 801	Inconel 801	312
N08810	alloy 800H	Inconel 800H	69, 80, 230, 289, 293, 302, 312, 313, 314, 319, 333, 347
N08811	alloy 800HT	Inconel 800HT	289, 293, 302, 312, 313, 314, 319, 325, 326, 338, 347

Nickel Alloys (Continued)

UNS No.	Common Name	Trade Name	Page No.
N08825	alloy 825	Incoloy 825	15, 248, 260, 263, 288, 293, 296, 298, 300, 303, 305, 306, 307, 308, 312, 313, 314, 319, 320, 324, 347, 379
N08904	alloy 904L	---	20, 232, 233, 237, 238, 240, 244, 248, 251, 252, 259, 262, 263, 264, 265, 269, 271, 273, 277, 372
N08925	---	---	21, 251, 252, 259
N08926	alloy 926, 1925hMo	INCO 25-6MO	21, 248, 251, 252, 259, 260, 278, 306, 307, 309
N08932	---	URSB-8	248, 251, 253, 259
N09901	alloy 901	Incoloy 901	318
N09908	alloy 908	Incoloy 908	321
N09925	alloy 925	Incoloy 925	289, 293, 296, 303, 304, 305, 306, 307, 310, 311, 347
N10001	alloy B, AWS ERNiMo-1, AWS ENiMo-1	Hastelloy B	8, 288, 379
N10002	alloy C	Hastelloy C	288
N10003	alloy N, AWS ERNiMo-2	Hastelloy N	324, 327, 379
N10004	alloy W, AWS ERNiMo-3, AWS ENiMo-3	Hastelloy W	379
N10008	AWS ERNiMo-8, AWS ENiMo-8	TGS-709S	379

Nickel Alloys (Continued)

UNS No.	Common Name	Trade Name	Page No.
N10009	AWS ERNiMo-9, AWS ENiMo-9	NITTETSU Filler 196	379
N10276	alloy C-276, AWS ERNiCrMo-4, AWS ENiCrMo-4	Hastelloy C-276	10, 261, 262, 263, 269, 275, 288, 293, 296, 298, 300, 302, 303, 305, 306, 307, 319, 320, 336, 347, 354, 372, 373, 379
N10629	alloy B-4	Hastelloy B-4	9, 300, 347
N10665	alloy B-2, AWS ERNiMo-7, AWS ENiMo-7	Hastelloy B-2	9, 288, 293, 300, 347, 379
N10675	alloy B-3, AWS ERNiMo-10, AWS ENiMo-10	Hastelloy B-3	9, 293, 300, 347, 379
N12160	---	HR-160	289, 293, 294, 324, 342, 343, 347
N19909	alloy 909	Incoloy 909	317, 318
N24025	A 494 M-25S	---	38
N24030	A 494 M-30H	---	38
N24130	A 494 M-30C	---	38
N24135	A 494 M-35-1	---	38
N26022	A 494 CX-2MW	---	10, 38, 289, 298
N26055	A 494 CY5SnBiM	---	38
N26455	A 494 CW-2M	---	10, 38
N26625	A 494 CW-6MC	---	38
N30002	A 494 CW-12MW	---	10, 38, 329
N30007	A 494 N-7M	---	9, 38
N30012	A 494 N-12MV	---	9, 38
N30107	A 494 CW-6M	---	38

Cast Stainless Steels

UNS No.	ACI Designation	Page No.
J29999	CG-3M	171
J91150	CA15	55
J91151	CA-15M	33, 55
J91153	CA-40	33, 55
J91154	CA40F	55
J91540	CA-15, CA-6NM	33, 55
J91803	CB-30	33, 53
J91804	CB-6	33
J92110	CB-7Cu-2	33, 56
J92180	CB-7Cu-1	33, 56
J92205	CD3MN	35, 213
J92500	CF-3	33, 57, 171, 176, 177
J92590	CF-10	34, 57, 171
J92600	CF-8	33, 57, 171, 176, 177
J92602	CF-20	34, 57, 59
J92603	HF	36
J92605	HC	36
J92613	HC30	36
J92615	CC-50	33, 53
J92630	---	171
J92660	CF-8C	171
J92700	CF-3MN	33
J92701	CF-16F	34, 57, 59, 171
J92710	CF-8C	33, 57
J92800	CF-3M	33, 40, 41, 171, 176, 177
J92803	HF30	36
J92804	CF3MN	57
J92805	CE-8N	33
J92900	CF3M, CF-8M	34, 57, 58, 171, 176, 177
J92901	CF-10M	34, 57
J92920	CF-8M	171
J92971	CF-10MC	34, 57
J92972	CF10SMnN	34, 57, 171
J93000	CG8M	57, 59
J93001	CG-12	34, 57
J93005	HD	36
J93015	HD50	36
J93254	CK-3MCuN	34

Cast Stainless Steels (Continued)

UNS No.	ACI Designation	Page No.
J93345	CE-8MN	33, 62, 213, 227
J93370	CD-4MCu	12, 17, 33, 56, 60, 213, 227
J93371	CD6MN	213
J93372	CD4MCuN	35, 61, 213
J93380	CD-3MWCuN	33, 213, 227
J93400	CH-8	34, 57
J93401	CH-10	34, 57
J93402	CH-20	34, 57, 171
J93403	HE	36
J93404	CE3MN	34, 62, 213
J93413	HE35	36
J93423	CE-30	33, 57
J93503	HH	36
J93513	HH30	36
J93633	HH33	36
J93790	CG-6MMN	34, 57
J94003	HI	36
J94013	HI35	36
J94202	CK-20	34, 57, 171
J94203	HK30	36
J94204	HK40	36, 41
J94213	HN	36
J94214	HN40	36
J94224	HK	36, 70, 77
J94650	CN-7MS	35
J94651	CN-3MN	34
J94652	CN-3M	34

Wrought Stainless Steels

UNS No.	Common Name	Trade Name	Page No.
S13800	PH 13-8 Mo	Armco PH 13-8 Mo	106, 108, 109, 111, 113, 118, 119, 121, 126, 127, 129, 130
S15500	15-5 PH	Armco 15-5 PH	106, 108, 109, 118, 119, 121, 126, 129, 130
S15700	PH 15-7 Mo	Armco PH 15-7 Mo	106, 108, 109, 118, 123, 127, 129, 130
S17400	17-4 PH	Armco 17-4 PH	91, 106, 108, 109, 110, 111, 113, 117, 118, 122, 126, 129, 130, 378
S17700	17-7 PH	Armco 17-7 PH	106, 108, 109, 110, 113, 118, 124, 127, 129, 130
S20100	type 201	---	162, 167, 168, 174, 176, 177, 178
S20161	---	Gall-Tough	162, 174, 178
S20200	type 202		162, 167, 168, 174, 178
S20400	---	Nitronic 30	162, 167, 168, 174, 176, 177, 178
S21800	---	Nitronic 60	162, 171, 174, 178
S30100	type 301	---	117, 118, 162, 167, 168, 174, 178
S30200	type 302	---	86, 162, 167, 168, 174, 78
S30300	type 303	---	162, 171, 174, 178
S30323	type 303Se	---	162, 171, 174, 178

Wrought Stainless Steels (Continued)

UNS No.	Common Name	Trade Name	Page No.
S30400	type 304	---	57, 110, 111, 135, 138, 140, 143, 144, 145, 146, 148, 156, 157, 160, 162, 165, 166, 167, 168, 171, 174, 176, 177, 178, 232, 237, 260, 261, 263, 266, 267, 271, 273, 354, 358, 363, 368, 369, 388
S30403	type 304L	---	13, 162, 166, 167, 168, 171, 174, 176, 177, 178, 214, 215, 222, 228, 230, 231, 236, 237, 239, 240, 265, 363, 369
S30409	type 304H	---	162, 169, 170, 171, 174, 176, 177, 178
S30430	type 302HQ	---	162, 167, 138, 174, 178
S30451	type 304N	---	162, 167, 168, 174, 176, 177, 178
S30453	type 304LN	---	369
S30500	type 305	---	162, 167, 168, 174, 178, 267
S30800	type 308	---	162, 178
S30815	253MA	---	163, 169, 170, 174, 176, 177, 178
S30883	type 308L	---	162, 178
S30900	type 309	---	163, 169, 170, 171, 174, 363

Wrought Stainless Steels (Continued)

UNS No.	Common Name	Trade Name	Page No.
S30908	type 309S	---	163, 169, 170, 174, 176, 177, 178
S30909	type 309H	---	163, 169, 170, 174, 176, 177, 179
S31000	type 310	---	41, 163, 169, 170, 171, 175, 300, 339, 354, 363, 366
S31002	type 310 low carbon	---	230, 236
S31008	type 310S	---	163, 169, 170, 175, 176, 177, 179, 277
S31009	type 310H	---	163, 169, 170, 175, 176, 177, 179, 331
S31050	type 310MoLN	---	248, 250, 252, 259, 266, 278
S31200	---	Avesta 44LN	212, 216, 227
S31254	---	Avesta SMO 254	20, 237, 238, 246, 248, 250, 252, 259, 278
S31260	---	DP-3	212, 216, 227
S31266	---	URB-66™	247, 248, 250, 252, 259, 260, 272
S31500	---	Sandvik 3RE60	17, 212, 214, 215, 216, 227
S31600	type 316	---	6, 148, 162, 166, 167, 168, 171, 174, 176, 177, 178, 217, 260, 263, 264, 265, 267, 271, 273, 277, 330, 363, 367, 371, 382

Wrought Stainless Steels (Continued)

UNS No.	Common Name	Trade Name	Page No.
S31603	type 316L	---	13, 40, 41, 162, 166, 167, 168, 171, 174, 176, 177, 178, 214, 215, 217, 222, 228, 230, 231, 232, 233, 235, 237, 238, 239, 240, 259, 261, 262, 278, 363, 371
S31609	type 316H	---	163, 169, 170, 171, 174, 176, 177, 178
S31653	type 316LN	---	232
S31700	type 317	---	217, 363, 371
S31703	type 317L	---	149, 162, 167, 168, 171, 174, 176, 177, 178, 217, 228, 230, 231, 232, 233, 240, 363, 371
S31726	type 317LMN	---	162, 167, 168, 171, 174, 178, 232, 274, 275
S31753	type 317LN	---	232
S31803	alloy 2205, alloy22/5	---	18, 211, 212, 215, 216, 217, 218, 220, 223, 224, 225, 227, 228, 229, 233, 235, 236, 237, 238, 239, 240, 354
S32001	---	19D	212
S32050	---	SR50A	248, 250, 252, 259

Wrought Stainless Steels (Continued)

UNS No.	Common Name	Trade Name	Page No.
S32100	type 321	---	163, 169, 170, 171, 174, 176, 177, 178, 314, 333, 363, 369
S32205	2205 high N, 2205 +	---	19, 211, 212, 215, 233
S32304	alloy 2304	---	211, 212, 216, 227, 236, 237
S32404	---	UR 50	17, 212
S32520	---	UR 52N+	212
S32550	alloy 255	Ferralium 255	18, 212, 216, 227, 235
S32654	alloy 654	Avesta SMO 654	21, 246, 248, 250, 252, 259, 260, 264, 266, 272, 274, 277, 284
S32750	alloy 2507	---	20, 212, 216, 227, 237, 238
S32760	---	Zeron 100	20, 212, 216, 227, 235
S32900	type 329	---	16, 212, 216, 227, 373
S32950		Carpenter 7-Mo Plus	212, 216, 227
S34565	4565S	---	247, 248, 250, 252, 259
S34700	type 347	---	163, 169, 170, 171, 174, 176, 177, 178, 363, 369, 382
S34709	type 347H	---	331
S35000	---	AM350	106, 108, 109, 110, 124, 128, 129, 130
S35045	alloy 803	Incoloy 803	291, 313
S35135	alloy 864	Incoloy 803	291, 334, 347

Wrought Stainless Steels (Continued)

UNS No.	Common Name	Trade Name	Page No.
S35500	---	AM355	106, 108, 109, 124, 128, 129, 130
S39274	DP-3W	---	212
S40300	type 403	---	90, 92, 93, 94, 95, 99, 100, 102, 360
S40500	type 405	---	99, 100, 133, 135, 138, 140, 142, 157, 360, 361
S40800	type 408	---	139
S40900	type 409	---	99, 100, 132, 133, 135, 138, 139, 140, 142, 144, 146, 148, 151, 156, 157, 333, 360, 361
S40910	---	---	146
S40920	---	---	146, 152
S40930	---	---	146
S40975	---	---	133, 142, 151
S41000	type 410	---	86, 88, 89, 90, 91, 92, 93, 94, 95, 97, 98, 99, 100, 102, 103, 104, 110
S41003	---	---	133
S41008	type 410S	---	133
S41040	XM-30	---	89, 92, 93, 94, 95, 102
S41045	---	---	133
S41400	---	---	91, 92, 93, 94, 96, 100, 102
S41500	---	---	92, 93, 103
S41600	type 416	---	90, 92, 94, 95, 97, 99, 100, 103, 360
S41610	XM-6, 416Plus X	---	92

Wrought Stainless Steels (Continued)

UNS No.	Common Name	Trade Name	Page No.
S41623	type 416Se	---	92, 94, 95, 100, 103
S42000	type 420	---	4, 85, 86, 88, 89, 90, 92, 93, 94, 96, 99, 100, 103, 360, 403
S42010	---	---	93, 104
S42020	type 420F	---	90, 92, 94, 100, 104
S42200	type 422	---	92, 94, 99, 100, 104
S42900	type 429	---	100, 135, 138
S43000	type 430	---	101, 132, 133, 135, 138, 139, 140, 142, 143, 144, 145, 146, 147, 354, 360, 361
S43020	type 430F	---	101, 133, 135, 138, 142
S43023	type 430F Se	---	101, 133, 142
S43035	type 439	---	101, 133, 135, 138, 139, 140, 142, 145, 147, 148, 151, 152
S43036	type 430Ti	---	136
S43100	type 431	---	91, 92, 93, 94, 96, 99, 101, 104, 138
S43400	type 434	---	101, 133, 136, 138, 140, 142, 143, 144, 145, 146, 147, 361
S43600	type 436	---	101, 133, 136, 138, 142, 147
S43932	type 439LT	---	151, 152, 154, 155, 156, 157

Wrought Stainless Steels (Continued)

UNS No.	Common Name	Trade Name	Page No.
S44002	type 440A	---	89, 90, 92, 93, 94, 96, 98, 99, 101, 105
S44003	type 440B	---	89, 90, 92, 93, 94, 96, 98, 99, 101, 105
S44004	type 440C	---	89, 90, 92, 93, 94, 96, 98, 99, 101, 105, 157
S44020	type 440F	---	90, 105
S44100	18CrCb	---	133, 151, 155
S44200	type 442	---	101, 136, 361
S44400	type 444	---	134, 136, 138, 141, 143, 144, 145, 146, 147, 148, 151
S44426	alloy 26-1, XM-33	---	22
S44500	type 430M	---	151
S44600	type 446	---	22, 101, 134, 136, 137, 138, 360, 361
S44625	XM-27	E-BRITE	22
S44626	26-1Ti	---	134, 138, 139, 141, 147, 148, 361
S44627	alloy 26-1, XM-27	E-BRITE 26-1	22, 134, 136, 149
S44635	---	Nu Monit	134, 136, 148, 149
S44660	---	Sea-Cure	23, 148, 149
S44700	---	AL 29-4	23, 139, 147, 148, 149, 361
S44800	---	AL 29-4-2	134, 137, 138, 141, 147, 148, 149, 157, 361
S44735	---	Al-29-4C	23, 134, 137, 148, 149
S44800	---	AL-29-4-2C	23